Biomathematics

Volume 9

Warren J. Ewens

Mathematical Population Genetics

Springer-Verlag
Berlin Heidelberg New York 1979

Warren John Ewens

Department of Mathematics
Monash University
Clayton 3168 Victoria
Australia

and

Department of Biology
University of Pennsylvania
Philadelphia, PA 19174
U.S.A.

AMS Subject Classification (1970): 92-02, 92A10, 92A15

ISBN 3-540-09577-2 Springer-Verlag Berlin Heidelberg New York
ISBN 0-387-09577-2 Springer-Verlag New York Heidelberg Berlin

Library of Congress Cataloging in Publication Data. Ewens, Warren John.
Mathematical population genetics. (Biomathematics ; v. 9). Bibliography: p.
Includes index. 1. Population genetics – Mathematics. I. Title. QH455.E93.
575.1. 79-18938

© by Springer-Verlag Berlin Heidelberg 1979
Printed in Germany

Typesetting: Elsner & Behrens, Oftersheim
Printing and binding: Konrad Triltsch, Würzburg
2141/3140-5 4 3 2 1 0

For Bronnie and Dave

Introduction

Population genetics occupies a central place in a variety of important biological and social undertakings. It has for many years been crucial to an understanding of evolutionary processes, of plant and animal breeding programs, and of various diseases of particular importance to man. While increased research in these areas naturally leads to a greater understanding of them, it also shows, particularly with the mathematical theory of population genetics, that previous arguments have sometimes been misleading, important points have been glossed over, and our knowledge of the genetic behavior of populations is not as firm as might previously have been thought. This observation is all the more important because much recent controversy on developments within or connected to population genetics has sometimes relied on now outdated population genetics theory. In this connection one might mention sociobiology, the effects of genetic manipulation with recombinant DNA, nature-nurture and heritability studies, and the knowledge of the detailed constitution of genetic material and the consequent possibility of its artificial creation. The importance of these developments is immense, as is the need to base controversies on them on firm population genetic and other scientific knowledge.

Population genetics embraces observational, experimental and theoretical components. While population genetics theory is in large measure quantitative, the complexities of Nature ensure that non-mathematical reasoning eventually outruns the purely mathematical aspects of the theory, which are necessarily based on simplified models of biological behavior. Nevertheless, the purely mathematical aspects of population genetics theory comprise a very large area of applied mathematical research, and the aim of this book is to give an account of this purely mathematical theory. Thus this book is not about population genetics theory, still less about population genetics itself. Indeed, the selection of material that must necessarily be made is biased towards that with the richest mathematical content, and this sometimes implies that topics of greater importance to population genetics generally are treated at shorter length than their real importance warrants. Given the number of books on population genetics and population genetics theory, I believe there is a place for an account of the purely mathematical theory, even if biased in this way. Despite this broad aim, the first chapter of this book is largely historical and considers more general questions on population genetics. This is so since I believe such a background is necessary even for a consideration of the purely mathematical theory.

The book has been aimed at the graduate or research level and should be supplemented by reading an introductory text. Perhaps the most useful for this purpose is C. C. Li's excellent *First Course in Population Genetics*. As indicated above, collateral reading in population genetics theory generally is also necessary to place the topics treated in this book in proper perspective.

What is the value of the mathematical side of population genetics theory? It may be argued that this merely makes quantitative arguments the general nature of which is already clear qualitatively. While in some measure this is true, there are many questions where common-sense qualitative arguments have led to quite incorrect conclusions on the genetic behavior of populations. This is true even for rather simple aspects of the theory and, of course, is increasingly true for more complex aspects and also aspects involving stochastic phenomena. This matter is discussed further in the concluding remarks of the book, to some extent in the light of examples of such questions treated in the preceding chapters.

The mathematical theory contributes in various degrees to the controversial areas mentioned in the opening paragraph. The theory of the correlation between relatives for a metrical trait, outlined in Chaps. 7 and 10, is the key ingredient in heritability studies and in nature-nurture allocations. The small but growing mathematical theory of altruistic traits concerns perhaps the central question of sociobiology. Detailed knowledge of the nature of genetic material has already led to considerable quantitative theory, particularly in the nature of evolutionary processes: it is perhaps in this area that, of those mentioned, mathematical population genetics theory will find its greatest application and from which, in turn, it will be most influenced so far as its nature and direction are concerned. The manipulation of genetic material now possible is perhaps, except in a negative sense, the area where mathematical theory is of least value. One population geneticist has claimed that the eventual goal of the study of evolution is to understand its processes quantitatively and thus be able to predict and control its course. The theory in this book, particularly that of Chaps. 6 and 7, should indicate the difficulty of achieving the first aim and the consequent great danger in an attempt to take control of evolutionary processes generally and in particular (as some enthusiasts would wish) of human evolution. The complexities of the genetic behavior of populations, as shown by the (still incomplete) mathematical theory, are far greater than our power to comprehend and control.

Various points concerning the presentation of this book should be mentioned. Aiming to concentrate on the mathematical theory, I have emphasized, particularly in Chap. 2, that such theory rests on models of biological reality which, no matter how simplistic, must be analysed on their own without the injection of extraneous assumptions during the analysis. If such assumptions are brought in, and the assumptions injected contradict those implicit in the model, in principle any result, no matter how incorrect, can arise. Of course the conclusions reached from a model must be treated with caution, depending on the reality of the initial assumptions made, but this is a different matter from interfering with the analysis of a model in mid-stream. Several incorrect conclusions in population genetics have arisen from such *ad hoc* interference.

So far as terminology is concerned I have followed the standard usage of the subject, even when this is perhaps unsatisfactory. Two unfortunate expressions, "gene frequency" (instead of the more logical "allele frequency") and "additive genetic variance" (instead of, perhaps, "genic variance") are entrenched in the literature, and

I have used them here except on specific occasions when a more precise usage seemed necessary.

The notation is not consistent throughout the book. Thus the symbol "x_i" might variously mean the frequency of an allele in generation i, in subpopulation i, the frequency of the allele A_i, and so on. Consistency would lead to cumbersome notation, and the context should always make clear, even if no explicit explanation is given, what any symbol stands for.

I have cited fewer rather than more references during this book, concentrating on those accounts that appear to be definitive, innovative, the most recent or in some other way important. Other references are easily sought out, particularly with the aid of the magnificent *Bibliography of Theoretical Population Genetics* by Felsenstein and Taylor.

This book has benefited greatly from the advice and criticism of many friends and colleagues, of whom I should mention Marc Feldman, Walter Fitch, Bob Griffiths, Sam Karlin, Ray Littler, Tom Nagylaki, Eugene Seneta, Richard Spielman and Glenys Thomson. I must thank in particular Frank Norman for many patient hours spent explaining to me the intricacies of mathematical diffusion theory, John Gillespie for his constant advice on biological, evolutionary and mathematical questions, and above all Geoff Watterson for his most careful and detailed reading of drafts of this book and for much discussion and guidance on the topics it considers. Naturally I am responsible for all errors and obscurities in the final version. Finally it is pleasure to thank Barbara Young in Melbourne for her careful and cheerful work in typing the manuscript.

Melbourne and Philadelphia   W. J. Ewens
December 1976 — October 1978

Contents

1. The Golden Age

1.1 Biometricians, Saltationists and Mendelians

During the early years of this century the field of evolutionary biology was in a state of controversy and disarray. Personality clashes exacerbated the already deep scientific divisions on fundamental questions of evolution and at times little hope could have been held for reconciliation between the various factions, much less that a new area of scientific thought would grow out of a synthesis of the several seemingly irreconcilable viewpoints. Such a synthesis, however, eventually did occur, and the subject of population genetics is in effect the outgrowth of that synthesis.

The immediate cause of the divisions on evolutionary questions was the rediscovery of Mendelism. This rediscovery reinforced an already existing division among biologists about the nature of evolution, and it is of some interest to review, very briefly, what this division concerned.

The *Origin of Species* was published in 1859. Apart from the controversies it brought about on a non-scientific level, it set biologists at odds as to various aspects of the theory. That evolution had occurred was not, on the whole, questioned. What was more controversial was the claim that the agency bringing about evolution was natural selection, and even among selectionists what was even more controversial was the nature of the evolutionary change brought about in this way. Darwin adhered to the "gradualist" point of view that changes in the nature of organisms in populations were gradual and incremental. Some of those who, in general, were his strongest supporters, for example T. H. Huxley and Francis Galton, were "saltationists": they believed that evolutionary changes most often occur in "jumps" of not inconsiderable magnitude. Two evolutionary schools of thought developed from these two points of view. Although any attempt to describe in brief terms the long and complex controversies that followed is bound to be incomplete, it is nevertheless possible to trace in general the threads of the arguments followed by members of both schools (for a much more detailed account see Provine, 1971).

Before doing so, it must be remembered that the mechanism of heredity was in effect unknown before 1900 and that, in so far as a common view of heredity existed, it would have been that the characteristics of an individual are, or tend to be, a blending of the corresponding characteristics of his parents. (It is, however, interesting to note that in a letter to Darwin in 1875 Galton came almost by pure reasoning to a proposition about the hereditary mechanism very close to the Mendelian one. Unfortunately this line of thought appears not to have been pursued: if it had been, the course of evolutionary thought during the last hundred years would have been different. For details of Galton's letter and comments on it see Olby, 1965.) This blending hypothesis brought perhaps the most substantial scientific objection to Darwin's theory. It is easy enough to see that, with random mating, the variance in a population for any characteristic would, under a blending theory, decrease by a factor of one-half in each generation. Thus uniformity of characteristics would rapidly be obtained so that no variation would exist upon which natural selection could act. This rapid rate would be modified only slightly under reasonable non-random mating. Since, of course, such uniformity is not observed, this argument is incomplete; but since variation of the degree observed could then only occur by postulating further factors of strong effect which cause the characteristics of offspring to deviate from those of their parents, it cannot then be reasonably argued that selectively favored parents produce offspring who closely resemble them and who are thus themselves selectively favored. This argument was recognized by Darwin as a major obstacle to his theory of evolution through selection and it is interesting to note that later versions of the *Origin* were somewhat influenced by this argument.

Galton played an ambiguous role in the controversy between the gradualists and the saltationists. On the one hand, he was himself a believer in the saltation theory, and this no doubt influenced him in advancing in 1875 the hereditary theory referred to above. On the other hand, he pursued a close intellectual and personal relationship with Darwin and, through this, attempted to quantify the gradualist evolutionary process. This led him to introduce the statistical concepts of correlation and regression which became the main tools of a group of scientists, later known as biometricians, who in a sense were the true inheritors of the gradualist Darwinian theory. This group's mathematical research in quantitative evolution began in the 1890s under the leadership of W. F. R. Weldon and Karl Pearson. At the same time the saltationists gained further adherents, notably William Bateson, and the struggle between the two groups moved in the closing years of the century towards a climax.

The year 1900 saw the rediscovery of Mendelism. The particulate nature of this theory was of course appealing to the saltationists. Rather soon many biologists believed in a non-Darwinian process of evolution through mutational jumps – the view that "Mendelism has destroyed Darwinism" was not uncommon. On the other hand, the biometricians continued to believe in the Darwinian theory of gradualist evolution through natural selection and were thus, in the main, disinclined to believe in the Mendelian mechanism, or at least that this mechanism was of fundamental importance in evolution.

It would be pointless to follow in detail the bitter acrimony that then followed. Even the inspired arguments of Yule (1902), based on a mathematical analysis of the Mendelian system, that Mendelism and Darwinism could be reconciled, were largely ignored. And yet, paradoxically, Darwinism and Mendelism are not incompatible; the former relies crucially on the latter, and further it would be difficult to conceive of a Mendelian system without some form of natural selection associated with it. To see why this should be so, it is now necessary to turn to the beginnings of the mathematical theory of population genetics.

1.2 The Hardy-Weinberg Law

We consider a random-mating monoecious population which is so large that gene frequency changes may be treated as deterministic and focus attention on a given gene locus at which two alleles may occur, namely A_1 and A_2. Suppose that in any generation the proportions of the three genotypes A_1A_1, A_1A_2 and A_2A_2 are $X, 2Y$ and Z, respectively. Since random mating obtains, the frequency of matings of the type $A_1A_1 \times A_1A_1$ is X^2, that of $A_1A_1 \times A_1A_2$ is $4XY$, and so on. We now consider the outcomes of each of these matings. If the very small probability of mutation is ignored, and if there are no fitness differentials between genotypes, elementary Mendelian rules indicate that the outcome of an $A_1A_1 \times A_1A_1$ mating must be A_1A_1 and that in an indefinitely large population, half the $A_1A_1 \times A_1A_2$ matings will produce A_1A_1 offspring, and the other half will produce A_1A_2 offspring with similar results for the remaining matings.

It follows that since A_1A_1 offspring can be obtained only from $A_1A_1 \times A_1A_1$ matings (with frequency 1 for such matings), from $A_1A_1 \times A_1A_2$ matings (with frequency $\frac{1}{2}$ for such matings), and from $A_1A_2 \times A_1A_2$ matings (with frequency $\frac{1}{4}$ for such matings), and since the frequencies of these matings are $X^2, 4XY, 4Y^2$, the frequency X' of A_1A_1 in the following generation is

$$X' = X^2 + \tfrac{1}{2}(4XY) + \tfrac{1}{4}(4Y^2) = (X + Y)^2. \tag{1.1}$$

Similar considerations give the frequencies $2Y'$ of A_1A_2 and Z' of A_2A_2 as

$$2Y' = \tfrac{1}{2}(4XY) + \tfrac{1}{2}(4Y^2) + 2XZ + \tfrac{1}{2}(4XZ) = 2(X + Y)(Y + Z), \tag{1.2}$$

$$Z' = \tfrac{1}{4}(4Y^2) + \tfrac{1}{2}(4YZ) + Z^2 = (Y + Z)^2. \tag{1.3}$$

The frequencies $X'', 2Y''$ and Z'' for the next generation are found by replacing $X', 2Y'$ and Z', by $X'', 2Y''$ and Z'' and $X, 2Y$ and Z by $X', 2Y'$ and Z' in Eqs. (1.1)–(1.3). Thus, for example,

$$X'' = (X' + Y')^2$$
$$= (X + Y)^2$$
$$= X', \qquad\qquad \text{[using (1.1) and (1.2)]}$$

and similarly it is found that $Y'' = Y', Z'' = Z'$. Thus, the genotypic frequencies established by the second generation are maintained in the third generation and consequently in all subsequent generations. Note that frequencies having this property can be characterized as those satisfying the relation

$$(Y')^2 = X'Z'. \tag{1.4}$$

Clearly if this relation holds in the first generation so that

$$Y^2 = XZ, \qquad (1.5)$$

then not only would there be no change in genotypic frequencies between the second and subsequent generations, but also these frequencies would be the same as those in the first generation. Populations for which Eq. (1.5) is true are said to have genotypic frequencies in Hardy-Weinberg form.

Since $X + 2Y + Z = 1$, only two of the frequencies X, $2Y$ and Z are independent. If, further, Eq. (1.5) holds, only one frequency is independent. Examination of Eqs. (1.1)–(1.3) shows that the most convenient quantity for independent consideration is $x = X + Y$, i.e. the frequency of the gene A_1.

The above results may be summarized in the form of a theorem:

Theorem (Hardy-Weinberg). Under the assumptions stated, a population having genotypic frequencies X (of A_1A_1), $2Y$ (of A_1A_2) and Z (of A_2A_2) achieves, after one generation of random mating, stable genotypic frequencies x^2, $2x(1 - x)$, $(1 - x)^2$ where $x = X + Y$ and $1 - x = Y + Z$. If the initial frequencies X, $2Y$, Z are already of the form x^2, $2x(1 - x)$, $(1 - x)^2$, then these frequencies are stable for all generations. Numerical examples of this theorem were given by Castle (1903), who possibly [cf. Keeler (1968)] knew the theorem in full generality, by Yule (1906), and by Pearson (1904). The first published general proof was by Hardy (1908) and Weinberg (1908), and it is after these authors that the theorem has become known.

Why is this rather simple theorem so important? Unfortunately it is important for two different reasons, one purely technical, and concentration on the technical reason has sometimes tended to obscure its truly basic value. The technical point is that if, as we may reasonably assume, Eq. (1.5) is true, the mathematical behavior of the population can be examined in terms of the single parameter x rather than the vector (X, Y); this is certainly a considerable convenience, but it is not fundamentally important. The really important part of the theorem lies in the stability behavior — if no external forces act, there is no intrinsic tendency for any variation present in the population (that is, variation caused by the existence of the three different genotypes) to disappear. This indicates immediately that the major earlier criticism of Darwinism, namely the fact that variation decreases rapidly under the blending theory, does not apply with Mendelian inheritance. It is clear from the Hardy-Weinberg Law that under a Mendelian system of inheritance, variation tends to be maintained. (Of course, the action of selection itself often tends to destroy variation; this qualification is of some importance and we will return to this point later and will find that the rate of loss of variation in any realistic Mendelian scheme is far less than the rate under any realistic blending scheme.)

Thus the Hardy-Weinberg Law shows that far from being incompatible, Darwinism and Mendelism are almost inseparable. It would be difficult to think of a hereditary process other than the "quantal" Mendelian scheme in which natural selection could act with such efficiency, while on the other hand fitness differential between genotypes will normally lead to changes in gene frequencies and thus ultimately to evolution. We will generalize the Hardy-Weinberg Law later in this book: for the moment we will be content with noting its historical significance.

It was thus beginning to become clear by the end of the first decade of the century that a reconciliation between Darwinism and Mendelism was not only possible but indeed inevitable. In 1911 this was already apparent to a young student of mathematics who read, during that year, a paper on "Heredity" to the Cambridge University Eugenics Society in which he stressed the necessity for this reconciliation. Such a reconciliation would carry with it a requirement to interpret on Mendelian principles the large bodies of data assembled by the biometricians on the correlations between relatives for various physical characteristics. Several years later R. A. Fisher, the young student in question, wrote a landmark paper in population genetics in which this reconciliation was achieved.

Special cases of such correlations had been treated earlier by Pearson (1904) and Yule (1906), but Fisher's paper was the first one to consider the problem in a rather complete degree of generality. We therefore consider the approach he used since several of the quantities which play a key role in his argument will appear subsequently to have considerable evolutionary importance.

1.3 The Correlation Between Relatives

Consider any character which is determined entirely by a locus A at which occur alleles A_1 and A_2. Suppose that all A_1A_1 individuals have measurement m_{11} for this character, that all A_1A_2 individuals have measurement m_{12} and that all A_2A_2 individuals have measurement m_{22}. Note that at the moment we assume no environmental contributions: once we know the genotype of any individual we know the value of his measurement.

Suppose random mating obtains with respect to this character and that the frequencies of A_1A_1, A_1A_2 and A_2A_2 are x^2, $2x(1-x)$ and $(1-x)^2$, respectively. Then the mean value of this measurement is

$$\bar{m} = x^2 m_{11} + 2x(1-x)m_{12} + (1-x)^2 m_{22}$$

and the variance in the measurement is

$$\sigma^2 = x^2(m_{11} - \bar{m})^2 + 2x(1-x)(m_{12} - \bar{m})^2 + (1-x)^2(m_{22} - \bar{m})^2. \tag{1.6}$$

Table 1.1.

	G		Son A_1A_1 m_{11}	A_1A_2 m_{12}	A_2A_2 m_{13}
		M			
	A_1A_1	m_{11}	x^3	$x^2(1-x)$	0
Father	A_1A_2	m_{12}	$x^2(1-x)$	$x(1-x)$	$x(1-x)^2$
	A_2A_2	m_{22}	0	$x(1-x)^2$	$(1-x)^3$

What is the covariance between father and son with respect to this measurement? Suppose the father is A_1A_1. Then the son will be A_1A_1 (with probability x) or A_1A_2 (with probability $1 - x$). The father himself will be A_1A_1 with probability x^2. Continuing in this way it is possible to draw up a table of the probabilities of the various father-son combinations in genotype and hence in the character measured. We find, using G for genotype and M for measurement, the values shown in Table 1.1.

The covariance between the measurement for the father and that for the son, assuming no change in the frequency of A_1 between the two generations, is thus

$$x^3 m_{11}^2 + 2x^2(1-x)m_{11}m_{12} + x(1-x)m_{12}^2 + 2x(1-x)^2 m_{12}m_{22}$$
$$+ (1-x)^3 \, m_{22}^2 - \bar{m}^2$$
$$= x(1-x)\{xm_{11} + (1-2x)m_{12} - (1-x)m_{22}\}^2. \tag{1.7}$$

The correlation between the two measurements, found by dividing the covariance by the variance (since the variance for sons is the same as that for fathers), is then

$$x(1-x)\{xm_{11} + (1-2x)m_{12} - (1-x)m_{22}\}^2/\sigma^2. \tag{1.8}$$

It is useful to write this expression in a different form. If we define

$$\left. \begin{aligned} \sigma_A^2 &= 2x(1-x)\{xm_{11} + (1-2x)m_{12} - (1-x)m_{22}\}^2, \\ \sigma_D^2 &= x^2(1-x)^2 \, \{2m_{12} - m_{11} - m_{22}\}^2, \end{aligned} \right\} \tag{1.9}$$

the expression (1.8) is clearly

$$\tfrac{1}{2}\sigma_A^2/\sigma^2. \tag{1.10}$$

Furthermore, it is simply a matter of algebra to show that

$$\sigma^2 = \sigma_A^2 + \sigma_D^2, \tag{1.11}$$

and in view of these relations it is of some interest to find interpretations for σ_A^2 and σ_D^2.

In order to find an interpretation for σ_A^2 we consider what changes are made in the measurement in question if we replace an A_1 allele by an A_2 allele. The effect of doing this depends on whether the replacement is made in an A_1A_1 individual or an A_1A_2 individual. The change is $m_{12} - m_{11}$ in the first case and $m_{22} - m_{11}$ in the second, and these will not generally be equal. We thus try to find some expression for this effect which in some sense is as close as possible to these two values.

Suppose we fit the measurements m_{11}, m_{12} and m_{22} as closely as possible (in the sense of least squares) by values of the form $\mu + 2\alpha_1, \mu + \alpha_1 + \alpha_2, \mu + 2\alpha_2$, where we make the constraint $x\alpha_1 + (1-x)\alpha_2 = 0$. If we differentiate

$$S = x^2(m_{11} - \mu - 2\alpha_1)^2 + 2x(1-x)(m_{12} - \mu - \alpha_1 - \alpha_2)^2$$
$$+ (1-x)^2(m_{22} - \mu - 2\alpha_2)^2$$

with respect to μ we find the minimizing value is $\mu = \bar{m}$. Differentiation of the Lagrangian expression

$$x^2(m_{11} - \bar{m} - 2\alpha_1)^2 + 2x(1-x)(m_{12} - \bar{m} - \alpha_1 - \alpha_2)^2$$
$$+ (1-x)^2(m_{22} - \bar{m} - 2\alpha_2)^2 - 4\lambda(x\alpha_1 + (1-x)\alpha_2)$$

with respect to α_1 and α_2 gives eventually

$$\left. \begin{aligned} \alpha_1 &= x(m_{11} - \bar{m}) + (1-x)(m_{12} - \bar{m}) \\ \alpha_2 &= x(m_{12} - \bar{m}) + (1-x)(m_{22} - \bar{m}) \end{aligned} \right\} \tag{1.12}$$

and we may then define the average effect of substituting A_2 for A_1 by

$$\alpha_2 - \alpha_1 = x(m_{12} - m_{11}) + (1-x)(m_{22} - m_{12}). \tag{1.13}$$

(When more than two alleles are involved we find it more convenient to adopt a slightly different usage and to call α_1 the average effect of A_1 and α_2 the average effect of A_2). Note that the value (1.13) could have been found almost immediately by taking a weighted average of $m_{12} - m_{11}$ and $m_{22} - m_{12}$. The present approach, while less direct, does on the other hand yield further information. The minimum value of the expression S is easily seen to be

$$x^2(1-x)^2(2m_{12} - m_{11} - m_{22})^2 \tag{1.14}$$

and the difference between this and σ^2, namely the sum of squares removed from σ^2 by the parameters α_1 and α_2, is

$$2x(1-x)\{xm_{11} + (1-2x)m_{12} - (1-x)m_{22}\}^2. \tag{1.15}$$

Now notice that the expression (1.15) is identical to the quantity σ_A^2 defined in (1.9), while the residual sum of squares (1.14) is identical to σ_D^2 defined in (1.10). Because σ_A^2 can be derived in the way just outlined, it can be called the genic or allelic variance: it is that part of the total variance in the character which can be accounted for by the average effects of the alleles A_1 and A_2. A more common name for σ_A^2 is the "additive genetic variance" in the character measured: this usage is perhaps unfortunate but because it is well established we follow it in this book. The residual variance σ_D^2 is called the dominance variance and, except for the trivial cases $x = 0$, $x = 1$, is zero only if $m_{12} = \frac{1}{2}(m_{11} + m_{22})$. We may thus express the result (1.10) as follows: under the conditions assumed, the correlation between father and son in the measurement considered is half the ratio of the additive genetic variance to the total variance in the measurement. If we denote this ratio by ρ^2, this result becomes

$$\text{corr (father, son)} = \tfrac{1}{2}\,\rho^2. \tag{1.16}$$

Note that this correlation is always non-negative, and will only take the value zero when $x = (m_{12} - m_{22})/(2m_{12} - m_{11} - m_{12})$, a possibility that can arise only if $m_{12} > m_{11}, m_{22}$ or $m_{12} < m_{11}, m_{22}$. We emphasize strongly the fact that this correlation has been found by basing all calculations on the Mendelian nature of the hereditary process.

A table analogous to Table 1.1, considering in this case full sibs, shows that under the same assumptions made above,

$$\text{corr (full sibs)} = \tfrac{1}{2}\,\rho^2 + \tfrac{1}{4}\,\delta^2, \tag{1.17}$$

where $\delta^2 = \sigma_D^2/\sigma^2$. Similar considerations, using tables of Mendelian association rather more complex than Table 1.1, show that under the same assumption

$$\text{corr (uncle, nephew)} \qquad = \tfrac{1}{4}\,\rho^2, \tag{1.18}$$

$$\text{corr (double first cousins)} \quad = \tfrac{1}{4}\,\rho^2 + \tfrac{1}{16}\,\delta^2, \tag{1.19}$$

and so on.

Having obtained these results, Fisher then considered more complex situations, in particular cases where more than two alleles are possible at each locus, where characters are determined by the alleles at many loci and where assortative mating obtains. We shall not pursue here the complexities associated with assortative mating: they are touched on briefly in Chap. 10. So far as multiple alleles are concerned, Fisher showed that the correlation formulae (1.16), (1.17), (1.18) and (1.19) remain unaltered provided that the additive and dominance variances are defined in the natural way through a generalization of the least-squares procedure just described.

The problem of characters determined by many loci is in principle more complex since interactive effects must now be taken into account. In the case of a character which is correlated with fitness it is very hard to determine how important these effects will be. If, however, the character is not correlated with fitness, we can reasonably assume that (cf. Sect. 7.7)

$$\text{freq } (A_i A_j B_k B_l \dots) = \text{freq } (A_i A_j) \times \text{freq } (B_k B_l) \times \dots \tag{1.20}$$

where $A_1, A_2, \dots$ are the alleles possible at locus A partially determining this character, $B_1, B_2, \dots$ are the alleles possible at a second locus B partially determining this character, etc. (Strictly speaking we assume that none of these alleles have any effect on fitness.) Equation (1.20) implies that the frequency of any chromosome can be written as the product of the frequencies of its constituent alleles. In this case the additive variance τ^2 can be found, as we show later (Sect. 7.4), by simply summing the additive variance at individual loci (i.e., $\tau^2 = \Sigma\,\sigma_A^2$), with a similar result for the total variance ($\omega^2 = \Sigma\,\sigma^2$: our notation here is different from Fisher's). Thus assuming that (1.20) is true, the correlation in the character measured between father and son becomes

$$\text{corr (father, son)} = \tfrac{1}{2}\,\tau^2/\omega^2, \tag{1.21}$$

which is the natural generalization of (1.16), with similar values for the other relationships. It is quite possible that, while these results are strictly true only when the character in question, and the alleles determining it, are not correlated with fitness, these values yield a satisfactory approximation even when there is some such correlation.

So far we have not taken any account of environmental variance. In practice this is extremely difficult because of the very high environmental correlation for father and son, for brother and brother, and so on. Ignoring the possibilities of such environmental correlation, Fisher used formulae such as those above, in conjunction with observed correlations, to estimate the various components of variance in any character. We do not pursue the details of this here (more will be said on this matter in Chap. 10). It is sufficient to note at this stage that, at least under simplified assumptions, the genetic component of the correlation between relatives is given in terms of two parameters and that the pattern of correlations so predicted agreed, for the data used by Fisher, reasonably well with those observed, so that Fisher had made a most significant beginning in reconciling biometry and Mendelism and for fusing these two into one discipline. From this point on theoretical population genetics could start on a firm quantitative basis, and, as we now see, the same variables used so effectively by Fisher in this reconciliation are, remarkably, central to a quantification of the evolutionary process.

1.4 Evolution

We turn now to the evolutionary consequences of Mendelism. The corner-stones of the Darwinian theory of evolution are variation and natural selection. Variation is provided, under a Mendelian system, ultimately by mutation: in all natural population mutation provides a continual source of genetic variation. Since the different genotypes thus created will often have different fitnesses (i.e. will differ in viability, mating success, and fertility), natural selection will occur. Our task is to quantify this process, and we now outline the work done during the 1920s and 1930s in this direction. Such a quantification amounts to a scientific description of the Darwinian theory in Mendelian terms.

It is necessary, at least as a first step, to make a number of assumptions and approximations about the evolutionary processes considered. Thus although, as we have seen, mutation is essential for evolution, mutation rates are normally so small that for certain specific problems we may ignore it. Further, although the fitness of an individual is determined in a complex way by his entire genetic make-up (and even then will often differ from one environment to another), we often assume as a first approximation that his fitness depends on his genotype at a single locus, or at least can be found by "summing" single locus contributions to fitness. It is also difficult to cope with that component of fitness which relates to fertility, and almost always special assumptions are made about this. More complete discussions of these problems will be given later in this book. If fitness relates solely to viability then much of the complexity is removed and for convenience we make this assumption, at least for the moment.

Suppose then that the fitnesses of the three genotypes A_1A_1, A_1A_2 and A_2A_2 at a certain locus "A" are as given below:

	A_1A_1	A_1A_2	A_2A_2
fitness	w_{11}	w_{12}	w_{22}
frequency	x^2	$2x(1-x)$	$(1-x)^2$

Note that we have written the frequencies of these genotypes in the form appropriate to random mating. Now Hardy-Weinberg frequencies apply only at the moment conception since from that time on differential viabilities alter genotype frequencies from the Hardy-Weinberg form: for this reason we count the population at the moment of conception of each generation. (Note that we are implicitly assuming a "non-overlapping" generation model: later we will consider several alternative models.)

Clearly the most interesting question to ask is: what is the behavior of the frequency x of the allele A_1 under natural selection? Since we take the fundamental unit of the evolutionary process to be the replacement in a population of an "inferior" allele by a "superior" allele, this question is essential to an understanding of the micro-evolutionary process.

This question was first attacked in certain specific cases by Norton (see Punnett, 1917), and later in much greater detail by Haldane (1924, 1926, 1927a, 1927b, 1930a, 1930b, 1932a) with a summary in Haldane (1932b). We consider here only the simplest of these cases. Before doing so, we should note that we are required to explain two seemingly contradictory phenomena. On the one hand, we must explain the dynamic process of the substitution of one allele for another and, on the other hand, we must explain the observed existence of considerable, apparently stable, genetic polymorphism.

The first concern is to find the frequency x' of A_1 in the following generation. By considering the fitnesses of each individual and all possible matings, we find rather easily that

$$x' - x = \frac{x(1-x)\{w_{11}x + w_{12}(1-2x) - w_{22}(1-x)\}}{w_{11}x^2 + 2w_{12}x(1-x) + w_{22}(1-x)^2} . \qquad (1.23)$$

Clearly continued iteration of the recurrence relation (1.23) yields the succesive values taken by the frequency of A_1. Unfortunately simple explicit expressions are not always possible, and resort must be made to approximation.

Before doing this it should be noted that x' depends on the ratios of the fitnesses w_{ij} rather than the absolute values so that x' is unchanged if we multiply each w_{ij} by some constant k. It is therefore possible to scale the w_{ij} in any convenient way. Different scalings are more convenient for different purposes. We indicate below two alternative scalings of the fitness values w_{ij}, and on different occasions either (1.24a), (1.24b), or (1.24c) will be most appropriate. It should be emphasized that nothing is involved here other than convenience of notation.

Fitness Values

A_1A_1	A_1A_2	A_2A_2	
w_{11}	w_{12}	w_{22}	(1.24a)
$1 + s$	$1 + sh$	1	(1.24b)
$1 - s_1$	1	$1 - s_2$	(1.24c)

We normally assume that except in extreme cases (perhaps involving lethality) the fitness differentials s, sh, s_1 and s_2 are small, perhaps of the order of 1%. In this case we ignore small-order terms in these parameters. Using the fitness scheme (1.24b), the recurrence relation (1.23) can be replaced, to a sufficient approximation, by

$$x' - x = sx(1 - x)\{x + h(1 - 2x)\}. \tag{1.25}$$

In turn, if we measure time in units of one generation, this equation may be approximated by

$$dx/dt = sx(1 - x)\{x + h(1 - 2x)\}. \tag{1.26}$$

If we denote the time required for the frequency of A_1 to move from some value x_1 to some other value x_2 by $t(x_1, x_2)$, then clearly

$$t(x_1, x_2) = \int_{x_1}^{x_2} [sx(1 - x)\{x + h(1 - 2x)\}]^{-1}dx. \tag{1.27}$$

(Naturally this equation applies only in cases where, starting from x_1, the frequency of A_1 will eventually reach x_2.)

While an explicit expression for $t(x_1, x_2)$ is possible, it is usually more convenient to use the expression (1.27) directly. Suppose first that $s > sh > 0$. Then it is clear from (1.26) that the frequency of A_1 steadily increases towards unity. However, as this frequency approaches unity, the time required for even small changes in it will be large, due to the small term $1 - x$ in the denominator of the integrand in (1.27). This behaviour is even more marked in the case $h = 1$ (A_1 dominant to A_2 in fitness), for then the integrand in (1.27) contains a term $(1 - x)^{-2}$. This very slow rate of increase is due to the fact that once x is close to unity, the frequency of the genotype A_2A_2 (against which selection is operating) is extremely low. In the important particular case $h = \frac{1}{2}$ (no dominance in fitness), Eq. (1.27) assumes the simple form

$$t(x_1, x_2) = \int_{x_1}^{x_2} \{\tfrac{1}{2} sx(1 - x)\}^{-1}dx. \tag{1.28}$$

From Eqs. (1.27) and (1.28) it is possible to evaluate the times required for any nominated changes in the frequency of A_1; some values are given in Table 1.2. Note that the times shown in the table support the conclusions just given and that, while selection acts so that variation is ultimately destroyed, the times required are usually very long and much longer than those required under any blending theory of in-

Table 1.2. Generations spent in various frequency ranges [calculated from (1.27): $s = 0.01$]

h	Range					
	0.001–0.01	0.01–0.1	0.1–0.5	0.5–0.9	0.9–0.99	0.99–0.999
$\frac{1}{2}$	462	480	439	439	480	462
1	232	250	309	1,020	9,240	90,231

heritance. We may therefore often expect to observe considerable genetic polymorphism in populations even though they are subject to directional natural selection. We shall find several uses later for this table and its various generalizations. The papers by Haldane referred to above provide values analogous to those in Table 1.2 in increasingly complex conditions (e.g. inbreeding, different sets of fitnesses in the two sexes). Clearly this procedure quantifies, at least approximately, the unit micro-evolutionary process of the replacement of an "inferior" allele by a "superior" allele.

It is clear that if $s < sh < 0$ a process parallel to the above, with A_2 steadily replacing A_1, will occur. This process is a mirror image of the one just considered and needs not further comment. An entirely different behavior arises when the fitness w_{12} of the heterozygote exceeds the fitnesses of the two homozygotes. This case is most conveniently treated by using the fitness parameters (1.24c) with $s_1, s_2 > 0$. Here the recurrence relation (1.23) may be rewritten, to a sufficient approximation, as

$$x' - x = x(1 - x)\{s_2 - x(s_1 + s_2)\}. \tag{1.29}$$

It is clear that there will be no change in the frequency x of A_1 if x takes the particular value

$$x = x^* = s_2/(s_1 + s_2) = (w_{22} - w_{12})/(w_{11} + w_{22} - 2w_{12}). \tag{1.30}$$

Note further that if $x < x^*$, then $x < x' < x^*$, while if $x > x^*$, then $x^* < x < x$. Thus x^* is a point of stable equilibrium and, whatever its initial value, the frequency x of A_1 will steadily approach x^*. It is not difficult to see that if $s_1 < 0, s_2 < 0$, then x^* is still an equilibrium point of the recurrence system (1.23), but in this case it is an unstable equilibrium and thus of little interest: the frequency of A_1 will steadily decrease to zero if its initial value is less than x^*, while it will steadily increase to unity if initially greater than x^*. All the above considerations show that a necessary and sufficient condition that there exist a stable equilibrium of the frequency of A_1 in (0, 1) is that the heterozygote have a larger fitness than both homozygotes. This most important fact was established by Fisher (1922) and gives one possible explanation for the occurrence of stable allelic frequencies in a population. Later we shall find a number of other possible explanations: for the moment we simply note that under the Mendelian system we can explain the occurrence of both dynamic substitutional processes and static equilibrium configurations. Thus, by the 1920s the first major steps were already being taken to explain in Mendelian terms and to quantify what are perhaps the two major properties of biological populations, namely their capacity to evolve and their capacity to maintain variation over long periods.

It is convenient now to consider briefly the effect of mutation. We suppose that A_1 mutates to A_2 at a rate u and that A_2 mutates to A_1 at rate v. Then it is easy to see that, if there is no selection,

$$x' = x(1 - u) + v(1 - x),\qquad(1.31)$$

and that a stable equilibrium is reached when

$$x = x^* = v/(u + v).\qquad(1.32)$$

Suppose now that both selection and mutation occur. We have in mind mainly the case where selective differences are of order 1% while mutation rates are of order 10^{-5} or 10^{-6}. Consider first the case where heterozygote selective advantage exists so that, under selection only, a stable equilibrium of the form (1.30) exists. It is clear under this assumption that if selection and mutation are now both taken into account there will exist a new stable equilibrium differing only trivially from that given by (1.30). We thus do not consider this case any further. Using the fitness scheme (1.24b) with $s > sh > 0$, it is clear that there will exist a stable equilibrium point for the frequency of A_1 close to unity. More exactly we find, for this equilibrium, the approximate formula

$$x = x^* = 1 - \{u/(s - sh)\},\qquad(1.33)$$

while if $s > 0, h = 1$ (A_1 dominant to A_2), the corresponding formula is

$$x = x^* = 1 - \sqrt{u/s}.\qquad(1.34)$$

Parallel formulae apply when $s < sh < 0$: here we find, at equilibrium,

$$x = x^* = v/sh,\qquad(1.35)$$

while when $s < 0, h = 1$,

$$x = x^* = \sqrt{v/s}.\qquad(1.36)$$

All these formulae were arrived at during the 1920s: note that they imply a second way in which genetic variation may be maintained in a population, i.e. by "mutation–selection balance." Note, however, that the frequency of one or other allele will be very small for any of the equilibria (1.33)–(1.36), although the frequency of the less frequent allele is less small where dominance is complete. Thus, when $s = 0.01, u = 10^{-6}$, the frequency of A_2 at equilibrium will be 0.01 when $h = 1$ (complete dominance) and 0.0002 when $h = \frac{1}{2}$ (no dominance).

We have noted earlier that the Mendelian system of heredity enables us to quantify, at least as a first approximation, the rate of allelic substitution in an evolutionary process. Is it possible to arrive at general principles, derived from the Mendelian system, which quantify the two main features of an evolutionary process through Darwinian natural selection, viz the requirement of variation for evolution to occur and the "improvement" brought about in a population through this evolution? In his Fundamental Theorem of Natural Selection, Fisher (1930a) attempted to find two such principles. His original derivation is not clearly based on an explicit model of the way in which an allele frequency changes in time, and thus we give here a derivation based on an explicit model which leads effectively to the same conclusions as were arrived at by Fisher.

Consider a population where the fitness of any individual depends only on his genetic constitution at a single locus "A". Suppose that two alleles, A_1 and A_2, are possible at this locus and that the fitnesses of the three possible genotypes are as given in (1.24a). The population is assumed to reproduce in non-overlapping generations so that Eq. (1.23) is applicable. In any generation we may define the mean fitness $\bar{w}$ of the population in that generation by

$$\bar{w} = w_{11}x^2 + 2w_{12}x(1-x) + w_{22}(1-x)^2, \tag{1.37}$$

where x is the frequency of A_1 in that generation. The frequency x' of A_1 in the following generation can be found from (1.23), and thus the mean fitness $\bar{w}'$ in that generation can be computed as

$$\bar{w}' = w_{11}(x')^2 + 2w_{12}x'(1-x') + w_{22}(1-x')^2. \tag{1.38}$$

Consider the change $\Delta\bar{w} = \bar{w}' - \bar{w}$ in mean fitness between these two generations. It is not difficult to show, using (1.23), that

$$\Delta\bar{w} = 2x(1-x)\{w_{11}x + w_{12}(1-2x) - w_{22}(1-x)\}^2$$
$$\times \{w_{11}x^2 + (w_{12} + \tfrac{1}{2}w_{11} + \tfrac{1}{2}w_{22})x(1-x) + w_{22}(1-x)^2\}\bar{w}^{-2}. \tag{1.39}$$

Clearly $\Delta\bar{w}$ is non-negative, so we may conclude that natural selection acts so as to increase, or at worst maintain, the mean fitness of the population. This is the first part of the Fundamental Theorem of Natural Selection, and is clearly a quantification in genetic terms of the Darwinian concept that an "improvement" in the population has been brought about by the action of natural selection.

We may also use (1.39) to quantify the second part of the Darwinian principle that variation (in our case genetic variation) is necessary for natural selection to operate. Clearly, $\Delta\bar{w}$ is zero if $x = 0$ or $x = 1$, i.e. if there is no genetic variation in the population. A much stronger statement than this is, however, possible. Note that if the w_{ij} are all close to unity we may write, to a sufficiently close approximation,

$$\Delta\bar{w} = 2x(1-x)\{w_{11}x + w_{12}(1-2x) - w_{22}(1-x)\}^2. \tag{1.40}$$

Now Eq. (1.9) shows immediately that the additive genetic variance in fitness is

$$\sigma_A^2 = 2x(1-x)\{xw_{11} + (1-2x)w_{12} - (1-x)w_{22}\}^2, \tag{1.41}$$

and we thus quantify in genetic terms the second major element of the Darwinian theory: the rate of increase of mean fitness is essentially equal to the additive component of the genetic variance in fitness. If fitness differentials are not small, a slightly different formula is found (Seneta, 1973).

One might initially have thought that the total variance in fitness, namely

$$\sigma^2 = w_{11}^2 x^2 + 2w_{12}^2 x(1-x) + w_{22}^2 (1-x)^2 - \overline{w}^2, \tag{1.42}$$

rather than the additive component of the variance, should be related to the increase in mean fitness. There are at least two arguments that show that this is not so. First, if the fitness values are of the form (1.24c) with $s_1, s_2 > 0$, and if the population is at the equilibrium point (1.30), then the total variance in fitness will be positive and yet, because the population is at equilibrium, there will be no increase in mean fitness. Second (and more importantly), the additive component of the genetic variance is that portion explained by the alleles themselves freed, as far as is possible, from deviations due to dominance. Since, in the model we consider, changes in allele frequencies are the fundamental components of evolution, the rate of increase of mean fitness can be expected to be related to that component of the total genetic variance which is accounted for by the alleles themselves.

Since our objective in this chapter is to review briefly the conclusions in theoretical population genetics up to about 1940, we do not mention here the various generalizations of the Fundamental Theorem of Natural Selection that are possible. We do, however, emphasize the great weight which Fisher placed on this theorem and the way in which it so beautifully quantifies in genetic terms the main theme of the Darwinian theory.

Before turning to the final major topic of this chapter, we consider briefly one question that will arise later in a generalized sense. We consider mutation–selection equilibria such as (1.33), where the less frequent allele is quite rare and is maintained only by recurrent mutation from the favored allele. Using the fitness scheme (1.24b), we note that if this mutation rate were zero the mean fitness of the population would be $1 + s$ and that the occurrence of the mutation causes the mean fitness to decrease somewhat from this value. What is the decrease in mean fitness? So long as $h < 1$, this decrease is found, to a close approximation, to be $2u$. For $h = 1$ a somewhat different calculation, using (1.34), gives a decrease of u, and for values of h close to 1 a value closer to u than to $2u$ is found. In other words, the population suffers a decrease in mean fitness proportional to the mutation rate (but not to fitness differentials): Haldane (1937), who first obtained this result, makes the most reasonable assertion that this situation has been reached by evolutionary modification of the mutation rate so that a small current decrease in mean fitness is traded off against an increase in genetic plasticity in the population suitable for possible future evolution. We term the loss is mean fitness the "mutational load" and later consider this and more general forms of genetic load.

We turn finally to an aspect of evolutionary behaviour which was considered at some length by Fisher, Haldane and Wright, namely the effect of the finite size of the population considered. This finiteness implies that changes in gene frequencies must be viewed as being part of a stochastic, rather than a deterministic, process. It is necessary, therefore, in order to arrive at a theoretical estimate of the importance of the stochastic factor, to set up a model which describes as well as possible the behaviour of a population in this circumstance. Perhaps more than any other part of the theory the choice of a model here is somewhat arbitrary, and we do not pretend that Nature necessarily follows at all closely the models we construct.

Although they did not use the terminology of Markov chain theory, the methods used by Fisher and Wright are in fact those of this theory and its close relative, diffusion theory. A brief summary of parts of this theory is given in Sect. 2.11: we anticipate here some of these results, and present the conclusion of Fisher and Wright stated in the terminology of Markov chains.

Consider, as the simplest possible case, a diploid population of fixed size N. Suppose that the individuals in this population are monoecious and that no selective differences exist between the two alleles A_1 and A_2 possible at a certain locus "A". There are $2N$ genes in the population in any generation, and it is sufficient to center our attention on the number X of A_1 genes. Clearly in any generation X takes one or other of the values $0, 1, ..., 2N$: denote the value assumed by X in generation t by $X(t)$.

We must now assume some specific model which describes the way in which the genes in generation $t + 1$ are derived from the genes in generation t. Clearly many reasonable models are possible and, for different purposes, different models might be preferable. We discuss various possible models later in this book: naturally, biological reality should be the main criterion in our choice of model, but we shall also consider mathematical convenience in this choice. The model used implicity by Fisher (1930a) and explicitly by Wright (1931), and which we call here the Wright-Fisher model, assumes that the genes in generation $t + 1$ are derived by sampling with replacement from the genes of generation t. This means that the number $X(t + 1)$ is a binomial random variable with index $2N$ and parameter $X(t)/2N$. More explicitly, given that $X(t) = i$, the probability p_{ij} that $X(t + 1) = j$ is given by

$$p_{ij} = \binom{2N}{j} (i/2N)^j \{1 - (i/2N)\}^{2N-j}, \quad i, j = 0, 1, 2, ..., 2N. \tag{1.43}$$

In this form, it is clear that $X(\cdot)$ is a Markovian random variable with transition matrix $P = \{p_{ij}\}$, so that in principle the entire probability behaviour of $X(\cdot)$ can be arrived at through knowledge of P and the initial value $X(0)$ of X. In practice, unfortunately, the matrix P does not lend itself readily to simple explicit answers to many of the questions one would like to ask, and we shall be forced, later, to consider alternative approaches to these questions.

On the other hand, Eq. (1.43) does enable us to make some comments more or less immediately. Perhaps the most important is that whatever the value $X(0)$, eventually $X(\cdot)$ will take either the value 0 or $2N$, and once this happens there will be no further change in the value of $X(\cdot)$. Genetically this corresponds, of course, to the fact that, since we do not allow mutation, once the population is pure A_2A_2 or pure A_1A_1, no variation exists, and no further evolution is possible at this locus. It was therefore natural for both Fisher and Wright to ask, assuming the model (1.43), what the probability is that eventual fixation of A_1 (rather than A_2) will occur and, perhaps more important, how much time might be expected to pass before fixation of one or other allele occurs.

It is easy enough to see that the answer to the first question is $X(0)/2N$. This conclusion may be arrived at by a variety of methods, the one most appropriate to Markov chain theory being that the solution $\pi_j = j/(2N)$ satisfies Eq. (2.118) and its boundary conditions. Setting $j = X(0)$ leads to the required solution. A second way of arriving at the value $X(0)/2N$ is to note that $X(\cdot)/2N$ is a martingale, i.e., satisfies the "invariant expectation" formula

$$E\{X(t+1)/2N \mid X(t)\} = X(t)/2N, \tag{1.44}$$

and then use either martingale theory or informal arguments to arrive at the desired value. A third approach, more informal and yet from a genetical point of view perhaps more useful, is to note that eventually every gene in the population is descended from one unique gene in generation zero. The probability that such a gene is A_1 is simply the initial fraction of A_1 genes, viz. $X(0)/2N$, and this must also be the fixation probability of A_1.

It is far more difficult to discuss the time until fixation occurs. The most obvious quantity to evaluate is the mean time $\bar{t}\{X(0)\}$ taken until $X(\cdot)$ reaches 0 or 2N, starting from $X(0)$. As it happens, the explicit formula for this is very complex, although, as we see later, some simple approximations are available. In fact, Fisher and Wright (no doubt noting this difficulty) paid comparatively little attention to the mean fixation time, concentrating on an approach centering around the leading non-unit eigenvalue of P. It is easy enough to see from (1.42) that if we put $x(t) = X(t)/2N$,

$$E[x(t+1)\{1-x(t+1)\}|x(t)] = \{1-(2N)^{-1}\}x(t)\{1-x(t)\}, \tag{1.45}$$

so that the expected value of the "heterozygosity measure" $2x(\cdot)\{1-x(\cdot)\}$ decreases by a factor of $1 - (2N)^{-1}$ each generation. It follows immediately that $1 - (2N)^{-1}$ is an eigenvalue of P, and the theory in Appendix B shows that it is the leading non-unit eigenvalue. Suppose the right and left eigenvectors corresponding to this eigenvalue are $\mathbf{r} = (r_0, r_1, r_2, \ldots, r_{2N})$, and $\mathbf{l}' = (l_0, l_1, l_2, \ldots, l_{2N})$ respectively. It is easy to see, from (1.45), that $\mathbf{r}'$ is proportional to the vector

$$\{0, 2N-1, 2(2N-2), 3(2N-3), \ldots, 2N-1, 0\}. \tag{1.46}$$

Unfortunately, no such simple formula exists for $\mathbf{l}$. If we suppose that $\mathbf{l}$ and $\mathbf{r}$ are normalized by the requirements

$$\sum_{k=1}^{2N-1} l_k = 1, \quad \sum_{k=0}^{2N} l_k r_k = 1, \tag{1.47}$$

then Eq. (2.117) shows that

$$p_{ij}(t) = \text{Prob}\{X(t) = j \mid X(0) = i\}$$
$$= r_i l_j \{1 - (2N)^{-1}\}^t + o\{1 - (2N)^{-1}\}^t \text{ for } t \text{ large.} \tag{1.48}$$

Equations (1.45) and (1.48) jointly provide much interesting information. It is clear that, especially in a large population, the mean heterozygosity of the population decreases extremely slowly with time as a result of the pure sampling drift implicit in the process under consideration, and we conclude that although, in the model (1.43), genetic variation must ultimately be lost, such loss is usually very slow. This slow rate of loss may be thought of as a stochastic analogue of the "variation-preserving" property of infinite genetic populations shown by the Hardy-Weinberg theorem, and we may quote Fisher (1958, p. 95) on this conclusion: "No result could bring out more forcibly the contrast between the conservation of the variance in particulate inheritance, and its dissipation in inheritance confirming to the blending theory". We shall generalize this conclusion later, taking into account not only mutation but also complications brought about through variation in the population size through geographical factors, through the existence of two sexes, etc.

What can be said about the distribution of $X(t)$ for large t, given $X(t) \neq 0, 2N$? Both Fisher (1958, pp. 90–96) and Wright (1931, pp. 111–116) paid considerable attention to this question. It is clear from (1.47) and (1.48) that

$$\lim_{t \to \infty} \Pr \{X(t) = j \mid X(t) \neq 0, 2N\} = l_j, \quad j = 1, 2, ..., 2N - 1. \tag{1.49}$$

Furthermore, both Fisher (1958, p. 94) and Wright (1931, p. 113) show that $l_j \approx (2N - 1)^{-1}$, so that the asymptotic distribution under consideration is essentially uniform. Although both Fisher and Wright devoted considerable attention to this distribution (and indeed to very accurate expressions for it, especially for very small and very large values of j), it is possibly of rather minor importance. The reason for this is that the complete spectral expression for $p_{ij}(t)$, of which (1.48) gives the leading terms and which was unknown to Fisher and Wright, shows that by the time this distribution becomes relevant it is almost certain that fixation or loss of A_1 will already have occured. This observation, due to Kimura (1955a), will be taken up in more detail later: for the moment we use it simply to justify passing over further discussion of the distribution l.

A more important question, also taken up by Fisher (1958, p. 96) and Wright (1931, p. 116) (although in a rather different form than that used later in this book) is the following. Suppose, in an otherwise pure $A_2 A_2$ population, a single new mutant A_1 gene arises. No further mutation occurs so we assume that from this point on the model (1.43) applies. How much time will pass before the mutant is lost (probability $1 - (2N)^{-1}$ or fixed (probability $(2N)^{-1}$)? The mean number of generations $\bar{t}_1$ for one or other of these events may be written in the form

$$\bar{t}_1 = \sum_{j=1}^{2N-1} \bar{t}_{1,j}, \tag{1.50}$$

where $\bar{t}_{1,j}$ is the mean number of generations that the number of A_1 genes takes the value j (before reaching 0 or $2N$). Both Fisher and Wright show (and we rederive this result later, see (5.23)) that

$$\bar{t}_{1,j} \approx 2j^{-1}, \quad j = 1, 2, ..., 2N - 1 \tag{1.51}$$

so that, using (1.50),

$$\bar{t}_1 \approx 2[\log(2N-1)+\gamma], \tag{1.52}$$

where γ is Euler's constant 0.5772 ... This expression is the C^{-1} of Wright (p. 117); Fisher gives an extremely accurate expression 2 $[\log(2N-1)+\gamma] + 0.200645 + O(N^{-1})$, which for large N is correct to at least 5 decimal places, as well as expressions for $\bar{t}_{1,j}$ more accurate, for small j, than are provided by (1.51).

There is an "ergodic" equivalent to (1.50) and (1.51) which is perhaps of more interest than (1.50) and (1.51) themselves, and which is indeed the route by which Fisher arrived at these formulae. Consider a sequence of independent loci, each initially pure "A_2A_2", and at which, in the kth member of the sequence, a unique mutation A_1 occurs in generation k. We may then ask how many such loci, after a long time has passed, will be segregating for A_1 and A_2, and at how many of these loci will there be exactly j "A_1" genes. It is clear that the mean values of these quantities are $\bar{t}_1$ and $\bar{t}_{1,j}$, respectively, and this gives us some idea, (at least insofar as the model (1.43) is realistic), of how much genetic variation we may expect to see in any population at a given time. This is a question we shall take up again at much greater length.

Wright (1931), p. 129) and Fisher (1958, p. 99) also considered the modifications to these results when selective differences exist. Again we do not pursue the details of their calculations since we arrive later at their results by other methods. Suppose we assume fitness values of the form (1.24b). Then it is reasonable to replace (1.43) by the model

$$P_{ij} = \binom{2N}{j}(\eta_i)^j(1-\eta_i)^{2N-j}, \quad i,j,=0,1,2,...,2N \tag{1.53}$$

where now

$$\eta_i = \frac{(1+s)i^2+(1+sh)i(2N-i)}{(1+s)i^2+2(1+sh)i(2N-i)+(2N-i)^2}. \tag{1.54}$$

We may again ask what values $\bar{t}_1$ and $\bar{t}_{1,j}$ assume. This problem was attacked by Fisher and Wright only in the case $h = \frac{1}{2}$. We shall show later, for general values of h, that

$$\bar{t}_{1,j} \approx 2\int_x^1 \psi(y)dy/[2Nx(1-x)\psi(x)\int_0^1 \psi(y)dy], \tag{1.55}$$

where $x = j/2N$,

$$\psi(x) = \exp\{-2\alpha hx + (2h-1)\alpha x^2\} \tag{1.56}$$

and $\alpha = 2Ns$. Note that when there is no selection ($\alpha = 0$) Eq. (1.55) reduces to Eq. (1.51), while for the zero dominance case ($h = \frac{1}{2}$), (1.55) reduces to

$$\bar{t}_{1,j} \approx 2[1 - \exp\{-\alpha(1-x)\}]/[2Nx(1-x)\{1 - \exp(-\alpha)\}], \tag{1.57}$$

agreeing with the value given by Fisher (his $4an$ is our α). For $h \neq \frac{1}{2}$, (1.55) cannot be evaluated explicitly although clearly numerical evaluation is possible. In all cases $\bar{t}_1 = \Sigma \bar{t}_{1,j}$ and to evaluate this, even for $h = \frac{1}{2}$, numerical methods will be required.

Both Fisher and Wright used (1.57) to find the probability that a new mutant A_1 will eventually become fixed in the population. Their method (which is quite different from the one we consider later) is as follows. Suppose in (1.57) we put $x = 1 - \delta$ and consider small values of δ. Then (1.57) reduces in effect to

$$2\alpha/[2N\{1 - \exp(-\alpha)\}] \tag{1.58}$$

which, as $\alpha \to 0$, approaches $2/2N$. We now argue that since the probability of fixation of A_1 for the neutral case ($\alpha = 0$) is known to be $(2N)^{-1}$, the probability of fixation in the case we are considering must be given by

$$\Pr(A_1 \text{ fixes}) = s/[1 - \exp(-2Ns)]. \tag{1.59}$$

This is identical to the value given by Fisher (p. 100) and Wright (p. 133) upon setting our s equal to Fisher's $2a$ and Wright's $2s$.

Equation (1.59) influenced Fisher considerably. He was accustomed to think in terms of very large populations; thus he gave a table of values of $\bar{t}_1$ [see Eq. (1.50)] for values of N ranging from 10^6 to 10^{12} and wrote later of populations of size of a thousand million as though they are typical. The ratio of the right-hand side in (1.59) to the value $(2N)^{-1}$ applying for the case $s = 0$ is

$$\alpha/[1 - \exp(-\alpha)], \tag{1.60}$$

and for the values $\alpha = -4, 0$ and 4 this ratio takes the values $0.08, 1$ and 4. Thus, as noted by Fisher, increasing α from -4 to $+4$ increases the probability of fixation of A_1 by a factor of 50. Thus, in a population of size 10^9, only a minute range of selective differences around zero lead effectively to the same fixation probability as for complete selective equivalence. As an alternative way of noting this, an increase in s from 0 to 10^{-6} increases the probability of fixation of A_1 by a factor of 2,000 in a population of this size. These considerations strongly influenced Fisher in arriving at the view that selective differentials are of paramount importance in determining the evolutionary behavior of populations and that the randomness in the behavior of gene frequencies brought about by the finite nature of the population size in no way seriously undermines the Darwinian theory. Wright was less influenced by formulae such as (1.60), since he was accustomed to think in terms of population sizes far smaller than 10^9. His view of the optimal circumstance under which evolution occurs (which we consider later) was rather different from Fisher's and fundamentally involves random changes in gene frequencies in populations of comparatively small size.

A further problem of an essentially stochastic nature, considered almost exclusively by Wright (pp. 133–134), concerns the stationary distribution of the frequency of A_1 when, in addition to the changes in frequencies brought about by selection and the random changes due to the finite nature of the population size, we allow mutation from A_1 to A_2 (at rate u) and from A_2 to A_1 (at rate v). In this case we may reasonably replace the transition probability (1.53) by

$$P_{ij} = \binom{2N}{j} (\eta_i^*)^j (1 - \eta_i^*)^{2N-j},$$

(1.61)

where η_i^* is given by

$$\eta_i^* = (1 - u)\eta_i + (1 - \eta_i)v,$$

(1.62)

η_i being defined by (1.54). If we put $x = X(\cdot)/2N$, Wright showed in effect that the stationary distribution of x is of the form

$$f(x) = \text{const } x^{4Nv-1}(1-x)^{4Nu-1} \exp \{2\alpha hx - (2h-1)\alpha x^2\}.$$

(1.63)

When the heterozygote is at a selective advantage it is perhaps better to use the fitness parameters (1.24c) to arrive at the equivalent formula

$$f(x) = \text{const } x^{4Nv-1}(1-x)^{4Nu-1} \exp \{2\alpha_2 x - (\alpha_1 + \alpha_2)x^2\},$$

(1.64)

where $\alpha_i = 2Ns_i$. In these formulae the relative effects of the population size, the selective coefficient and the mutation rate on the form of the distribution can be ascertained. Thus, if mutation rates are sufficiently small ($4Nu, 4Nv < 1$) some accumulation of probability occurs near $x = 0, x = 1$. This does not, however, nenessarily mean that most of the mass of the probability distribution is near these points, and it is quite possible that the most likely values for x are determined more by selection than by mutation.

An example is given by Ewens (1969c). We consider the case $N = \frac{1}{4} \times 10^5, u = v = 5 \times 10^{-6}$, and, using the notation (1.24c), $s_1 = s_2 = 2 \times 10^{-3}$. Inserting these values in (1.64) we arrive at the density function

$$f(x) = Cx^{-1/2} (1-x)^{-1/2} \exp 200x(1-x).$$

(1.65)

It is clear that the mean value of x is $\frac{1}{2}$, but this gives us no indication of whether the mutation rates or the selective coefficients are more important in determining the form of the distribution since in this example both mutation and selection are symmetric. To make progress on this point we compare the integral of the density function over two small sub-intervals, one near 0 and the other near $\frac{1}{2}$. Thus we find, for example, that the probability that the frequency x of A_1 is less than 0.0001 or greater than 0.9999 is approximately

$$2C \int_0^{0.0001} x^{-1/2} dx = 0.04C,$$

(1.66)

while the probability that x is between 0.4999 and 0.5001 is approximately

$$0.0004\,C\exp{(50)}. \tag{1.67}$$

This is about 10^{22} times larger than the value given in (1.66) and indicates that in this case the selective forces have a far greater influence on the likely values that x will assume than have the mutation rates. Although the above example has a high degree of symmetry implicit in it, a parallel result will hold for asymmetric cases where the selective coefficients and the mutation rates are of the same order of magnitude as those just considered. Thus if $u = 5 \times 10^{-6}, v = 10^{-5}, s_1 = 10^{-3}, s_2 = 2 \times 10^{-3}$, selection is again far more important than mutation in determining the likely values of x. In general this conclusion will hold so long as the selective differentials are at least 100 times larger than the mutation rate: thus if in the above example $s_1 = s_2 = 2 \times 10^{-4}$, the probability of a value of x less than 0.0001 or greater than 0.9999 is of the same order of magnitude as the probability of a value between 0.4999 and 0.5001, while if $s_1 = s_2 = 2 \times 10^{-5}$, the former probability is rather larger than the latter.

Before leaving the topic of stochastic processes we note that Fisher (1922), Haldane (1927b) and Wright (1931) all considered the specific problem of the probability of survival of a single new favourable mutant allele. This probability has already been computed, for the case of selection without dominance, in Eq. (1.59). A rather different approach, using the theory of branching processes, may also be used to find this probability, and it is of some interest to outline the elements of this method. To do this we follow the treatment of Fisher (1930a).

We consider a population with non-overlapping generations, the various generations existing at a sequence of time points, 0, 1, 2, 3, ..., and suppose X_n genes (or "individuals") at time point n. Each of these X_n individuals gives rise to a number of offspring individuals and then dies. At time $n + 1$ each of these offspring in turn produces offspring, and so on. We suppose a given fixed distribution for the number of offspring for each individual and that the distributions of the numbers of offspring to those individuals alive at any given time point are independent. The values $X_0, X_1, X_2, \ldots$, form a Markov chain: in this case no fixed upper limit can be set to the values X_i. We suppose that each individual leaves i offspring with probability p_i and introduce the generating function

$$p(z) = p_0 + p_1 z + p_2 z^2 + \ldots, \tag{1.68}$$

where z is a dummy variable. Clearly, if the mean and variance of the distribution $\{p_i\}$ are μ and σ^2, we have

$$p(1) = 1, \quad p'(1) = \mu, \quad p''(1) = \sigma^2 - \mu + \mu^2. \tag{1.69}$$

We suppose the branching process starts with one individual in generation zero (i.e. $X_0 \equiv 1$). Then the generating function of the number of individuals in generation one is $p(z)$, and for generation two can be found in the following way. We have

$$\text{Prob } \{X_2 = i\} = \sum_j \text{Prob } \{X_2 = i \,|\, X_1 = j\} \times \text{Prob } \{X_1 = j\}. \tag{1.70}$$

Given that $X_1 = j$, the probability that $X_2 = i$ is evidently the coefficient of z^i in $\{p(z)\}^j$. Thus from (1.70)

$$\text{Prob } \{X_2 = i\} = \text{coeff } z^i \text{ in } \sum_j \{p(z)\}^j p_j$$
$$= \text{coeff } z^i \text{ in } p(p(z)).$$

It follows that the generating function of the distribution of X_2 is $p(p(z))$ and, more generally, the generating function of the distribution of X_n is the nth functional iterate $p_n(z)$, defined by

$$p_{n+1}(z) = p(p_n(z)) = p_n(p(z)). \tag{1.71}$$

Fisher (1930a) was interested in three quantities: first, the probability π_n that $X_n = 0$; second, the limiting value of π_n as $n \to \infty$; and finally the conditional distribution of X_n for n large, given $X_n \neq 0$. By setting $z = 0$ in (1.71) we see immediately that π_n satisfies the functional relation

$$\pi_{n+1} = p(\pi_n), \quad n = 1, 2, 3, \ldots \tag{1.72}$$

with $\pi_0 = 0$. By letting $n \to \infty$ in (1.72), we see that limiting value π of π_n satisfies

$$\pi = p(\pi), \tag{1.73}$$

and it is not hard to show that the required value π is the smallest positive root of (1.73). Putting $\pi = 1 - \delta$ (δ small), a Taylor series expansion in (1.73) yields

$$1 - \delta \approx 1 - \delta p'(1) + \tfrac{1}{2} \delta^2 p'(1), \tag{1.74}$$

and if $\mu = 1 + \epsilon$ (ϵ small, positive), (1.69) and (1.74) show that

$$\delta \approx 2\epsilon/\sigma^2. \tag{1.75}$$

We shall defer consideration of the conditional distribution of $X_n(X_n \neq 0)$ for a moment and examine it only in a case of particular genetic interest.

We turn now to the application of these results in genetics, following the approach used by Fisher (1930a). Consider the case of a non-recessive A_1 mutant gene introduced into a previously pure A_2A_2 population. Homozygotes A_1A_1 will not usually appear until the number of A_1 genes is comparatively large (of order $\sqrt{N}$, where N is the population size) and by this time the fate of the new mutant, i.e., whether it will die out or not, is usually settled. Thus although it is clear that the assumptions made in the theory of branching processes are not exactly met for populations of fixed size, rather close approximations, increasing in accuracy as $N \to \infty$, should be possible by using this theory. In practice, the expression "survival of a new mutant" is then taken to mean the increase in the frequency of a mutant to a point where the probability of loss of the mutant by accidents of sampling in anything other than a very long time can be ignored. We have in mind in particular either the fixation of the mutant in the population or the attainment of a quasi-stable equilibrium point determined, for example, by heterozygote selective advantage.

We are mainly interested in establishing results for populations of stable size, and by convention we do this by using the fitness scheme (1.24b) where the values are now taken as absolute fitnesses. We thus identify the unit fitness of the prevailing genotype A_2A_2 with stable population size. Assuming the model (1.53), we may reasonably suppose each mutant A_1 gene produces a random number of A_1 "offspring" according to the binomial distribution with index $2N$ and parameter $(1 + sh)/2N$. To a sufficient approximation we may replace this distribution by a Poisson distribution with parameter $1 + sh$. In this case the generating function (1.68) becomes

$$p(z) = \exp \{(z - 1)(1 + sh)\} \tag{1.76}$$

and the approximation (1.75) yields

$$\delta \approx 2sh. \tag{1.77}$$

For $h = \frac{1}{2}$ this agrees with the value (1.59) found by diffusion methods, at least for values of N sufficiently large so that $\exp(-Ns)$ can be ignored. (This confirms our view that the branching process approximation is most accurate for large N.) Equation (1.72) becomes

$$\pi_{n+1} = \exp \{(\pi_n - 1)(1 + sh)\}, \tag{1.78}$$

an equation which may be iterated numerically to provide values of π_n for any value of n. This was done by Fisher for $s = 0$ and for $sh = 0.01$ and the numerical values found confirm the approximation (1.77) (which Fisher did not use explicitly).

The case $s = 0$ is of particular interest. Here

$$\pi_{n+1} = \exp(\pi_n - 1). \tag{1.79}$$

Since $\pi_n \to 1$ as $n \to \infty$ it is interesting to attempt an approximate solution of (1.79) in the form

$$\pi_n \approx 1 - cn^{-1}.$$

Insertion of this trial value into (1.79) gives $c = 2$ and hence

$$\pi_n \approx 1 - 2n^{-1}. \tag{1.80}$$

This value was given by Fisher from inspection of the numerical iteration (1.79).

We turn finally to the conditional distribution of X_n, given $X_n \neq 0$ for the case $s = 0$. Here we merely outline Fisher's conclusion since we will give more general formula (embracing his as a particular case) in Chap. 10. It is clear that the unconditional mean of X_n is unity, and hence from (1.80) the conditional mean of X_n (given

$X_n > 0$) is approximately $\frac{1}{2}n$. It is thus reasonable to consider the normalized variable $y_n = X_n/n$ which, given $X_n > 0$, we may hope will possess a limiting distribution as $n \to \infty$. By using generating function techniques, Fisher showed that the limiting $(n \to \infty)$ distribution of y_n is

$$f(y) = 2 \exp(-2y), \quad y > 0 \tag{1.81}$$

so that in particular

$$\text{Prob}(X_n > kn) = \text{Prob}(y > k) \approx \exp(-2k). \tag{1.82}$$

In the case $sh > 0$, Fisher proved that the conditional distribution of $y = X_n/(1 + sh)^n$, given $X_n > 0$, is asymptotically

$$f(y) = 2sh \exp(-2shy), \quad y > 0. \tag{1.83}$$

Thus,

$$\text{Prob}(X_n > X(1 + sh)^n \,|X_n > 0) = \text{Prob}(y > X|y > 0) \approx \exp(-2Xsh). \tag{1.84}$$

It should be emphasized that these conclusions, while they are arrived at by considering indefinitely large values of n, nevertheless apply only if the numbers of mutants involved is far less than the population size N, for it is only for such values that branching process approximations are legitimate. This is true particularly of equation (1.84).

What evolutionary conclusions can be drawn from the above? The first, and perhaps most important, is that while the survival probability (1.77) is small, it is nevertheless positive. Thus while the lines initiated by most favorable mutations will die out (and usually rather rapidly), the eventual survival of a favorable mutant is certain if mutation is recurrent. Thus taking the case $s = 0.01$, $h = \frac{1}{2}$, a mutation rate of 10^{-6} in a population of size 10^8 will produce 200 mutations per generation, and the probability that none of the mutational lines initiated in just the first generation survive is only $(0.99)^{200} \approx 0.14$. In a larger population (or with a larger mutation rate) the probability is diminished even further. It follows that in large populations a favorable new mutant will begin to establish itself rather soon after mutation to it commences. (We may then use equations such as (1.27) to consider how long various degrees of establishment will require). On the other hand, in small populations, and even more important with unique mutational events, the small individual probability of survival of a single mutant line is a factor which must be incorporated in evolutionary considerations.

A second observation concerns the origin and potential selective advantage of a mutant which has spread to large numbers in a population. We may take as a numerical example a population of size 10^7 with 10^5 A_1 genes. If these genes enjoy no selective advantage and arose from a single mutational event, Eq. (1.82) shows that the mutation most likely occurred at least 10^5 generations in the past. However, if mutation to the allele in question is recurrent, the average time required for the current frequency

10^5 is rather less, while if the mutant possesses a selective advantage its present frequency can be explained by a comparatively rapid recent increase in numbers.

A final comment concerns populations whose sizes are not stationary. Any mutant in a population of uniformly increasing size will have its survival probability increased. Consider as an example a new mutant having selective advantage 0.01 arising in a population of 10^4. Suppose now that the population doubles in size for eight generations and stabilizes at a size of 256×10^4. If the doubling in population size were to continue indefinitely, the new mutant would have a probability π of loss satisfying

$$\pi = \exp\left(2.02\left(\pi - 1\right)\right),$$

the solution of which is $\pi = 0.1978$. When doubling stops after eight generations the probability of loss of the mutant is rather greater than this, being approximately 0.3. A converse comment applies for mutants in decreasing populations. Thus increasing populations should exhibit some variety of forms compared to stable or decreasing populations. The variety will perhaps diminish once stability of population size is reached (since some unfavorable mutants, which increased in numbers because of the increase in population size, will now die out). In practice, of course, any protracted increase in size must occur at a rather low rate, and thus this argument applies most to mutations whose selective advantage or disadvantage is rather small.

1.5 Evolved Genetic Phenomena

In the previous section we have asked the question, "assuming the Mendelian genetic scheme and given the numerical values of various genetic parameters (e.g. mutation rates, degree of dominance), what conclusions can be drawn about evolutionary processes?" Fisher, Wright and Haldane also asked a converse question, namely "given that evolution has occurred, what purely genetic characteristics can be explained as a result of this evolution?" Perhaps the most interesting such questions concern mutation rates, dominance, linkage intensities, and the sex-ratio, while on a broader level the existence of sexual dimorphism (a Mendelian phenomenon) and even the pervasiveness of the Mendelian scheme itself can be considered. Here we limit attention to brief comments on the first four topics, again restricting attention to the work done in the period we are considering.

We have alluded already to the question of observed rates of mutation and the possibility that these are the result of evolutionary processes whereby the contrasting requirements of a low mutation rate (to preserve such favorable gene complexes as have been built up) and high mutation rates (so that a large number of potentially or actually favorable new mutations will arise) are optimally balanced. It is difficult to quantify this argument, and no real attempt to do so was made during the time we are considering. (For a more recent quantitative discussion see Sect. 10.2.) Of course one must avoid the assumption that all genetic phenomena are purely the result of evolutionary processes: it is certainly possible to argue that current mutation rates are partly the result of extrinsic factors having nothing to do with evolution and that, while they no doubt vary from locus to locus and time to time and are capable of some evolutionary modification, are not presently at optimal evolutionary values.

We turn next to the question of dominance. Fisher argued that dominance is the outcome of an evolutionary process through the (induced) selection of modifier genes at loci other than the primary one under consideration. He was strongly influenced in this view by the observation that it is normally the prevailing wild-type allele that is dominant, so that in the course of its becoming the prevalent type it presumably acquired this property. We consider the details of this argument later (Sect. 6.5): for the moment we merely introduce the elements of the analysis. Consider two alleles A_1 and A_2 at a locus and suppose the fitness scheme of the form (1.24b). If A_1 mutates to A_2 at rate u we may suppose that the frequency of A_1 is at the mutation-selection equilibrium point (1.33). Suppose now that at a locus M, at which the allele M_2 was previously fixed, a mutant allele M_1 arises with the effect that those $A_1 A_2$ individuals carrying the allele M_1 are altered in phenotypic expression towards that of the prevailing homozygote $A_1 A_1$. We assume that fitness is determined by the phenotype so that the fitness scheme takes the following form:

	$A_1 A_1$	$A_1 A_2$	$A_2 A_2$	
$M_1 M_1$	1	1	$1 - s$	
$M_1 M_2$	1	$1 - sk$	$1 - s$	(1.85)
$M_2 M_2$	1	$1 - sh$	$1 - s$	

Here $s > 0$ and $0 \leqslant k \leqslant h \leqslant 1$. Clearly M_1 is at an induced selective advantage to M_2 and will steadily increase in frequency to unity, bringing about dominance of A_1 over A_2.

Several qualifications should be made about this argument. Perhaps the most important is that we have ignored any possible selective differences between M_1 and M_2 which might arise for reasons quite separate from dominance modification at the A locus. Clearly the rate of change in the frequency of M_1 through dominance modification is very small, since the selective superiority of M_1 over M_2 through this agency arises only in the comparatively rare heterozygotes $A_1 A_2$. It would require only a minute selective advantage of M_2 over M_1 for other reasons to overcome this. Wright (1929a, b) was strongly influenced by this argument in forming his doubts about Fisher's theory. Wright's view on evolution, which we shall examine more closely in the next section, was centered around the assumption of almost universal interactive effects of genes, so that the fate of any allele is determined by the net selective force acting on it, the direction of this force being normally determined by factors more important than dominance modification. Fisher, on the other hand, believed that the selective advantage due to dominance modification would ultimately be effective. This view also we examine in the next section.

Wright (1934) put forward the more purely physiological view that dominance is a natural pristine characteristic, rather than an evolved characteristic, of an allele. We do not go into the details of this argument here: it is sufficient to note that the theory recognizes the role of genes in controlling the production of enzymes, which act as catalysts in physiological processes, and that one gene may well produce sufficient enzyme for a certain process so that no further effect is produced by a second gene. The reader is encouraged to read Wright (1934 and the references therein) for in no other way than by reading the original papers can the flavor of the long dispute on this matter, and its bearing on the general evolutionary principles of Fisher and Wright, be appreciated.

We consider next the question of linkage modification. The circumstances under which Fisher envisaged the evolution of close linkage between two loci (see for example Fisher, 1958, p. 116) occur when, at two loci A and B, the allele A_1 is favored in the presence of B_1 while A_2 is favored in the presence of B_2. This will imply that the double heterozygote A_1B_1/A_2B_2 will occur more frequently than the double heterozygote A_1B_2/A_2B_1 and that a recombination between A and B loci will break down the former in greater absolute numbers than they are formed by recombination from the latter. Thus a higher recombination fraction will lead to a greater breakdown of the "favored" gametes A_1B_1 and A_2B_2 and hence to a decrease in the mean fitness of the population. It is convenient to give an example of the form of fitness scheme envisaged for such a process. One set of fitnesses having the desired characteristics is of the form

$$
\begin{array}{cccc}
 & B_1B_1 & B_1B_2 & B_2B_2 \\
A_1A_1 & 1 & 1-a & 1-4a \\
A_1A_2 & 1-a & 1 & 1-a \\
A_2A_2 & 1-4a & 1-a & 1
\end{array}
\tag{1.86}
$$

This fitness scheme was introduced by Wright (1952) and considered in some detail by him for purposes other than that of present interest. The behaviour of this model is not so simple as one initially expects. Thus while it can be shown that whatever the value of the recombination fraction R between A and B loci $(0 < R \leqslant \frac{1}{2})$, there is an equilibrium point of gamete frequencies with all frequencies positive, this equilibrium is never stable. In other words, a fitness scheme of the form (1.86) cannot maintain a stable genetic polymorphism at either A or B locus and is thus of no use for considering the argument in question.

Another scheme which, more approximately, yields the fitnesses desired is

$$
\begin{array}{cccc}
 & B_1B_1 & B_1B_2 & B_2B_2 \\
A_1A_1 & 1 & 1-a & 1-2a \\
A_1A_2 & 1-a & 1+2a & 1-a \\
A_2A_2 & 1-2a & 1-a & 1
\end{array}
\tag{1.87}
$$

This fitness scheme leads to an equilibrium point with

$$
\begin{aligned}
\text{freq}\,(A_1B_1) &= \text{freq}\,(A_2B_2) = c^* \\
\text{freq}\,(A_1B_2) &= \text{freq}\,(A_2B_1) = \tfrac{1}{2} - c^*
\end{aligned}
\tag{1.88}
$$

where c^* is the unique solution in $(\frac{1}{4}, \frac{1}{2})$ of the equation

$$12ac^3 - 8ac^2 + ac + R(1 + 2a)(c - \tfrac{1}{4}) = 0. \qquad (1.89)$$

It is easy to verify geometrically that c^* increases as R decreases and that $c^* \to \frac{1}{2}$ as $R \to 0$.

We turn next to the equilibrium value of the mean fitness $\overline{w}$, considered as a function of c^*. This turns out to be

$$\overline{w} = 1 - 4ac^* + 12a(c^*)^2, \qquad (1.90)$$

and since this increases as c^* increases for $\frac{1}{4} < c^* < \frac{1}{2}$, we conclude immediately that $\overline{w}$ *decreases* as R increases.

We will show later that the equilibrium (1.90) is stable, at least for small R, and thus we have shown that, for small R at least, the stable equilibrium mean fitness decreases as the recombination fraction between the loci increases.

Fisher now argued that if "different strains" have different recombination fractions, the strain with the smallest value will, because of its higher mean fitness, tend to replace the others, so that tight linkage will have evolved in the population. This argument, involving the new concept of interpopulation selection, well be considered further (with arguments not involving such selection) in Sect. 6.5.

The final characteristic we consider is sex-ratio. Fisher's argument on this is curiously non-genetic in the sense that it could well have been made in pre-Mendelian times. The argument involves the introduction of the concept of "parental expenditure", which does not initially appear to be a necessary, or indeed the most obviously appropriate, vehicle for explaining sex-ratio. The argument is that each offspring receives, while young, a certain expenditure on the part of its parents. Consider now a cohort of such offspring about to embark on reproduction. The males in this cohort will supply exactly half the ancestry of the descendents of this cohort, as will of course the females. Suppose now that the total parental expenditure on behalf of males is less than that for females. Then parents having the tendency to produce male offspring in excess will, for the same expenditure, tend to contribute disproportionally to the ancestry of subsequent generations. Since the same argument in reverse would apply if the expenditure on females were less, selection will tend to change the sex-ratio to the point where an equal expenditure is made on female and male offspring. If now males suffer a heavier pre-adult mortality, then as compared to females more of this expenditure will take place for males who die early and do not participate in reproduction. It follows that the sex-ratio of males to females should exceed unity at birth but be lower than unity at the age of reproduction.

This argument leads to an evolutionary adjustment of sex-ratio. Whether the various assumptions implicit in it are valid is uncertain, and what appears to be a superior verbal argument, the consequence of which is that sex-ratio should be unity at the age of reproduction, is given in Crow and Kimura (1970, pp. 288–289). We examine an argument parallel to Fisher's, but based more firmly on genetic concepts, in Chap. 10.

1.6 Overall Evolutionary Theories

We now outline the two contrasting views of evolution arrived at by Fisher and Wright.

Fisher's view was in a way a simple one. Populations are considered to be very large — the numerical values of Fisher (1930a) for population size are often of order 10^9 or larger — so that, apart from the particular case of the probability of survival of an individual new mutant, stochastic effects are not important and deterministic analyses are sufficient to describe the essence of evolutionary behaviour. (Even in the case of new mutants, we have seen that essentially deterministic behaviour arises for recurrent mutations in large populations.) Thus, once the genetic raw material has been furnished by mutation, natural selection remains as the sole important agency in shaping genetic evolution.

The nature of this selection is also seen as being rather straightforward. In the first place, since complexes of genes at various loci, even if harmonious, tend to be broken up by recombination, strong emphasis is placed on single alleles at particular loci. (This is not to deny the fact that, as we have just seen, Fisher viewed interactive systems as being important: but, for example, so far as evolutionary processes are concerned, the effect of an interactive system such as (1.87) simply has the effect of yielding a selective advantage to M_1 over M_2, and the primary emphasis is placed on this fact.) This leads to the point of view (Fisher, 1953) that "it is often convenient to consider a natural population not so much as an aggregate of living individuals but as an aggregate of gene ratios". Fisher would have regarded this view as an approximation, but one which is nevertheless sufficient to describe the main characteristics of evolution. This view pervades, directly of indirectly, his work not only in population genetics but also, interestingly enough, in the statistical theory of experimental design (see, for example, Fisher, 1926, p. 511), which was strongly influenced, if indeed not suggested, by his research in genetics.

In population genetics a corollary of this view is that frequencies of gametes can be found, at least to a sufficient approximation, as the product of the frequencies of the constituent alleles. This approximation is implicit in his pioneer work in both quantitative and evolutionary genetics, except in special cases involving, for example, assortative mating. Thus, for example, in both fields the total additive genetic variance, a quantity of central importance, appears to be defined as the sum of the constituent one-locus marginal values (Fisher, 1918, p. 405; 1958, p. 37). We shall see later that while this is correct if indeed gamete frequencies can be so calculated, it is not generally so. Furthermore the Fundamental Theorem of Natural Selection, perhaps the apotheosis of Fisher's evolutionary theory, is true under these circumstances but not necessarily otherwise.

A further characteristic of Fisher's evolutionary view, arising from the above considerations and the assumed very large sizes of populations, is that an allele having a net selective advantage, no matter how small, is destined for fixation, at least while the selective advantage persists. Thus, for example, one of his main objectives in putting forward his theory of the evolution of dominance through the natural selection of modifiers was to show that even such a minute selective force would have evolutionary consequences. This was seen as being so even if the modifiers are subject to selective forces other than through dominance modification. Fisher's reasoning on this point, (in particular, Fisher, 1934, pp. 372–373) is not clear to this writer, who shares Wright's (1934) doubts on its acceptability.

To summarize, Fisher's view on the nature of evolution involves large population sizes, an emphasis on the main effects of single loci (as contrasted with complexes of loci), and a steady and essentially deterministic increase in the frequency of each allele having a selective advantage, no matter how small, with regard to the various alternative alleles at its locus. Evolution can be viewed on a locus-by-locus basis, and the net evolutionary pattern can be found by "adding" together such single-locus events.

Fisher's view has a grand simplicity to it. Is it, however, simplistic? The evolutionary theory reached by Wright (1931, 1956, 1960, 1965b, 1969b) appears, at least at first sight, to be more subtle. Wright reaches his view of evolution by discussing in turn several modes of the way in which gene substitution by selection can occur. He considers first selection in a very large random-mating population in a stable environment. The rates of change of gene frequency can be assumed to follow, at least to a reasonable approximation, differential equations of the form (1.26). Successive substitutional processes depend on the occurrence of favorable new mutations, and these are seen as arising sufficiently rarely so that evolution in this manner takes place too slowly to be effective. (Note that despite this point of view, this seems to be precisely the circumstance in which Fisher saw most of evolution taking place.)

A second circumstance in which gene replacement procedures occur is that where the environment changes, often deteriorating from the point of view of the population, so that a previously deleterious allele becomes favored and increases in frequency. Perhaps the best-known example is that of industrial melanism in moths during the nineteenth century. While such evolution can be rather more rapid than that just outlined (since there is no delay in waiting for a favorable mutation to arise, the store of previously deleterious alleles, maintained perhaps by a mutation-selection balance characterized by (1.35), being assumed large enough to provide immediately alleles favored in the new environment), at least one difficulty arises with it. This is that the evolution resembles a treadmill in that a population may do little more than merely keep abreast of the environmental deterioration. It is thus more difficult than under the first mode of evolution to account for the marvellous adaptations which populations have made to their environments and for the amazing intricacy and efficiency of observed physiological systems.

Despite these difficulties, Wright ultimately placed some weight on this second mode of evolution. From the very beginning, however, Wright's major view of evolution was more complex and of a different nature. He envisaged difficulties in building up favorable interactive genetic systems at least under the first mode described, and proposed a three-phase process under which the evolution of such systems could most easily occur. This view assumes that large populations are normally split up into semi-isolated subpopulations, or "demes", each of which is comparatively small in size. Within each deme there exists a "surface" of mean fitness, depending on the genetic constitution at many loci, and in conformity with the fundamental theorem of natural selection gene frequencies tend to move so that local peaks in this surface are approached. The surface of mean fitness is assumed to be very complex with a multiplicity of local maxima, some higher than others. If a fully deterministic behavior obtains the system simply moves to the nearest selective peak and remains there. The importance of the comparatively small deme size is that such strict deterministic behavior does not occur: random drift can move gene frequencies across a saddle and possibly under the control of a higher selective peak. (Random changes in selective values can also perform the same function.) In this way a succession of peaks can be reached, each one higher than the previous one. Interpopulational selection, arising from migration of individuals from demes which have higher selective peaks than have other demes, allows the favorable gene complex to spread ultimately throughout the entire population. Note that the unit of selection here is the entire gene complex and not individual alleles. Indeed the latter are viewed as often having no absolute selective advantage, being perhaps favorable in some gene combinations but unfavorable in others.

A case where evolution can more easily take place under this mode compared to that of Fisher is that of two alleles, one at each of two loci, which individually are deleterious but together are favorable. Calling the alleles in question A_1 and B_1, one selective scheme where this might occur is the following:

	B_1B_1	B_1B_2	B_1B_2	B_2B_2
A_1A_1	$1+r$	$1+s$	$1-t$	
A_1A_2	$1+s$	1	$1-u$	
A_2A_2	$1-t$	$1-u$	1	

$$(1.91)$$

Here $r > s > 0$ and $t > u > 0$. Under a deterministic scheme the frequencies of A_1 and B_1, if initially small, will be kept small (because of the selective disadvantage of $A_1A_2B_2B_2$ and $A_2A_2B_1B_2$ to $A_2A_2B_2B_2$). If however in one deme the frequencies of A_1 and B_1 can reach a sufficiently high value, the selective advantage of $A_1A_1B_1B_1$, and to a lesser extent of $A_1A_1B_1B_2$ and $A_1A_2B_1B_1$, will lead to fixation of A_1 and B_1. In terms of the previous discussion, this implies passing across a saddle from a selective peak at frequency (A_1) = frequency (B_1) = 0 to a higher selective peak at frequency (A_1) = frequency (B_1) = 1. By migration the favored complex, involving the gamete A_1B_1 in high frequency, is now assumed to spread to all demes.

It will be clear that Wright's emphasis, at least compared to Fisher's, was on interactive genetic systems in which most characters are affected by the genes at many loci and most genes have pleiotropic effects (i.e., influence several characters). In this view he was no doubt strongly influenced by his early experimental work on the coat color of guinea pigs, which revealed the importance of these effects. From the very first (see, in particular, Wright, 1935, 1952 and a summary in Wright, 1969b) his theoretical work involved multi-locus analysis and in particular an examination of the "optimum" model (1.88) and its various generalizations for more than two loci. We examine such models in more detail in Chap. 6.

Just as we asked whether Fisher's view of evolution in Mendelian populations was too simplistic, it is equally reasonable to ask whether Wright's overall views, particularly those involving population subdivision with migration between partially isolated demes, are not too complex. His picture of evolution may well rely on an equipoise of migration rates, fitness differentials and deme sizes of an unrealistically finely-tuned nature. We shall examine this point later when assessing the roles of these various factors, and of linkage, in evolution. It should in conclusion be mentioned that the facile criticism of Wright's evolutionary theory, that random drift is seen as an alternative to selection, has no basis in reality. Random drift acts merely as a trigger mechanism in the first phase of the process, changing gene frequencies within each deme before the more permanent and important changes brought about by selection.

1.7 An End and a Beginning

The alternative theories of evolution favored by Wright and Fisher should not obscure the fact that there was a great measure of agreement between them, and with Haldane, on evolution in Mendelian populations. The work of these three was decisive in reconciling biometry with genetics, in describing the features of Mendelian evolution and in part in quantifying the process. To this day any serious student of population genetics must read their seminal research papers and books, the flavor of which is almost impossible to recapture. They brought to a glorious climax a major field of characteristically twentieth-century science.

It may then seem paradoxical to state that population genetics theory is presently flourishing as never before. There are several reasons why this should be so. In the first place, a resolution of the differences between Wright and Fisher will require not only more empirical information on the sizes of populations, on migration rates, on selective values, on linkage arrangements, and on inbreeding coefficients than was available to them or is indeed presently available, but also a more detailed theory than they used, for a satisfactory analysis. In many instances the verbal discussion of Wright and Fisher far outran the mathematical basis of their arguments. Of course at some level this is inevitable and desirable, but it is possible that firmer and more detailed theoretical guidelines would have led to the avoidance of several dubious lines of argument. We amplify this statement later. Second, and least important, several purely mathematical errors in the early work have led to incomplete or sometimes misleading conclusions. But of far larger importance in the revitalization of theoretical population genetics are the advances which have been made in genetics, particularly molecular genetics, in ecology and ecological genetics, and also in mathematical and statistical techniques and in computers during the last quarter century. One example will serve to illustrate the importance of these advances. The work of Fisher, Wright and Haldane on evolutionary theory, with some important exceptions (notably on the rates of evolution in moths through industrial melanism, and the selective values necessary to explain these, as calculated by Haldane) were abstract and were not matched with real data. Thus if selective parameters took certain values, and mutation took place at given rates, evolution could and would occur, and at a given rate. But could evolution take place at the rate we have observed, given our observed or estimated values of these and other parameters? The enormous increase in evolutionary knowledge provided by observations on the structure of the gene as revealed by molecular genetics at least allows us to attempt to answer these questions, a condition unfortunately not granted to the pioneers. The rich knowledge of population structures and of inter- and intra-specific genetic variation now available to us from a variety of sources greatly stimulates our research in population genetics and provides us with a basis for far more detailed investigations than was hitherto possible. The following chapters attempt to describe the mathematical theory upon which much of the research on these and other questions is presently based.

2. Technicalities and Generalizations

2.1 Introduction

This chapter is largely technical in nature. Its aim in part is to consider further some of the theoretical points raised in Chap. 1 and to put these in a setting so that a more detailed and up-to-date discussion of them can be made in later chapters. A second aim is to introduce some further techniques not discussed in the previous chapter. Some rather straightforward generalizations of the theory are also made.

Population genetics models often make a number of simplifying assumptions, for example that random mating obtains, that fitnesses are fixed constants, that the population size is effectively infinite, and so on. In this chapter we consider what happens when some of these assumptions are relaxed or even dropped altogether. It is difficult enough to consider the effect of relaxing two or three of these assumptions simultaneously and quite impossible to consider the effect of relaxing them all. In the various sections of this chapter we therefore consider one or other generalization of the theory brought about by relaxing one or other of these assumptions. No systematic attempt will be made to assess mathematically the effect of simultaneous relaxation of two or more assumptions: such an assessment at the moment must largely be non-quantitative.

2.2 Random Union of Gametes

In elementary textbooks the way in which the frequencies of the various genotypes in a daughter generation are derived from those in the parent generation is by means of a two-way table. All the various matings are enumerated, their frequencies and the relative frequencies with which they produce various offspring genotypes are noted, and thus the frequencies of the daughter generation genotypes are calculated. This procedure was outlined in the previous chapter. It is more efficient, however, at least for random mating populations, to proceed in a different way. Restricting attention to autosomal loci, we observe that each individual transmits, for each locus, one gene to each of his offspring: the union of two such genes, one from each parent, defines at that locus the genotype of the offspring individual. Random mating of parents is equivalent to random union of genes. Thus, for example, using the notation of Sect. 1.2, since the frequency of A_1 in the parent generation is $X + Y$, the frequency of A_1A_1 in the daughter generation, being the probability that two genes drawn at random from the parent generation are both A_1, is $(X + Y)^2$. This and parallel arguments for the other genotypes gives Eqs. (1.1)–(1.3) immediately. Only slight extensions of this argument are needed for more complex cases such as sex-linked loci, multiple alleles, dioecious populations, etc., and we use this form of argument below in developing the properties of these more complex models.

We will emphasize later the importance of setting up a well-defined model in the framework of which a particular genetical process is analysed. A model is a mathematical abstraction which, although it cannot usually embody all the complexities of the real world, nevertheless aims to capture the essential features of reality in a sufficiently simplified form so that a mathematical analysis of the model is possible. Model-building is a delicate art as is the assessment, knowing the inadequacies of the model in describing the real world, of the biological value of its conclusions. It is nevertheless imperative that well-defined models be set up and all derivations be made ultimately from the assumptions of the model. The injection of *ad hoc* assumptions (which can possibly contradict the implicit properties of the model) can lead to "proving" any absurd conclusion whatsoever.

Having said this, we should state more explicitly the model assumed in the above argument. It has been assumed that the population is monoecious, of effectively infinite size, that any daughter-generation individual is formed by the mating of two randomly chosen individuals of the parent generation and, perhaps most important, that distinct generations can be recognised so that matings occur only between individuals of the same generation who do not participate in further mating once the daughter generation is formed. These assumptions imply no age structure, no geographical effects, no mating success differentials, and so on. Later, models with assumptions rather different from these will be introduced.

2.3. Dioecious Populations

In this section we drop the assumption that the population is monoecious and suppose instead that it is dioecious (i.e. admits two sexes). The other assumptions of the previous section are maintained. Suppose first there is no selection and that in a given generation the genotypic frequencies are as given in (2.1) below:

	A_1A_1	A_1A_2	A_2A_2	
males:	X_M	$2Y_M$	Z_M	(2.1)
females:	X_F	$2Y_F$	Z_F	

The argument of the random union of gametes, suitably modified to the dioecious case, shows that the frequency of A_1A_1 among both males and females of the daughter generation is $(X_M + Y_M)(X_F + Y_F)$, with parallel formulas for A_1A_2 and A_2A_2. This implies that after one further generation of random mating the frequencies in both sexes are in the Hardy-Weinberg form

$$A_1A_1 \qquad\qquad A_1A_2 \qquad\qquad\qquad A_2A_2$$
$$x^2 \qquad\qquad 2x(1-x) \qquad\qquad (1-x)^2 \qquad\qquad\qquad\qquad (2.2)$$

where

$$x = \tfrac{1}{2}\,(X_M + X_F + Y_M + Y_F). \qquad\qquad\qquad\qquad (2.3)$$

Suppose now that viability selection exists so that the relative fitnesses (i.e. relative rates of survival from birth to maturity) of A_1A_1, A_1A_2 and A_2A_2 males are w_{11}, w_{12}, and w_{22}, with corresponding values v_{11}, v_{12} and v_{22} in females. Consider genotypic frequencies immediately after formation of the zygotes of any generation, and suppose that in a given generation the males produce A_1 gametes with frequency x and A_2 gametes with frequency $1 - x$. Let the corresponding frequencies for females be y and $1 - y$. Then at the time of conception of the zygotes in the daughter generation the genotypic frequencies are, in both sexes,

$$A_1A_1 \qquad\qquad A_1A_2 \qquad\qquad\qquad A_2A_2$$
$$xy \qquad\qquad x(1-y)+y(1-x) \qquad\qquad (1-x)(1-y)$$

By the age of maturity these frequencies will have been altered by differential viability to the relative values

	A_1A_1	A_1A_2	A_2A_2
males:	$w_{11}xy$	$w_{12}\{x(1-y)+y(1-x)\}$	$w_{22}(1-x)(1-y)$
females:	$v_{11}xy$	$v_{12}\{x(1-y)+y(1-x)\}$	$v_{22}(1-x)(1-y)$

The frequencies x' and y' of A_1 gametes produced by males and females of the daughter generation are thus

$$x' = \frac{w_{11}xy + \tfrac{1}{2}w_{12}\{x(1-y)+y(1-x)\}}{w_{11}xy + w_{12}\{x(1-y)+y(1-x)\} + w_{22}(1-x)(1-y)}, \qquad (2.4a)$$

$$y' = \frac{v_{11}xy + \tfrac{1}{2}v_{12}\{x(1-y)+y(1-x)\}}{v_{11}xy + v_{12}\{x(1-y)+y(1-x)\} + v_{22}(1-x)(1-y)}. \qquad (2.4b)$$

These recurrence relations cannot in general be solved explicitly. It is nevertheless possible to arrive at certain important properties of their equilibrium points. It is clear that if selection favours the same allele in both males and females there will be no internal equilibrium, so the two cases of real interest are where different genes are favoured in the two sexes and where overdominance is involved. Our analysis of these two cases follows that of Kidwell et al. (1977).

Suppose first there is no (additive) dominance in fitness for each sex and that selection acts in opposite directions. We may assume the fitnesses are

	A_1A_1	A_1A_2	A_2A_2
males	1	$1 - \frac{1}{2} s_m$	$1 - s_m$
females	$1 - s_f$	$1 - \frac{1}{2} s_f$	1

where s_m, $s_f > 0$. Solution of the equilibrium equations $x = x'$, $y = y'$ gives, as the only possible equilibrium,

$$x = 1 - s_m^{-1} + \{(s_m s_f - s_m - s_f + 2)(2 s_m s_f)^{-1}\}^{1/2},$$
$$y = 1 - s_f^{-1} + \{s_m s_f - s_m - s_f + 2)(2 s_m s_f)^{-1}\}^{1/2}.$$

However, this equilibrium will be admissible $(0 < x < 1, 0 < y < 1)$ only if

$$\frac{s_m}{1 + s_m} < s_f < \frac{s_m}{1 - s_m} \tag{2.5a}$$

or equivalently

$$\frac{s_f}{1 + s_f} < s_m < \frac{s_f}{1 - s_f}. \tag{2.5b}$$

When these conditions apply the equilibrium can be shown to be stable. We conclude, especially if s_m and s_f are small, that additive selection acting in opposite directions in the two sexes will maintain a stable equilibrium only if the selective differences in the two sexes are fairly close.

Suppose now that dominance is introduced so that the fitness scheme becomes

	A_1A_1	A_1A_2	A_2A_2
males	1	$1 - h_m s_m$	$1 - s_m$
females	$1 - s_f$	$1 - h_f s_f$	1

An interesting special case occurs when $h_f + h_m = 1$. Here the conditions (2.5) that there exist a single stable internal equilibrium point continue to apply. When $h_f + h_m < 1$ there will be at most one equilibrium point, and the conditions on s_m and s_f for this to occur are rather less stringent than (2.5). Thus, roughly, for smaller h_f and h_m values a larger range of s_m and s_f values will lead to an equilibrium point. When $h_f + h_m > 1$ it is possible that more than one internal equilibrium point can arise, but the conditions for this are not given here.

When directional selection obtains for one sex and overdominance in the other, one suspects that a stable polymorphic equilibrium is possible provided the directional selection is not too strong. We quantify this statement in a moment when considering conditions for a stable polymorphic equilibrium to exist.

It is of considerable interest to ask how effective different selective schemes in the two sexes are, compared to identical selective schemes, in maintaining genetic variation.

We attack this question quantitatively by considering the conditions for the existence of an internal polymorphism in the two cases. For practical purposes we may suppose that such a polymorphism exists when the two equilibria freq $(A_1) = 0$ in males and females, freq $(A_2) = 0$ in males and females, are both unstable. If we linearize the recurrence relations (2.4) around $x = y = 0$ and around $x = y = 1$ we find that the condition for such a polymorphism is that both the inequalities

$$(w_{12}/w_{22}) + (v_{12}/v_{22}) > 2, \qquad\qquad\qquad (2.6a)$$

$$(w_{12}/w_{11}) + (v_{12}/v_{11}) > 2 \qquad\qquad\qquad (2.6b)$$

should hold. These requirements are the natural extensions to the monoecious population requirement that the heterozygote be more fit than both homozygotes for a stable internal polymorphism to occur.

When A_1 is at a selective advantage in males (so that $w_{11} > w_{12} > w_{22}$) but overdominance applies in females (so that $v_{12} > v_{11}, v_{22}$), condition (2.6a) holds automatically. However, condition (2.6b) will hold only if the overdominance in females is sufficiently strong compared to the directional selection in males. Thus, (2.6b) quantifies our earlier discussion of this point.

How stringent are the conditions (2.6)? Suppose we normalize so that $w_{12} = v_{12} = 1$. The conditions (2.6) then reduce to the requirements that the harmonic means of v_{11} and w_{22} and of v_{22} and w_{22} should both be less than unity. Since harmonic means are less than arithmetic means, this is a less stringent requirement than that the arithmetic means both be less than unity. In other words, the existence of different selective parameters in the two sexes provides a stronger mechanism for maintaining genetic polymorphism than taking average selective values over sexes would suggest.

The above analysis concerns autosomal loci and clearly a special analysis is needed in the sex-linked case. Taking the males as the heterogamatic sex the frequencies of the various genotypes in this case can be written

male		female		
A_1	A_2	A_1A_1	A_1A_2	A_2A_2
x	$1 - x$	Y_{11}	$2Y_{12}$	Y_{22}

If there is no selection, following the model outlined in the previous section it is evident that the frequencies in the following generation are

$$x' = Y_{11} + Y_{12},$$
$$Y'_{11} = x(Y_{11} + Y_{12}),$$
$$2Y'_{12} = x(Y_{12} + Y_{22}) + (1 - x)(Y_{11} + Y_{12}),$$
$$Y'_{22} = (1 - x)(Y_{12} + Y_{22}).$$

In contrast to the autosomal case, one generation of random mating is not sufficient to yield equal frequencies of A_1 in the two sexes. Nor does one further generation of random mating produce female genotypic frequencies in Hardy-Weinberg form. On the other hand, since

$$x' - (Y'_{11} + Y'_{12}) = -\tfrac{1}{2}\{x - (Y_{11} + Y_{12})\},$$

the absolute value of the difference between male and female frequencies of A_1 decreases by $\tfrac{1}{2}$ in each generation. For practical purposes we may thus assume that after a short time these frequencies are equal: one further generation of random mating then yields frequencies in the form

males		females		
A_1	A_2	$A_1 A_1$	$A_1 A_2$	$A_2 A_2$
z	$(1-z)$	z^2	$2z(1-z)$	$(1-z)^2$

where

$$z = \tfrac{1}{3}x + \tfrac{2}{3}(Y_{11} + Y_{12}).$$

When selection operates the behavior is clearly more complex (see Sprott, 1957; Bennett, 1957; Cannings, 1967, 1968). We do not go into details here and in this book pay little attention (perhaps less than is deserved) to sex-linked genes under the assumption that properties of autosomal loci are normally mirrored, perhaps with minor alterations, in the sex-linked case.

Note that while, in both autosomal and the sex-linked cases, the evolutionary behavior of two-sex systems is slightly more complex than in the monoecious case, the important Mendelian properties of conservation of genetic variation and suitability for evolutionary processes continue to apply.

2.4 Multiple Alleles

We turn now to the case of multiple alleles. Suppose at an autosomal locus A alleles $A_1, A_2, ..., A_k$ can occur. We consider a model identical to that of Sect. 2.2 and suppose there is no selection. If the frequency of A_i in any generation is x_i, the concept of the random union of gametes shows that in the next generation the frequency of $A_i A_i$ will be x_i^2 and that of $A_i A_j$ $(i \neq j)$ will be $2x_i x_j$. These frequencies are in generalized Hardy-Weinberg form and are maintained through future generations.

Suppose now that viability differentials exist and that the fitness of $A_i A_j$ is w_{ij}. It is clear that, if we count individuals at the moment of conception of each generation, the genotypic frequencies are in Hardy-Weinberg form. The gene frequencies will normally change from one generation to another, and the appropriate recurrence relations are

$$x_i' = x_i \, \Sigma \, w_{ij} x_j / \bar{w}, \tag{2.7}$$

$$= x_i w_i / \bar{w}, \tag{2.8}$$

where w_i, the "marginal fitness of the allele A_i", is defined as

$$w_i = \Sigma \, w_{ij} x_j \tag{2.9}$$

and $\bar{w}$, the mean fitness of the population, is defined by

$$\bar{w} = \Sigma \, x_i w_i = \Sigma \, \Sigma \, w_{ij} x_i x_j. \tag{2.10}$$

In view of the questions raised in Chap. 1 it is natural to ask whether the mean fitness increases from one generation to another and to seek the conditions on the w_{ij} that ensure a stable internal equilibrium point (i.e. each $x_i > 0$) of gene frequencies.

The most efficient proof that mean fitness increases was given by Kingman (1961a) and is reproduced here. The daughter generation mean fitness $\bar{w}'$ is defined by $\bar{w}' = \Sigma \, \Sigma \, w_{ij} x_i' x_j'$, and we are required to prove that, with this definition, $\bar{w}' - \bar{w} \geqslant 0$. Using (2.7),

$$\bar{w}' = \bar{w}^{-2} \, \{ \underset{i \; j}{\Sigma \, \Sigma} \, w_{ij} (x_i w_i)(x_j w_j) \}$$

$$= \bar{w}^{-2} \, \{ \underset{i \; j \; k}{\Sigma \, \Sigma \, \Sigma} \, w_{ij} w_{ik} x_i x_j x_k w_j \}.$$

By interchanging the roles of j and k we also have

$$\bar{w}' = \bar{w}^{-2} \, \{ \underset{i \; j \; k}{\Sigma \, \Sigma \, \Sigma} \, w_{ij} w_{ik} x_i x_j x_k w_k \}$$

and thus by averaging

$$\bar{w}' = \tfrac{1}{2} \, \bar{w}^{-2} \, \{ \underset{i \; j \; k}{\Sigma \, \Sigma \, \Sigma} \, w_{ij} w_{ik} x_i x_j x_k (w_j + w_k) \}$$

$$\geqslant \bar{w}^{-2} \, \{ \underset{i \; j \; k}{\Sigma \, \Sigma \, \Sigma} \, w_{ij} w_{ik} (w_j w_k)^{1/2} x_i x_j x_k \} \tag{2.11}$$

$$= \bar{w}^{-2} \, \underset{i}{\Sigma} x_i \, \{ \underset{j}{\Sigma} x_j w_{ij} (w_j)^{1/2} \}^2$$

$$\geqslant \bar{w}^{-2} \, [\underset{i}{\Sigma} x_i \, \underset{j}{\Sigma} x_j w_{ij} (w_j)^{1/2}]^2 \tag{2.12}$$

$$= \bar{w}^{-2} \, [\underset{j}{\Sigma} x_j (w_j)^{1/2} \, \underset{i}{\Sigma} x_i w_{ij}]^2$$

$$= \bar{w}^{-2} \, [\underset{j}{\Sigma} x_j (w_j)^{3/2}]^2$$

$$\geqslant \bar{w}^{-2}\left[\{\Sigma\,(x_j w_j)\}^{3/2}\right]^2 \tag{2.13}$$

$$= \bar{w}^{-2}\,(\Sigma\,x_j w_j)^3$$

$$= \bar{w}.$$

In this sequence of steps the inequality (2.11) is justified by the inequality $\frac{1}{2}\,(a+b) \geqslant (ab)^{1/2}$ for positive quantities a and b, and the inequalities (2.12) and (2.13) are justified by the convexity property $\Sigma\,x_i a_i^n \geqslant (\Sigma\,x_i a_i)^n$ for non-negative a_i and $n \geqslant 1$. If we assume each $x_i > 0$, this proof also shows that $\bar{w}' = \bar{w}$ if and only if $w_1 = w_2 = \ldots = w_k$, and when this is so it follows that

$$w_i = \bar{w}, \quad i = 1, 2, \ldots, k. \tag{2.14}$$

But this equation, from (2.8), implies $x_i' = x_i$, so that the system is at an equilibrium point. We thus conclude that in the system (2.7) the mean fitness always increases except when the system has reached an equilibrium point (where of course it remains unchanged). This conclusion also applies when some of the x_i are zero, although here of course (2.14) is true only for those values of i for which x_i is positive at the equilibrium point.

In view of the discussion in Chap. 1 it is natural to ask whether the change in mean fitness can be approximated by σ_A^2, the additive genetic variance in fitness. The natural generalization of the procedure that led to (1.15) is to define σ_A^2 as the maximum sum of squares removed by $\alpha_1, \ldots, \alpha_k$ in the expression

$$S = \Sigma\,\Sigma\,x_i x_j (w_{ij} - \bar{w} - \alpha_i - \alpha_j)^2,$$

where the α_i are subject to the constraint $\Sigma\,x_i \alpha_i = 0$. This is a standard problem in population genetics and is discussed in Appendix A. The theory there shows that the minimizing values are

$$\alpha_i = w_i - \bar{w} \tag{2.15}$$

and thus that

$$\sigma_A^2 = 2\Sigma_i\,x_i (w_i - \bar{w})^2. \tag{2.16}$$

When $k = 2$ this reduces to the value given by Eq. (1.41). We now wish to compare (2.16) with $\bar{w}' - \bar{w}$, which we write as

$$\bar{w}' - \bar{w} = \bar{w}^{-2}\left[\Sigma\,\Sigma\,w_{ij} x_i x_j w_i w_j - \bar{w}^3\right].$$

If $w_{ij} = \bar{w} + \delta_{ij},\ w_i = \bar{w} + \delta_i$, where the δ_{ij} are assumed small, this becomes, on ignoring terms of order δ_{ij}^3,

$$\bar{w}' - \bar{w} \approx \sum \sum \{\delta_{ij}\delta_i + \delta_{ij}\delta_j + \delta_i\delta_i\}x_ix_j$$

$$= 2 \sum_i x_i\delta_i^2 + \sum_i x_i\delta_i\sum_j x_j\delta_j$$

$$= 2 \sum_i x_i\delta_i^2. \qquad (2.17)$$

This is identical to (2.16), and we conclude that for small fitness differentials the increase in mean fitness is very closely approximated by the additive genetic variance in fitness. When fitness differentials are not small a rather different conclusion is found (Seneta, 1973).

Suppose each $x_i > 0$. Then Eq. (2.16) shows that σ_A^2 is zero if and only if $w_1 = w_2 \ldots = w_k = \bar{w}$. If some x_i are zero σ_A^2 is zero if (2.14) applies for those values of i for which x_i is positive. In both cases the discussion above shows that σ_A^2 is zero if and only if the system is at an equilibrium point. We see later that in multi-locus systems the identification

$$\sigma_A^2 = 0 \Leftrightarrow \text{population in equilibrium} \qquad (2.18)$$

no longer holds, although a restricted version of this conclusion can be found.

We consider now the evolution of a metrical character (not necessarily fitness) under the system (2.7). Consider some character which for A_iA_j individuals takes the measurement m_{ij}. The mean value $\bar{m}$ of this character is given by $\bar{m} = \sum \sum x_ix_jm_{ij}$, and we wish to compute the change in this mean after one generation. To a first order of approximation,

$$\Delta(\bar{m}) = 2 \sum \sum (\Delta x_i)x_jm_{ij}$$

$$= 2 \sum (\Delta x_i)m_i$$

$$= 2 \sum (\Delta x_i)(m_i - \bar{m}),$$

$$\approx 2 \sum x_i(w_i - \bar{w})(m_i - \bar{m}) \qquad (2.19)$$

where we have defined m_i, the marginal measurement for the allele A_i, by

$$m_i = \sum x_j m_{ij}. \qquad (2.20)$$

A verbal description of this conclusion is that the change in the character is twice the covariance between marginal allelic values of the character itself and fitness. (For further details see Robertson 1966, 1968.) When the character is fitness itself this conclusion reduces to that obtained in (2.17).

We turn now to the conditions under which a stable equilibrium of gene frequencies exists. Assume first that each x_i is positive at the equilibrium. The equilibrium conditions (2.14) can be written

$$w_i - w_1 = 0, \quad i = 2, 3, \ldots, k$$

$$x_1 + x_2 + \ldots + x_k = 1, \qquad (2.21)$$

and this is just a system of k linear equations in k unknowns. It thus possesses no solution, one solution or an infinity of solutions. The first and third cases arise only for special values of the w_{ij} (such as, for example, when all fitnesses are equal). In practice it is perhaps of most interest to ignore these cases and suppose there is a unique solution to (2.21). Unfortunately this solution might be inadmissible (i.e. the condition $0 < x_i < 1$, $i = 1, \ldots, k$, might not be met), and even if the equilibrium is admissible it need not be stable. Fortunately the stability criteria have been obtained (Kingman, 1961b). A unique admissible solution to (2.21) will be stable if and only if the matrix $W = \{w_{ij}\}$ has exactly one positive eigenvalue and at least one negative eigenvalue. In such a case the system moves, for any initial frequency point for which each x_i is positive, to this equilibrium. If the equilibrium (2.21) is inadmissible or unstable the system (2.7) evolves in such a way that one or more alleles become eliminated. The behaviour then becomes considerably more complicated, and in practice perhaps the best procedure is to note that the system always moves so that $\bar{w}$ is maximized, so that finding the maximum value of $\bar{w}$ subject to the constraints $0 \leqslant x_i \leqslant 1$, $\Sigma x_i = 1$, via the Kuhn-Tucker theory for quadratic programming, will provide the stable equilibrium point. A further result of Kingman (1961b) is that if W has j positive eigenvalues at most $k - j + 1$ alleles will exist with positive frequencies at this equilibrium.

As the simplest possible example of this theory suppose all homozygotes have fitness $1 - s$ $(0 < s < 1)$ and all heterozygotes have fitness 1. Clearly there is an admissible equilibrium at $x_i = k^{-1}$. This will be stable if the matrix

$$W = \begin{pmatrix} 1-s & 1 & 1 & \cdots & 1 \\ 1 & 1-s & 1 & \cdots & 1 \\ 1 & 1 & 1-s & \cdots & 1 \\ \vdots & \vdots & \vdots & & \\ 1 & 1 & 1 & \cdots & 1-s \end{pmatrix}$$

has exactly one positive eigenvalue and at least one negative eigenvalue. But standard theory shows that the eigenvalues of this matrix are $k - s$, $-s$, $-s$, $\ldots$, $-s$, and thus the stability conditions are met.

We turn finally to the correlation between relatives in the k-allele system and take as an example the correlation between father and son. Suppose the father has genotype A_iA_i (and thus measurement m_{ii}). The son will be A_iA_j (and have measurement m_{ij}) with probability x_j, and since the frequency of A_iA_i fathers is x_i^2 this will make a contribution to the covariance of

$$x_i^2 \Sigma x_j m_{ii} m_{ij} = x_i^2 m_i m_{ii}. \tag{2.22}$$

If the father is A_iA_j (frequency $2\,x_ix_j$) the son will be A_iA_i (probability $\frac{1}{2}x_i$) or A_jA_j (probability $\frac{1}{2}x_j$), A_iA_j (probability $\frac{1}{2}(x_i + x_j)$), A_iA_l (probability $\frac{1}{2}x_l$) or A_jA_l (probability $\frac{1}{2}x_l$). The contribution to the covariance is thus

$$2x_ix_jm_{ij}[\tfrac{1}{2}(x_1m_{i1} + \ldots + x_km_{ik}) + \tfrac{1}{2}(x_1m_{j1} + \ldots + x_km_{jk})]$$

$$= x_ix_jm_{ij}(m_i + m_j). \tag{2.23}$$

Adding (2.22) over all i and (2.23) over all $i, j(i < j)$ we arrive at the covariance

$$\sum_i x_i^2 m_i m_{ii} + \sum_{i < j} \sum x_i x_j m_{ij}(m_i + m_j) - \bar{m}^2 = \sum_i x_i(m_i - \bar{m}^2).$$

This is just half the expression (2.16) (if we replace w_{ij} by the more general m_{ij}), and in this way we recover expression (1.10) for the correlation in the measurement between father and son, where now both variance terms have the more general k-allele interpretation. Identical conclusions apply for other relationships, and we conclude that correlation questions are not affected by the number of alleles at the locus in question.

2.5 Frequency-Dependent Selection

In all of the above constant fitness values for each genotype have been assumed. It is likely in reality that many fitness values are not constant but depend on the number of individuals in the population, on the frequencies of the various alleles or on both. In this short section we consider briefly some aspects of frequency-dependent selection. We assume the model of Sect. 2.2 with two alleles at the locus considered.

Using the fitness scheme (1.24a) we arrived at the equation

$$\Delta x = x(1 - x)\{w_{11}x + w_{12}(1 - 2x) - w_{22}(1 - x)\}/\bar{w},$$

and this equation continues to hold if the w_{ij} are functions of the allele frequency x. Clearly there are equilibria when $x = 0$, $x = 1$, or when

$$w_{11}x + w_{12}(1 - 2x) - w_{22}(1 - x) = 0. \tag{2.24}$$

If the functions w_{ij} are sufficiently complex functions of x, (2.24) can have a number of solutions, several of which can be stable. There is little point in considering special cases. Further, $\bar{w}$ need not be maximized at such an equilibrium: Eq. (2.24) and the equation $d\bar{w}/dx = 0$ show that means fitness will not be maximized at an equilibrium if, at that equilibrium,

$$x^2 dw_{11}/dx + 2x(1 - x)dw_{12}/dw + (1 - x)^2 dw_{22}/dx \neq 0.$$

Thus evolution can cause a steady decrease in mean fitness. In a classical example due to Wright (1948) it is supposed that the fitnesses of A_1A_1, A_1A_2, and A_2A_2 individuals are $1 - s + t(1 - x)$, 1, and $1 - s - t(1 - x)$ where $s, t > 0$. (There are reasonable real-world situations where such fitnesses might obtain.) If $s < t$ there is a point of stable equilibrium where $x = x^* = 1 - st^{-1}$, whereas the mean fitness is maximized at $\frac{1}{2}(\frac{1}{2} + x^*)$, halfway between x^* and $\frac{1}{2}$. Clearly, for suitable initial frequencies of A_1, the mean fitness can steadily decrease during the course of evolution.

2.6 Fertility Selection

Until now we have assumed that selection operates through viability differentials. (Note that we can also incorporate differential genotypic mating success into our models, but we do not consider this possibility in this book.) This assumption was made for mathematical convenience, and we now suppose that further selective differences between genotypes arise through differential fertility as well as through viability differences. The analysis now becomes more complex since fertility relates to mating combinations rather than single genotypes. Our discussion assumes the natural generalizations of the model of Sect. 2.2 and closely follows the work of Bodmer (1965) and Kempthorne and Pollak (1970). Using the natural generalization of (1.24a), suppose that the viability of an A_iA_j genotype is w_{ij} $(i, j = 1, ..., k)$ (assumed the same in both sexes) and that the fertility of an $A_iA_j \times A_mA_n$ mating is f_{ijmn}. (We adopt some standard ordering convention such that A_iA_j is the male and A_mA_n the female.) It is clear that male and female genotypic frequencies will be equal: let X_{ij} be the frequency of A_iA_j just before the conception of a new generation. Consider those matings leading to A_iA_i offspring: these must be of the form $A_iA_j \times A_iA_m$ for some j and m. Consideration of the genotypic products of such matings shows that the frequency of A_iA_i at the birth of the next generation will be proportional to

$$Z_{ii} = f_{iiii}X_{ii}^2 + \frac{1}{2}\sum_{m \neq i} f_{iiim}X_{ii}X_{im} + \frac{1}{2}\sum_{j \neq i} f_{ijii}X_{ij}X_{ii}$$

$$+\frac{1}{4}\sum_{j \neq i}\sum_{m \neq i} f_{ijim}X_{ij}X_{im}. \tag{2.25}$$

These A_iA_i individuals are now subject to viability selection between birth and the age of maturity, and it follows that the frequency X'_{ii} of A_iA_i just before the birth of the next following generation is given by

$$\mu X'_{ii} = w_{ii}Z_{ii}, \tag{2.26a}$$

where μ is a normalizing constant to be discussed later. Similar considerations for A_iA_j individuals yield

$$\mu X'_{ij} = w_{ij}Z_{ij}, \quad (i \neq j) \tag{2.26b}$$

where

$$Z_{ij} = (f_{iijj} + f_{jjii})X_{ii}X_{jj} + \frac{1}{2}\sum_{m \neq j} f_{iijm} X_{ii}X_{jm}$$

$$+ \sum_{m \neq i} f_{imjj}X_{im}X_{jj} + \frac{1}{4}\sum_{m \neq i}\sum_{n \neq j} f_{imjn}X_{im}X_{jn}.$$

The constant μ in Eq. (2.26) is now chosen so that $\sum\sum X'_{ij} = 1$.

These recurrence relations are far too complex to solve in general and we make no attempt to do so. The existence and stability of equilibrium points of the system (2.26) have been discussed by Hadeler and Liberman (1975); we do not pursue these here. Some simplification is possible if it is supposed that the fertilities f_{ijmn} are of the multiplicative form

$$f_{ijmn} = a_{ij}b_{mn}, \quad (a_{ij} = a_{ji}, b_{mn} = b_{nm}). \tag{2.27}$$

Introducing the new variables

$$x_i = (a_{ii}X_{ii} + \tfrac{1}{2} \sum_{j \neq i} a_{ij}X_{ij})/ \sum_{j \leqslant i} \sum a_{ij}X_{ij}, \tag{2.28}$$

$$y_i = (b_{ii}X_{ii} + \tfrac{1}{2} \sum_{j \neq i} b_{ij}X_{ij})/ \sum_{j \leqslant i} \sum b_{ij}X_{ij},$$

the recurrence relations (2.26) become

$$\mu^* X'_{ii} = w_{ii}x_iy_i,$$

$$\mu^* X'_{ij} = w_{ij}(x_iy_j + x_jy_i), \quad i \neq j, \tag{2.29}$$

where μ^* is a new normalizing constant ensuring that the sum of genotypic frequencies is unity. Use of (2.28) and (2.29) shows that

$$x'_i = \{a_{ii}w_{ii}x_iy_i + \tfrac{1}{2} \sum_{j \neq i} a_{ij}w_{ij}(x_iy_j + x_jy_i)\}/ \sum \sum a_{ij}w_{ij}x_iy_j,$$

$$y'_i = \{b_{ii}w_{ii}x_iy_i + \tfrac{1}{2} \sum_{j \neq i} b_{ij}w_{ij}(x_iy_i + x_jy_i)\}/ \sum \sum b_{ij}w_{ij}x_iy_j. \tag{2.30}$$

These recurrence relations are identical in form to (2.4) and thus the latter system, once appropriate changes in fitnesses have been made to include the viability parameters, continue to apply: some specific examples are given by Bodmer (1965). One question of particular interest is whether the mean fitness of the system increases with time. Unfortunately it is not at all evident that a natural definition for mean fitness exists. Using (2.29) and the analogy with previous recurrence systems it would be reasonable to define mean fitness as

$$\sum_i w_{ii}x_iy_i + \sum_{i<j} \sum w_{ij}(x_iy_j + x_jy_i), \tag{2.31}$$

but with this definition it is possible for mean fitness to decrease with time. Thus (Kempthorne and Pollak, 1970) if $k = 2$, $w_{11} = w_{12} = 1$, $w_{22} = 0.5$, $a_{11} = a_{12} = 1$, $a_{22} = 2$, $b_{11} = 0.25$, $b_{12} = b_{22} = 1$, $X_{11} = X_{22} = 0$, $X_{12} = 1$, then $x_i = y_i = 0.5$, and the mean fitness, as defined by (2.31), is 0.875. From (2.30), $x'_1 = x'_2 = \tfrac{1}{2}$, $y'_1 = 5/11$, $y'_2 = 6/11$ and using these values in (2.31) the daughter generation mean fitness is $19/22 \approx 0.864$. It is clear that this decrease is caused essentially because the genotype with highest fecundity has lowest viability.

Suppose now, in Eq. (2.27), it is assumed that $a_{ij} = b_{ij}$. Then immediately $x_i = y_i$ and that the birth of the new generation genotypic frequencies are in Hardy-Weinberg form. Further the recurrence relations (2.30) are of the form (2.7), and therefore the conclusions deriving from that system, including in particular the result that the mean fitness, defined now as $\Sigma \Sigma a_{ij} w_{ij} x_i x_j$, cannot decrease, continue to hold. The change in mean fitness again is approximately equal to the additive genetic variance when the latter is suitably defined so as to include both viability and fertility parameters.

Despite this it is possible that (2.31) is not a natural definition of the mean fitness of the infant population. The classical definition is that the fitness of any genotype is proportional to half the number of offspring individuals (of whatever genotype) from individuals of the genotype in question, counting being performed at the same stage of the life cycle. We now attempt to find an algebraic definition of mean infant fitness along these lines.

Consider infants of genotype $A_i A_j$: these survive to adulthood with probability w_{ij}. An $A_i A_j$ individual mating with an $A_m A_n$ individual has $a_{ij} a_{mn}$ offspring and crediting half of these to the $A_i A_j$ individual and averaging over all $A_m A_n$, the $A_i A_j$ individuals are credited with a proportionate amount

$$\tfrac{1}{2} w_{ij} a_{ij} \sum_m \sum_n x_m x_n w_{mn} a_{mn} / \overline{w} = w_{ij} a_{ij} \overline{m} / 2\overline{w}$$

of offspring, where $m_{ij} = a_{ij} w_{ij}$ and $\overline{m} = \Sigma \Sigma x_i x_j a_{ij} w_{ij}$. The mean fitness of the infant population may then reasonably be defined as the weighted average of these quantities, or

$$\Sigma \Sigma x_i x_j w_{ij} a_{ij} \overline{m} / 2\overline{w} = (\overline{m})^2 / 2\overline{w}. \tag{2.32}$$

In a parallel fashion the mean fitness of the adult population may be defined: details are given by Kempthorne and Pollak (1970). Curiously neither the infant mean fitness, defined by (2.32), nor the adult mean fitness must necessarily increase with time, decreases again possibly occurring when those genotypes with high fertility have low viability. We do not pursue this matter further and simply note the great complexity in general of fertility selection models. During most of the rest of this book selection will be taken to mean viability selection. This is no more than a reflection of the fact that, simply because the mathematics is easier, more is known about viability selection models, and it certainly does not necessarily represent the true state of nature.

2.7 Continuous-Time Models

In all of this book so far it has been assumed that populations reproduce at discrete time points. There are certainly some real-world populations for which this is a reasonable assumption. On the other hand, it is sometimes more appropriate biologically, or simpler mathematically, to use continuous-time models in which births and deaths can take place at any instant. This normally leads to mathematical systems where changes in gene frequency are described by differential equations or differential equation systems. In this section we outline some of these mathematical models and discuss their properties, relying heavily on the definitive work of Nagylaki (1974c, 1976), Nagylaki and Crow (1974) and Kimura (1958).

Much emphasis was laid in the Introduction and in Sect. 2.2 on the need to set up well-defined mathematical models of biological processes. It is perhaps in continuous-time models that the lack of such an approach has most sorely been felt. Thus, for example, in much of Fisher's work the extrinsic assumption (not proven as a consequence of the model used) was made that genotypic frequencies are in Hardy-Weinberg form for these models. As we see below there is possibly no continuous-time model where this is the case, and there are some examples when it is not even approximately true or true at equilibrium. As noted earlier such a procedure can in theory lead to the "proof" of an absurdity, and we attempt here to set up from the start a well-defined model and to remain within the confines of the model during the entire analysis.

Consider a locus "A" in a monoecious population and let this locus admit alleles $A_1, \ldots, A_k$. At a given time let the number of A_iA_j individuals be n_{ij}, where we adopt an ordering notation such that the A_i gene has derived from the male parent. Define n_i by $n_i = \frac{1}{2} \Sigma(n_{ij} + n_{ji})$: then $2n_i$ is the number of A_i genes in the population. If $N = \Sigma n_i$ is the population size we may write

$$x_i = n_i/N, \quad X_{ij} = n_{ij}/N \tag{2.33}$$

as the frequencies of A_i and the (ordered) genotype A_iA_j, respectively. Consider a continuous-time deterministic process of population change in which, if terms of order $(\delta t)^2$ are ignored throughout, $NX_{ij}d_{ij}\delta t$ individuals of genotype A_iA_j die in the time interval $(t, t + \delta t)$. Let $M\delta t$ be the number of matings during this time interval, $X_{im,nj}$ be the fraction of these matings which are of the (ordered) type $A_iA_m \times A_nA_j$, and $\tilde{a}_{im,nj}$ the number of offspring from such a mating. We introduce the standardized parameter $a_{im,nj} = M\tilde{a}_{im,nj}/N$ so that $NX_{im,nj}a_{im,nj}\delta t$ is the number of offspring from all (ordered) $A_iA_m \times A_nA_j$ matings in the time interval $(t, t + \delta t)$. Defining $n_{ij}(t)$ as the number of A_iA_j individuals in the population at time t and noting that A_iA_j individuals can arise from various ordered matings in various frequencies, we get

$$n_{ij}(t + \delta t) = n_{ij}(t) + \delta t[\sum_{m,n} NX_{im,nj}a_{im,nj} - d_{ij}n_{ij}(t)]$$

or, letting $\delta t \to 0$ in the usual way,

$$\dot{n}_{ij} = \sum_{m,n} NX_{im,nj}a_{im,nj} - d_{ij}n_{ij}, \tag{2.34}$$

where the time derivative, here and below, is denoted by a superior dot. This equation and the verbal description leading to it form the basis of the model we will consider.

It is convenient to define a "birth-rate" for A_iA_m individuals. Noting that the number of offspring (of whatever genotype) to such individuals acting as "first" partner in an $A_iA_m \times A_nA_j$ mating during $(t, t + \delta t)$ is $N \sum_{n,j} X_{im,nj}a_{im,nj}\delta t$ and that the number of A_iA_m individuals available to act as parents is n_{im}, it is reasonable to define th birth-rate b_{im} for such individuals by the equation

$$n_{im}b_{im} = N \sum_{n,j} X_{im,nj}a_{im,nj}. \tag{2.35}$$

From this the fecundity b_i, mortality d_i, and "Malthusian parameter" m_i of the allele A_i are defined by

$$x_i b_i = \sum_j X_{ij} b_{ij}, \quad x_i d_i = \sum_j X_{ij} d_{ij}, \quad m_i = b_i - d_i \tag{2.36}$$

and the mean fecundity $\bar{b}$, mortality $\bar{d}$, and Malthusian parameter $\bar{m}$ are then given by

$$\bar{b} = \sum x_i b_i, \quad \bar{d} = \sum x_i d_i, \quad \bar{m} = \bar{b} - \bar{d}. \tag{2.37}$$

Equations (2.34)–(2.37) jointly yield

$$\dot{N} = \bar{m}N \tag{2.38}$$

and

$$\dot{X}_{ij} = \sum_{mn} X_{im,nj} a_{im,nj} - (d_{ij} + \bar{m})X_{ij}, \tag{2.39}$$

together with the "classical" equation (sometimes extrinsically assumed on the basis of no particular model)

$$\dot{x}_i = x_i(m_i - \bar{m}). \tag{2.40}$$

To make further progress it is necessary to make certain assumptions. We here assume first that random mating obtains, so that

$$X_{im,nj} = X_{im} X_{nj} \tag{2.41}$$

and that $a_{in,nj}$ can be expressed in the additive form

$$a_{im,nj} = \beta_{im} + \beta_{nj} \tag{2.42}$$

for some parameters β_{ij}. Equation (2.42) is the natural analogue for continuous-time models of an equation like (2.27) for discrete-time models. Equations (2.36)–(2.42) lead to

$$a_{im,nj} = \bar{b} + (b_{im} - \bar{b}) + (b_{nj} - \bar{b})$$

so that

$$\dot{X}_{ij} = x_i x_j(b_i + b_j - \bar{b}) - (d_{ij} + \bar{m})X_{ij}. \tag{2.43}$$

Perhaps the most important question to ask is whether Hardy-Weinberg frequencies hold in this model. Defining $Q_{ij} = X_{ij} - x_i x_j$ as a measure of departure from Hardy-Weinberg, Eqs. (2.40) and (2.43) yield

$$\dot{Q}_{ij} = x_i x_j(d_i + d_j - d_{ij} - \bar{d}) - (d_{ij} + \bar{m})Q_{ij}. \tag{2.44}$$

Suppose that $d_i + d_j - d_{ij} - \bar{d} \neq 0$. Then whether or not Hardy-Weinberg proportions exist initially Eq. (2.44) shows that they do not persist and do not even hold at an equilibrium of the system (2.34). One particular consequence of this is that, despite the frequent assertions of Fisher, the rate of change of mean fitness is *not* in general equal to the additive genetic variance in fitness. It is of some interest to determine the relationship between the two quantities and we now do this in the simple special case where the quantities $a_{im,nj}$ and d_{ij} (which are functions of the $X_{im,nj}$ and of time) are adjusted so that the Malthusian parameter m_{ij} $(= b_{ij} - d_{ij})$ of the genotype A_iA_j is constant in time.

To find the additive genetic variance we minimize

$$S = \Sigma \Sigma X_{ij}(m_{ij} - \bar{m} - \alpha_i - \alpha_j)^2 \tag{2.45}$$

subject to $\Sigma x_i \alpha_i = 0$. If Hardy-Weinberg holds, so that $X_{ij} = x_i x_j$, this would be done as in Appendix A. To measure the effect of departure from Hardy-Weinberg we introduce the parameters θ_{ij}, defined by

$$X_{ij} = x_i x_j \theta_{ij}. \tag{2.46}$$

Clearly $\theta_{ij} \equiv 1$ implies Hardy-Weinberg. Inserting (2.46) in (2.45), the minimization equations yield

$$x_i \alpha_i + \sum_j x_i x_j \theta_{ij} \alpha_j = \sum_j x_i x_j \theta_{ij}(m_{ij} - \bar{m}) \tag{2.47}$$

or

$$\alpha_i + \sum_j x_j \theta_{ij} \alpha_j = \sum_j x_j \theta_{ij} a_{ij} = a_i, \tag{2.48}$$

where we define

$$a_{ij} = m_{ij} - \bar{m}, \quad a_i = x_i^{-1} \sum_j X_{ij} a_{ij}. \tag{2.49}$$

Further, the additive genetic variance, being the sum of squares removed by this procedure, is

$$\sigma_A^2 = 2 \sum_i x_i a_i \alpha_i, \tag{2.50}$$

where a_i is defined explicity by (2.49) and α_i implicitly by (2.48). Note that in view of (2.40) this may also be written

$$\sigma_A^2 = 2 \sum_i \dot{x}_i \alpha_i. \tag{2.51}$$

We turn now to the rate of change of mean fitness $\bar{m}$. By definition

$$\bar{m} = \Sigma \Sigma m_{ij} X_{ij}$$

and since under our assumptions the m_{ij} are constant,

$$\dot{m} = \Sigma \Sigma m_{ij} \dot{X}_{ij}$$

$$= \Sigma \Sigma a_{ij} \dot{X}_{ij}$$

$$= \Sigma \Sigma \alpha_{ij}(\dot{x}_i x_j \theta_{ij} + x_i \dot{x}_j \theta_{ij} + x_i x_j \dot{\theta}_{ij})$$

$$= 2 \Sigma \Sigma a_{ij} \dot{x}_i x_j \theta_{ij} + \Sigma \Sigma a_{ij} x_i x_j \dot{\theta}_{ij} \qquad (2.52)$$

$$= 2 \Sigma_i \dot{x}_i \Sigma_j a_{ij} x_j \theta_{ij} + \Sigma \Sigma a_{ij} x_i x_j \dot{\theta}_{ij}$$

$$= 2 \Sigma_i \dot{x}_i \{\alpha_i + \Sigma_j x_j \theta_{ij} \alpha_j\} + \Sigma \Sigma a_{ij} x_i x_j \dot{\theta}_{ij}$$

$$= \sigma_A^2 + 2 \Sigma \Sigma \dot{x}_i x_{ij} \theta_j + \Sigma \Sigma a_{ij} x_i x_j \dot{\theta}_{ij}. \qquad (2.53)$$

We wish to simplify the final two terms in (2.53). Now

$$x_j = \Sigma_i X_{ij} = \Sigma_i x_i x_j \theta_{ij}$$

so that

$$\Sigma_i x_i \theta_{ij} \equiv 1.$$

Differentiating with respect to t,

$$\Sigma_i \dot{x}_i \theta_{ij} + \Sigma_i x_i \dot{\theta}_{ij} \equiv 0 \quad \text{for each } j.$$

Thus the second term in (2.53) can be written

$$- 2 \Sigma \Sigma x_i x_j \alpha_j \dot{\theta}_{ij} = - \Sigma \Sigma x_i x_j (\alpha_i + \alpha_j) \dot{\theta}_{ij}.$$

The final two terms in (2.53) thus become

$$\Sigma \Sigma (a_{ij} - \alpha_i - \alpha_j) x_i x_j \dot{\theta}_{ij} = \Sigma \Sigma \delta_{ij} X_{ij} (\dot{\theta}_{ij}/\theta_{ij}),$$

where $\delta_{ij} = a_{ij} - a_i - a_j$ is a measure of non-additivity in the Malthusian parameters m_{ij}. We conclude that

$$\dot{m} = \sigma_A^2 + \Sigma \Sigma X_{ij} \delta_{ij} d(\log \theta_{ij})/dt. \qquad (2.54)$$

Thus the rate of increase of mean fitness is equal to the additive genetic variance in general only if Hardy-Weinberg proportions hold (which, as we have seen in our model at least, they do not) or if the Malthusian parameter is additive ($m_{ij} = \alpha_i + \alpha_j$). A more general and more important conclusion, with m_{ij} no longer kept constant, is given by Kimura (1958).

How important then are departures from Hardy-Weinberg? In our model Eq. (2.44) shows that departures will be negligible after some time has passed if $d_i + d_j - d_{ij} - \bar{d} = 0$. But there is another circumstance under which departures will also be negligible. Suppose that the deviations $b_{ij} - \bar{b}$ and $d_{ij} - \bar{d}$ are all of order s, where s is a small parameter. Then Nagylaki (1976) has shown that the deviation Q_{ij} defined above changes in time (according to Eq. (2.44)) in such a way that after a small time period t_1 (an explicit formula for which is given), Q_{ij} differs from zero only by a term of order s even though at that time the gene frequencies themselves may be far from their equilibrium values. After time $2t$, the rate of change of Q_{ij} is of order s^2. When this occurs a state of "quasi-Hardy-Weinberg" (QHW) is said to obtain. In this case departures from Hardy-Weinberg may be trivial, and as a consequence the Fundamental Theorem should hold to an excellent approximation. More exactly, under the assumptions we have made, the term σ_A^2 in (2.54) is of order s^2, and when QHW obtains the final term is of order s^3. Thus the first term on the right-hand side will dominate the second, leading, as noted, to the essential accuracy of the Fundamental Theorem. The only exception to this rule occurs close to equilibrium points: since $\sigma_A^2 = 0$ at equilibrium it is possible that near equilibrium σ_A^2 is smaller than the final term in (2.54). This is probably of minor importance and during the period of substantial change in gene frequencies the Fundamental Theorem is effectively true.

2.8 A Remark on Average Effect and Average Excess

We consider in this section certain properties of discrete-time models where, for one reason or another, Hardy-Weinberg proportions do not hold.

Denote the frequency of the (ordered) genotype A_iA_j by $X_{ij}(= X_{ji})$ and assume that this genotype has (viability) fitness w_{ij}, so that the mean fitness $\bar{w}$ of the population is $\bar{w} = \Sigma\,\Sigma\, w_{ij}X_{ij}$. The additive genetic variance in fitness will be defined as in (2.50) if m_{ij} and $\bar{m}$ in (2.45)–(2.49) are replaced by w_{ij} and $\bar{w}$. More precisely, if we denote

$$b_i = x_i^{-1} \sum_j X_{ij}(w_{ij} - \bar{w}) \tag{2.55}$$

and $\beta_1 \ldots \beta_k$ as the solutions of the equations

$$\beta_i + \sum_j x_j \theta_{ij}\beta_j = b_i, \quad \Sigma\, x_i\beta_i = 0, \tag{2.56}$$

then the additive genetic variance in fitness is

$$\sigma_A^2 = 2 \sum_i x_i b_i \beta_i. \tag{2.57}$$

If we write

$$w_{ij} = \bar{w} + \beta_i + \beta_j + \epsilon_{ij}$$

it is easy enough to see that

$$\Sigma\,\Sigma\, X_{ij}\epsilon_{ij} = 0. \tag{2.58}$$

We turn now to the change in mean fitness. If the frequency of A_i at the birth of a given generation is $\sum_j X_{ij}$, at the birth of the next generation it will be $\sum_j X_{ij}w_{ij}/\overline{w}$. The change in frequency is thus

$$\Delta x_i = x_i b_i / \overline{w}. \tag{2.59}$$

Then

$$
\begin{aligned}
\Delta\overline{w} &= \sum\sum X'_{ij}w_{ij} - \overline{w} \\
&= \sum\sum X'_{ij}(\beta_i + \beta_j + \epsilon_{ij}) \\
&= 2\sum_i \beta_i x'_i + \sum\sum (X_{ij} + \Delta X_{ij})\epsilon_{ij} \\
&= 2\sum_i \beta_i(\Delta x_i) + \sum\sum (\Delta X_{ij})\epsilon_{ij} \quad [\text{Using (2.56) and (2.58)}] \\
&= \sigma_A^2/\overline{w} + \sum\sum (\Delta X_{ij})\epsilon_{ij}, \quad [\text{Using (2.57) and (2.59)}].
\end{aligned}
\tag{2.60}
$$

This equation has some parallels with (2.54). Our main interest in it, however, is that in deriving it we have generalized to k alleles the definition of average effect (β_i) and average excess (b_i) of the allele A_i, defined implicitly by (2.56) and explicitly by (2.55), respectively. Note that when $\theta_{ij} \equiv 1$, $b_i = \beta_i$. In the two allele case explicit solution of (2.56) gives

$$\beta_i = b_i x_1 x_2 / \{ X_{11} X_{12} + 2X_{11} X_{22} + X_{12}X_{22} \}, \quad (i = 1, 2), \tag{2.61}$$

so that b_i and β_i have the same sign and are zero or non-zero together. Fisher often described $\beta_2 - \beta_1$ as the average effect of replacing A_1 by A_2, but in the k allele case the definition of β_i simply as the average effect of A_i is rather more flexible.

As an example of the use of these expressions, consider the change under natural selection of some character determined by the locus under consideration. Let individuals of genotype $A_i A_j$ have measurement m_{ij} for this character and put

$$\overline{m} = \sum\sum m_{ij} X_{ij}.$$

Then

$$\Delta(\overline{m}) = \sum\sum m_{ij} \Delta(X_{ij}).$$

Suppose it happens that $m_{ij} = \overline{m} + \gamma_i + \gamma_j$, where γ_i is the average effect (in the character measured) of allele A_i. Then

$$
\begin{aligned}
\Delta(\overline{m}) &= \sum\sum (\gamma_i + \gamma_j)\Delta(X_{ij}) \\
&= 2\sum_i \gamma_i \sum_j \Delta(X_{ij}) \\
&= 2\sum_i \gamma_i \Delta(x_i) \\
&= 2\sum_i x_i \gamma_i b_i/\overline{w}, \quad \text{from (2.59)},
\end{aligned}
$$

and this may be defined as twice the covariance between the average effect of the character and the average excess in fitness. A more complex formula obtains when the representation $m_{ij} = \bar{m} + \gamma_i + \gamma_j$ is not possible, but we do not pursue the matter here (see, for example, Nagylaki and Crow, 1974).

We make a final remark on the Fundamental Theorem, taken to be the statement that the rate of increase of mean fitness is equal to the additive genetic variance in fitness. Over the years many purported general "proofs" of this theorem have been given, some claiming the status of an authorized version, many mutually contradictory. In the previous section we have shown that for continuous-time models the theorem is exactly true only in certain special cases. For the discrete-time models we have so far considered the increase in mean fitness is not generally identical to the additive genetic variance. We show in the next section that for discrete-time models involving two loci the mean fitness can decrease. The value of such "proofs" is thus somewhat in doubt. A more fruitful and biologically useful approach, initiated by Nagylaki (1974c), is to introduce concepts such as QHW (discussed above) or quasi-linkage equilibrium (QLE, Kimura, (1965)), together possibly with numerical simulations, to delimit the circumstances under which the theorem is approximately true. It is indeed possible that population genetics theory generally will soon emphasize such approximative and sometimes qualitative approaches rather than more formal mathematical methods.

2.9 Two Loci

So far in this chapter we have assumed that the fitness of any individual depends on his genetic constitution at a single locus. This is of course only an initial simplification: we have already noted in Chap. 1 that for some questions (for example, the evolution of recombination rate) a more complicated theory is required. We now introduce briefly the case where fitness depends on the genetic constitution at two loci, deferring a more complete treatment to Chap. 6. Although such a "two-locus" theory may often be little more realistic than "single-locus" theory, it does allow at least two advances to be made. First, some assessment can be made of the accuracy of approximating two-locus behavior and measurements by combining two single-locus results. Second, no assessment of the evolutionary importance of linkage between loci can be made without at least a two-locus analysis.

For convenience we assume viability selection only, random mating and a discrete-time parameter. Consider two loci "A" and "B" at which occur alleles A_1, A_2 and B_1, B_2, respectively, and let the recombination fraction between the loci be $R(0 \leqslant R \leqslant 0.5)$. It is convenient conceptually to suppose that these loci are on the same chromosome: the unlinked case ($R = 0.5$) may be treated by imagining the distance along the chromosome between the two loci to be so long that the recombination fraction between them is 0.5. We then use the words gamete and chromosome interchangeably in what follows.

It is possible to write down recurrence relations connecting the (ten) zygotic frequencies (of A_1B_1/A_1B_1, A_1B_2/A_1B_1, ..., A_2B_2/A_2B_2). These relations show that a simpler set of recurrence relations can be found for the frequencies of the four gametes

A_1B_1, A_1B_2, A_2B_1 and A_2B_2. (We will sometimes call these gametes 1, 2, 3, 4 respectively.) This simplification arises through the concept of the random union of gametes and is parallel to treating gene frequencies rather than genotypic frequencies at a single locus. To illustrate this we consider first the case of no selection. The gametes forming the zygotes of any generation may be thought of as being drawn randomly from a pool containing gametes of type 1–4 in certain proportions. These gametes will not, however, be subsequently passed on in the same proportions since, for example, there will be a decrease in the frequency of A_1B_1 gametes through recombination in A_1B_1/A_2B_2 individuals which might not be exactly counterbalanced by an increase through recombination in A_1B_2/A_2B_1 individuals. Writing the frequency of gamete i as c_i ($i = 1, \ldots, 4$), these arguments show that the frequencies c_i' in the next generation are

$$
\begin{aligned}
c_1' &= c_1 + R(c_2c_3 - c_1c_4), \\
c_2' &= c_2 - R(c_2c_3 - c_1c_4), \\
c_3' &= c_3 - R(c_2c_3 - c_1c_4), \\
c_4' &= c_4 + R(c_2c_3 - c_2c_4),
\end{aligned}
\tag{2.62}
$$

or more economically

$$
c_i' = c_i + \eta_i R(c_2c_3 - c_1c_4),
\tag{2.63}
$$

where

$$
\eta_1 = \eta_4 = 1, \quad \eta_2 = \eta_3 = -1.
\tag{2.64}
$$

Several conclusions can be drawn immediately from these equations. First, since $c_1' + c_2' = c_1 + c_2$ and $c_1' + c_3' = c_1 + c_3$ there is no change in the frequencies of A_1 and B_1, confirming the one-locus analysis of Chap. 1. Second, elementary algebra shows that

$$
c_1'c_4' - c_2'c_3' = (1 - R)(c_1c_4 - c_2c_3)
\tag{2.65}
$$

so that if $R > 0$ (as we reasonably assume)

$$
c_1(t)c_4(t) - c_2(t)c_3(t) \to 0 \quad \text{as} \quad t \to \infty.
\tag{2.66}
$$

It follows that, under the assumptions (in particular that of no selection) we have made, if the population has evolved for some time we may reasonably assume

$$
c_1c_4 - c_2c_3 = 0.
\tag{2.67}
$$

It is of considerable interest to establish what this equation means in genetical terms. A little algebra shows that (2.67) is equivalent to

$$
\text{freq}(A_iB_j) = \text{freq}(A_i) \times \text{freq}(B_j)
\tag{2.68}
$$

for all i, j. When Eq. (2.68), or equivalently (2.67), holds the population is said to be in a state of linkage equilibrium with respect to these loci. The quantity $c_1 c_4 - c_2 c_3$, which we denote by D, is often called the "coefficient of linkage disequilibrium" although, as we see in a moment, this can be a rather misleading expression for this quantity.

We turn now to the case where selective differences between genotypes exist. In the previous chapter we used a fitness display, such as (1.87), which focusses attention on the genotypes at each of the two loci. For theoretical purposes, however, it is usually more convenient to adopt a notation focussed around the two gametes making up each individual: (we have just noted that gametic frequencies are the most natural vehicle for studying evolutionary behavior in two-locus systems under random mating). We thus adopt the fitness scheme shown in (2.69) below:

	$A_1 B_1$	$A_1 B_2$	$A_2 B_1$	$A_2 B_2$	
$A_1 B_1$	w_{11}	w_{12}	w_{13}	w_{14}	
$A_1 B_2$	w_{21}	w_{22}	w_{23}	w_{24}	(2.69)
$A_2 B_1$	w_{31}	w_{32}	w_{33}	w_{34}	
$A_2 B_2$	w_{41}	w_{42}	w_{43}	w_{44}	

Thus the fitness of zygotes made up of gametes i and j is w_{ij} (which we assume equal to w_{ji}). If coupling and repulsion double heterozygotes are assumed to have the same fitness then also $w_{23} = w_{14}$. We make this assumption throughout. If, for specific purposes, we wish to adopt a fitness display emphasizing single-locus genotypes, (2.69) becomes

	$B_1 B_1$	$B_1 B_2$	$B_2 B_2$	
$A_1 A_1$	w_{11}	w_{12}	w_{22}	
$A_1 A_2$	w_{13}	w_{14}	w_{24}	(2.70)
$A_2 A_2$	w_{33}	w_{34}	w_{44}	

The marginal fitness w_i of gamete i may be defined by

$$w_i = \sum_j c_j w_{ij}, \tag{2.71}$$

and the mean fitness $\bar{w}$ of the population then becomes

$$\bar{w} = \sum \sum c_i c_j w_{ij} = \sum c_i w_i. \tag{2.72}$$

Consideration of all possible matings, their frequencies and genetic outputs, as well as the fitnesses of the various genotypes, shows that the gametic frequencies c_i' in the following generation are

$$c_i' = \bar{w}^{-1}[c_i w_i + \eta_i R w_{14}(c_2 c_3 - c_1 c_4)], \quad i = 1, 2, 3, 4. \tag{2.73}$$

Here η_i is defined in (2.64) and if the w_{ij} are all equal, these recurrence relations reduce to (2.63). These important equations are due in this form to Lewontin and Kojima (1960) but were essentially derived earlier, for a continuous-time model, by Kimura (1956b). Our present aim is to discuss some of the more immediate consequences of these equations.

First, the mean fitness, as defined in (2.72), is similar in form to the definition (2.10) with $k = 4$. It follows from the discussion in Sect. 2.4 that if we assume that mean fitness is maximized at a unique internal ($c_i > 0$) point, than at this point $w_i = \bar{w}$, where now w_i and $\bar{w}$ are defined by (2.71) and (2.72). What is the connection between this point and equilibrium points of the system (2.73)? The equations $c_i' = c_i$ show that this system is in equilibrium when

$$\bar{w} = w_i + c_i^{-1}\eta_i R\, w_{14}(c_2 c_3 - c_1 c_4), \quad i = 1 \ldots 4. \tag{2.74}$$

Unless linkage equilibrium holds at the equilibrium point the latter then cannot be a point of maximum fitness. We show later that linkage equilibrium holds at equilibrium only in special cases: it follows that mean fitness can decrease in the system (2.73). The Fundamental Theorem cannot then be true in general in two-locus selection systems. We now demonstrate this by a numerical example. Suppose, using the notation (2.70), that the fitness scheme is

	$B_1 B_1$	$B_1 B_2$	$B_2 B_2$	
$A_1 A_1$	1.000	1.024	1.021	
$A_1 A_2$	1.025	1.066	1.026	(2.75)
$A_2 A_2$	1.018	1.019	1.007	

and let $R = \frac{1}{2}$ (A and B loci unlinked). If initially

$$c_1 = 0.168, \quad c_2 = 0.362, \quad c_3 = 0.292, \quad c_4 = 0.178 \tag{2.76}$$

the population mean fitness is 1.033106. The mean fitness now *decreases* for about 14 generations and after that steadily increases, reaching a value of 1.031212 at the equilibrium point

$$c_1 = 0.24136, \quad c_2 = 0.28164, \quad c_3 = 0.22192, \quad c_4 = 0.25508. \tag{2.77}$$

The net effect of the evolution of the population from the starting point (2.76) to the equilibrium point (2.77) is to decrease mean fitness by 0.001894. Note that at the equilibrium point $D = c_1 c_4 - c_2 c_3 = -0.000935$.

Second, the equilibrium point (or points) of (2.73) will depend on the recombination fraction R between the loci if linkage equilibrium does not obtain at equilibrium. Thus various values of R can be considered and the equilibrium mean fitnesses computed for each. When $R = 0$ the "equilibrium" Eq. (2.74) and the "maximization" equation $\bar{w} = w_i$ ($i = 1, \ldots, 4$) agree so that if each $c_i > 0$ at equilibrium, the value of R for which the greatest equilibrium mean fitness is achieved is for $R = 0$. This conclusion remains true if some of the c_i are zero at equilibrium but strangely, as we see later, it is not necessarily true that equilibrium mean fitness decreases as R increases. To the extent that equilibrium mean fitness is maximized for extremely tight linkage the argument of Fisher given in Chapt. 1 concerning the evolution of tight linkage between epistatic loci is justified. Note, however, that this argument can be made only when $D \neq 0$ at equilibrium for all R values: if $D = 0$ at equilibrium for all R the equilibrium mean fitness is independent of R. It is perhaps paradoxical that Fisher's argument on linkage modification relies on non-zero values of D, while it is precisely when D is non-zero that the Fundamental Theorem cannot remain valid.

The third topic we treat (at rather greater length) concerns the additive genetic variance in fitness. We are particularly interested in the relationship between this and the two marginal single-locus values, and we begin by defining the latter. Using the fitness scheme (2.70), we may define the marginal fitnesses of the various single-locus genotypes as follows:

Genotype	Frequency	Marginal fitness
A_1A_1	$(c_1 + c_2)^2$	$(w_{11}c_1^2 + 2w_{12}c_1c_2 + w_{22}c_2^2)/(c_1 + c_2)^2 = u_{11}$
A_1A_2	$2(c_1 + x_2)(c_3 + c_4)$	$(w_{13}c_1c_3 + w_{14}c_1c_4 + w_{14}c_2c_3 + w_{24}c_2c_4)/$ $(c_1 + c_2)(c_3 + c_4) = u_{12}$
A_2A_2	$(c_3 + c_4)^2$	$(w_{33}c_3^2 + 2w_{34}c_3c_4 + w_{44}c_4^2)/(c_3 + c_4)^2 = u_{22}$

$$(2.78)$$

B_1B_1	$(c_1 + c_3)^2$	$(w_{11}c_1^2 + 2w_{13}c_1c_3 + w_{33}c_3^2)/(c_1 + c_3)^2 = v_{11}$
B_1B_2	$2(c_1 + c_3)^2(c_2 + c_4)$	$(w_{12}c_1c_2 + w_{14}c_1c_4 + w_{14}c_2c_3 + w_{34}c_3c_4)/$ $(c_1 + c_3)(c_2 + c_4) = v_{12}$
B_2B_2	$(c_2 + c_4)^2$	$(w_{22}c_2^2 + 2w_{24}c_2c_4 + w_{44}c_4^2)/(c_2 + c_4)^2 = v_{22}$

From Eq. (1.41) the marginal additive genetic variance at the A locus may be defined as

$$2(c_1 + c_2)(c_3 + c_4)G_A^2, \tag{2.79}$$

where

$$G_A = u_{11}(c_1 + c_2) + u_{12}(1 - 2c_1 - 2c_2) - u_{22}(c_3 + c_4), \tag{2.80}$$

and similarly at the B locus the additive genetic variance is

$$2(c_1 + c_3)(c_2 + c_4)G_B^2, \tag{2.81}$$

where

$$G_B = v_{11}(c_1 + c_3) + v_{12}(1 - 2c_1 - 2c_3) - v_{22}(c_2 + c_4). \tag{2.82}$$

We now find the two-locus additive genetic variance. To do this we assign additive parameters α_1 and α_2 to A_1 and A_2 and β_1 and β_2 to B_1 and B_2 and then minimize the expression

$$S = c_1^2(w_{11} - \bar{w} - 2\alpha_1 - 2\beta_1)^2 + 2c_1c_2(w_{12} - \bar{w} - 2\alpha_1 - \beta_1 - \beta_2)^2$$
$$+ \ldots + c_4^2(w_{44} - \bar{w} - 2\alpha_2 - 2\beta_2)^2$$

subject to the constraints

$$(c_1 + c_2)\alpha_1 + (c_3 + c_4)\alpha_2 = 0, \quad (c_1 + c_3)\beta_1 + (c_2 + c_4)\beta_2 = 0. \tag{2.83}$$

Details of this minimization procedure are given by Kojima and Kelleher (1961) and Kimura (1965) and are not pursued here. It turns out that the additive genetic variance can be written

$$2\{(c_1 + c_2)(c_3 + c_4)H_A^2 + 2H_A H_B D + (c_1 + c_3)(c_2 + c_4)H_B^2\} \tag{2.84}$$

where H_A and H_B are the solutions of the equations

$$H_A + \{c_1 + c_2)(c_3 + c_4)\}^{-1} D H_B = G_A,$$
$$H_B + \{(c_1 + c_3)(c_2 + c_4)\}^{-1} D H_A = G_B, \tag{2.85}$$

G_A and G_B being given by (2.80) and (2.82).

Several interesting conclusions follow from these equations. Perhaps the most important is that if $D = 0$ (i.e. linkage equilibrium between the two loci) then $H_A = G_A$, $H_B = G_B$, and the true two-locus additive genetic variance is the sum of the two single-locus marginal values. When $D \neq 0$ this is no longer true, and there is no simple relationship between this sum and the true value. This is an important conclusion since it seems to be widely assumed in the classical literature (see for example Fisher, 1918, p. 405; 1958, p. 37; Wright, 1969, p. 439) that in a multi-locus system the true additive genetic variance can be found by simply summing single-locus marginal values. Since we have shown above that changes in mean fitness can be negative in two-locus systems (and thus cannot be equal to any form of genetic variance) it follows that

$$\Delta\bar{w}, \quad \sigma_A^2 \text{ (two-locus)}, \quad \Sigma \sigma_A^2 \text{ (single-locus marginals)} \tag{2.86}$$

have in general no clear and obvious connection with each other. This conclusion is generalized later in this book (Sect. 7. 4).

These conclusions may also be associated with properties of changes in gene frequency. Equations (2.79), (2.81), and (2.84) show that

$$\sigma_A^2 \text{ (two-locus)} - \Sigma \sigma_A^2 \text{ (single-locus marginals)} = 2D(G_A H_B + H_A G_B), \tag{2.87}$$

and if D is small this may be approximated by $-4D G_A G_B$.

Now since

$$\Delta(\text{frequency } A_1) = (c_1 + c_2)(c_3 + c_4)G_A/\bar{w}$$

with a corresponding expression for $\Delta(\text{frequency } B)$, it is found, if terms of order D^2 are ignored, that the left-hand side in (2.87) may be written

$$\frac{-4D\,\bar{w}^2\,\Delta(\text{frequency } A_1)\,\Delta(\text{frequency } B_1)}{(c_1 + c_2)(c_3 + c_4)(c_1 + c_3)(c_2 + c_4)}.$$

This gives an interesting relationship between the various additive genetic variances, the linkage disequilibrium, and the gene frequency changes in a two-locus system. Note also that if in a certain generation $\Delta(\text{frequency } A_1) = 0$ then $G_A = 0$ to the order of accuracy we use, and the total additive genetic variance is simply the marginal B locus value. Note, however, that this is true only as an approximation and that, more precisely, if there is linkage disequilibrium between A and B loci there is a small perturbation from the A locus to the total additive variance, even though gene frequencies are not changing at that locus.

We expect the additive genetic variance to be of importance in discussing the correlation between relatives. Before exploring this, we recall that gene frequencies alone are not sufficient to describe the evolution of two-locus systems so that it is plausible that the additive genetic variance (which involves gene frequencies) is not the appropriate component of variance for evolutionary considerations. We thus consider a variance defined by gamete frequencies which, since gamete frequencies do describe the evolutionary behavior, is possibly of greater evolutionary significance.

The marginal fitnesses w_i of the four gametes have been defined in Eq. (2.71). The total chromosomal (or gametic) variance in fitness, denoted σ_G^2, may be defined by

$$\sigma_G^2 = 2 \sum_{i=1}^{4} (w_i - \bar{w})^2 c_i, \tag{2.88}$$

(the factor 2 being inserted because there are two gametes per zygote). Suppose now we attempt to fit the marginal gametic fitnesses by additive components depending on the genes on each gamete. This is done by minimizing

$$c_1(w_1 - \bar{w} - \alpha_1 - \beta_1)^2 + c_2(w_2 - \bar{w} - \alpha_1 - \beta_2)^2 + c_3(w_3 - \bar{w} - \alpha_2 - \beta_1)^2$$
$$+ c_4(w_4 - \bar{w} - \alpha_2 - \beta_2)^2$$

with respect to $\alpha_1, \alpha_2, \beta_1$ and β_2, subject to (2.83). The sum of squares so removed may be described as being due to the additive effects of genes in gametes and for short may be called the additive gametic variance. It is found (see Kimura, 1965) that this is identical to the additive genetic variance (2.84) and thus the latter, perhaps unexpectedly, is of use in evolutionary and other considerations. This conclusion is generalized later (Sect. 7.4). The total gametic variance (2.88) has three degrees of freedom and the additive component, two. The remaining degree of freedom is taken up by the epistatic gametic variance σ_{EG}^2 which turns out to be defined by

$$\sigma_{EG}^2 = 2(w_1 - w_2 - w_3 + w_4)^2/(c_1^{-1} + c_2^{-1} + c_3^{-1} + c_4^{-1}). \qquad (2.89)$$

This is zero if and only if an additive genetic fitness scheme exactly fits the marginal gametic fitnesses.

We turn now to the correlation between relatives, restricting attention to the case where (2.68) holds, i.e. that the two loci are in linkage equilibrium. (This assumption was also made by Fisher, 1918). We consider both linked and unlinked loci: note that Fisher's 1918 analysis is concerned only with the unlinked case. Our treatment is based on Cockerham (1954, 1956) and Kempthorne (1954).

We first isolate various components of the total variance of the character measured. Suppose that the measurements for the various genotypes are

	B_1B_1	B_1B_2	B_2B_2	
A_1A_1	m_{11}	m_{12}	m_{13}	
A_1A_2	m_{21}	m_{22}	m_{23}	(2.90)
A_2A_2	m_{31}	m_{32}	m_{33}	

We form these measurements into a single vector $\mathbf{m} = (m_{11}, m_{12}, ..., m_{33})'$. If the frequency of A_1 is x and of B_1 is y, then since linkage equilibrium is assumed the frequency of $A_1A_1B_1B_1$ is x^2y^2, of $A_1A_1B_1B_2$ is $2x^2y(1-y)$ and so on. It is convenient to write these frequencies as the entries in a diagonal matrix F, so that

$$F = \begin{pmatrix} x^2y^2 & & & \\ & 2x^2y(1-y) & & 0 \\ & & \ddots & \\ & & & \ddots \\ 0 & & (1-x)^2(1-y)^2 \end{pmatrix}. \qquad (2.91)$$

Evidently the mean value $\bar{m}$ in the measurement is

$$\bar{m} = x^2y^2m_{11} + 2x^2y(1-y)m_{12} + ... + (1-x)^2(1-y)^2m_{33}, \qquad (2.92)$$

and further, adopting the notation of (2.78), the marginal means of A_1A_1, A_1A_2 and A_2A_2 are

$$u_{11} = y^2m_{11} + 2y(1-y)m_{12} + (1-y)^2m_{13},$$
$$u_{12} = y^2m_{21} + 2y(1-y)m_{22} + (1-y)^2m_{23}, \qquad (2.93)$$
$$u_{22} = y^2m_{31} + 2y(1-y)m_{32} + (1-y)^2m_{33}.$$

Similarly the marginal means at the B locus are

$$
\begin{aligned}
v_{11} &= x^2 m_{11} + 2x(1-x)m_{21} + (1-x)^2 m_{31}, \\
v_{12} &= x^2 m_{12} + 2x(1-x)m_{22} + (1-x)^2 m_{32}, \\
v_{22} &= x^2 m_{13} + 2x(1-x)m_{23} + (1-x)^2 m_{33}.
\end{aligned}
\tag{2.94}
$$

Finally the total variance σ^2 in the character measured is

$$
\sigma^2 = x^2 y^2 m_{11}^2 + \dots + (1-x)^2 (1-y)^2 m_{33}^2 - \bar{m}^2 = \mathbf{m}'F\mathbf{m} - \bar{m}^2.
\tag{2.95}
$$

This total variance has eight degrees of freedom, and our aim is to break it down into the sum of eight components, each of genetical significance. These components will measure the additive effects at each of the two loci, two dominance effects and the four interactions between these.

Suppose a matrix T exists such that $TFT' = I$ (or equivalently $(T')^{-1}F^{-1}T^{-1} = I$), where I is the unit 9 x 9 matrix, and define a vector $\mathbf{z}$ by $\mathbf{z} = TF\mathbf{m}$. Then

$$
\begin{aligned}
\mathbf{m}'F\mathbf{m} &= \mathbf{z}'(T')^{-1}F^{-1}FF^{-1}T^{-1}\mathbf{z} \\
&= \mathbf{z}'\mathbf{z} \\
&= z_1^2 + z_2^2 + \dots + z_9^2
\end{aligned}
\tag{2.96}
$$

and if the last row in T can be chosen to be $(1, 1, \dots, 1)$ then $z_9 = \bar{m}$ and

$$
\sigma^2 = z_1^2 + z_2^2 + \dots + z_8^2.
\tag{2.97}
$$

The equation $TFT' = I$ reduces to the requirement

$$
x^2 y^2 t_{i1} t_{j1} + 2x^2 y(1-y)t_{i2} t_{j2} + \dots + (1-x)^2(1-y)^2 t_{i9} t_{j9} = \delta_{ij}
\tag{2.98}
$$

where $\delta_{ij} = 1$ if $i = j$ and $\delta_{ij} = 0$ otherwise. The choice $t_{91} = t_{92} = \dots = t_{99} = 1$ does satisfy (2.98) with $i = j = 9$. Thus σ^2 can indeed be broken down into the sum (2.97), where

$$
z_i = x^2 y^2 t_{i1} m_{11} + 2x^2 y(1-y)t_{i2} m_{12} + \dots + (1-x)^2(1-y)^2 t_{i9} m_{33},
\tag{2.99}
$$

provided that the t_{ij} satisfy (2.98) and the further requirement

$$
x^2 y^2 t_{i1} + 2x^2 y(1-y)t_{i2} + \dots + (1-x)^2(1-y)^2 t_{i9} = 0, \quad i = 1 \dots 8.
\tag{2.100}
$$

Apart from these purely mathematical requirements we wish to choose the t_{ij} so that the z_i have the genetical interpretations described above.

Suppose z_1^2 and z_2^2 are to represent the additive and dominance components of the character from the A locus. Recalling equations (1.9) and using the marginal fitness values (2.93) we would like to have

$$z_1^2 = 2x(1-x)\{xu_{11} + (1-2x)u_{12} - (1-x)u_{22}\}^2$$
$$z_2^2 = x^2(1-x)^2\{2u_{12} - u_{11} - u_{22}\}^2. \tag{2.101}$$

Such a representation is in fact possible if in (2.99) we choose

$$t_{11} = t_{12} = t_{13} = x^{-1}\{2x(1-x)\}^{1/2}$$
$$t_{14} = t_{15} = t_{16} = (1-2x)\{2x(1-x)\}^{-1/2} \tag{2.102}$$
$$t_{17} = t_{18} = t_{19} = -(1-x)^{-1}\{2x(1-x)\}^{1/2}$$

and

$$t_{21} = t_{22} = t_{23} = -x^{-1}(1-x)$$
$$t_{24} = t_{25} = t_{26} = 1 \tag{2.103}$$
$$t_{27} = t_{28} = t_{29} = (1-x)^{-1}x.$$

These choices do satisfy the requirements (2.98) and (2.100), and thus our desired representation (2.101) is allowable. A parallel procedure for the B locus gives additive and dominance B locus contributions as

$$z_3^2 = 2y(1-y)\{yv_{11} + (1-2y)v_{12} - (1-y)v_{22}\}^2 \tag{2.103}$$

and

$$z_4^2 = y^2(1-y)^2\{2v_{12} - v_{11} - v_{22}\}^2,$$

and once more, with the choice of the t_{ij} implicit in these definitions, the orthogonality conditions are met. If z_5^2 is to represent the additive-by-additive component of the total variance it would be natural to choose $t_{5i} = t_{1i} \times t_{3i}$, and the remaining three interactive components would naturally be chosen by similar multiplications. If this is done it is found that all the orthogonality conditions are met, and this also implies that the representation (2.97) is completed. We do not go into details here and note only that the various components can be expressed as

(add x add): $z_5^2 = 4xy(1-x)(1-y)\{xye_{11} + x(1-y)e_{12} + (1-x)ye_{21}$
$\qquad\qquad + (1-x)(1-y)e_{22}\}^2,$ (2.104)

(add x dom): $z_6^2 = 2x(1-x)y^2(1-y)^2\{x(e_{11} - e_{12}) + (1-x)(e_{21} - e_{22})\}^2,$

(dom x add): $z_7^2 = 2x^2(1-x)^2y(1-y)\{y(e_{11} - e_{21}) + (1-y)(e_{12} - e_{22})\}^2,$

(dom x dom): $z_8^2 = x^2y^2(1-x)^2(1-y)^2\{e_{11} - e_{12} - e_{21} + e_{22}\}^2,$

where

$$e_{11} = m_{11} - m_{12} - m_{21} + m_{22},$$

$$e_{12} = m_{12} - m_{13} - m_{22} + m_{23},$$

$$e_{21} = m_{21} - m_{22} - m_{31} + m_{32},$$

$$e_{22} = m_{22} - m_{23} - m_{32} + m_{33}.$$

These expressions, given more generally to include the effect of inbreeding, were derived by Cockerham (1954). It is sometimes convenient to write

$$\sigma_A^2 = z_1^2 + z_3^2, \quad \sigma_D^2 = z_2^2 + z_4^2, \quad \sigma_{AA}^2 = z_5^2, \quad \sigma_{AD}^2 = z_6^2 + z_7^2, \quad \sigma_{DD}^2 = z_8^2,$$

so that

$$\sigma^2 = \sigma_A^2 + \sigma_D^2 + \sigma_{AA}^2 + \sigma_{AD}^2 + \sigma_{DD}^2. \tag{2.105}$$

A slightly shorter representation collects the final three terms as a single term σ_E^2 (epistatic variance), but for our purposes this is not useful as the final three terms in (2.105) are involved differently in the correlation between relatives and are therefore best kept separate.

Consider now the father-son and the brother-brother correlations in the measurement. It is possible to write down all 81 father-son genotypic combinations and, using a table extending Table 1.1, arrive at a father-son covariance. By doing this and a parallel procedure for full brothers it is found that if the A and B loci are unlinked,

$$\text{corr (father-son)} = (\tfrac{1}{2}\sigma_A^2 + \tfrac{1}{4}\sigma_{AA}^2)/\sigma^2, \tag{2.105a}$$

$$\text{corr (full brothers)} = (\tfrac{1}{2}\sigma_A^2 + \tfrac{1}{4}\sigma_D^2 + \tfrac{1}{4}\sigma_{AA}^2 + \tfrac{1}{8}\sigma_{AD}^2 + \tfrac{1}{16}\sigma_{DD}^2)/\sigma^2. \tag{2.105b}$$

When the two loci are linked, Cockerham (1956) has shown that the former expression remains unchanged but that the latter must be replaced by

$$\text{corr (full brothers)} = \{\tfrac{1}{2}\sigma_A^2 + \tfrac{1}{4}\sigma_D^2 + \tfrac{1}{8}(3 - 4R + 4R^2)\sigma_{AA}^2$$

$$+ \tfrac{1}{4}(1 - 2R + 2R^2)\sigma_{AD}^2 + \tfrac{1}{4}(1 - 2R + 2R^2)^2\sigma_{DD}^2\}/\sigma^2. \tag{2.106}$$

Note that the effect of linkage is always to increase the full brother correlation. We will rederive these formulae later (Chap. 7) as particular cases of correlations where the trait in question depends on an arbitrary number of loci, using a more efficient approach. The analysis in this section has assumed a discrete-time model, and it is expected that qualitatively similar conclusions would hold for a continuous model. One possible complication for such models does, however, occur. In the discrete models the frequency of any genotype is found immediately from the frequencies of the gametes making up this genotype, so that, for example,

$$\text{freq } (A_1A_1B_1B_2) = 2 \text{ freq } (A_1B_1) \text{ freq } (A_1B_2). \tag{2.107}$$

In the continuous-time model of Nagylaki and Crow (1974) the existence of linkage disequilibrium between the two loci implies that "Hardy-Weinberg" equations such as (2.107) are no longer true. This is of some interest since many theoretical analyses of continuous-time two-locus models have assumed the truth of equations like (2.107). However, Nagylaki (1976) has shown that when fitness differentials are small a state of "quasi-Hardy-Weinberg" soon emerges when genotypic frequencies can, to a very close approximation, be found from the constituent gametic frequencies.

2.10 Genetic Loads

A genetic load may be said to arise if the population mean fitness is less than that of some optimal value which in some idealized sense it could take. Thus in Sect. 1.4 the concept of a mutational load was introduced: with the fitness scheme (1.24b) the equilibrium mean fitness in an idealized case without mutation is $1 + s$, while with mutation it is $1 + s - 2u$. We then say that there is a mutational load of $2u$. (Of course, while this load represents a real hardship for a population because it partly represents, for example, genetically handicapped individuals, it is necessary for long-term evolutionary prospects that such a load exist.) A second form of load arises through recombination. Thus, for example, in the model (1.88) the mean fitness decreases as the recombination rate increases. As with the mutational load there are long-term reasons why a recombinational load is desirable.

The two forms of genetic load we now consider in detail are the substitutional load and the segregational load, where in both cases we define the load l by

$$l = (w_{\max} - \bar{w})/\bar{w}, \tag{2.108}$$

where $w_{\max}$ is the fitness of the most fit genotype and $\bar{w}$ is the mean fitness. If the latter is close to unity we approximate (2.108) by

$$l \approx w_{\max} - \bar{w}. \tag{2.109}$$

It is easier to describe the segregational load, and the principles can be largely carried over for the substitutional load. Consider then two alleles A_1 and A_2 at a locus A for which the fitness scheme (1.24c) applies with $s_1, s_2 > 0$. The population mean fitness at the stable internal equilibrium is $1 - s_1 s_2/(s_1 + s_2)$, and if $s_1 = s_2 = s$ this is $1 - \frac{1}{2}s$. Using the approximation (2.109), the segregational load is $\frac{1}{2} s$. This load arises because, through genetic segregation, it is impossible that all individuals be of the optimal genotype A_1A_2.

What happens if fitness depends on two loci? Perhaps the simplest case of two-locus fitnesses arises when the scheme (2.70) is in the multiplicative form

	B_1B_1	B_1B_2	B_2B_2
A_1A_1	$(1-s)^2$	$1-s$	$(1-s)^2$
A_1A_2	$1-s$	1	$1-s$
A_2A_2	$(1-s)^2$	$1-s$	$(1-s)^2$

where we assume $s > 0$. We will see later that when $R > \frac{1}{4} s^2$ the equilibrium mean fitness is $(1 - \frac{1}{2} s)^2$ as a straightforward multiplication of single-locus values would suggest. When $R < \frac{1}{4} s^2$ the picture is a little more complicated: here the equilibrium mean fitness is $(1 - \frac{1}{2} s)^2 + \frac{1}{4} s^2 - R$ and this is slightly larger than $(1 - \frac{1}{2} s)^2$. In both cases the segregational load is approximately s and is thus approximately twice the single-locus value. One might expect that a natural continuation of this scheme to n loci would produce a load of $(1 - \frac{1}{2}s)^{-n} - 1$, at least for loose linkage and that this load can be substantial for large values of n. This argument was once used as an argument in favor of the "neutral" theory, and we consider it at greater length when discussing this theory in Chap. 9.

For fitness schemes more complicated than the above these simple arguments are no longer satisfactory. Consider for example the fitness scheme (2.75). When the recombination fraction R is positive there will be a recombinational load as well as a segregational load. To consider the segregational load more or less in isolation we take a small value of R, say $R = 0.001$. Numerical computation shows that for this value there are two stable equilibria of gametic frequencies, namely (2.77) and

$$c_1 = 0.013, \quad c_2 = 0.469, \quad c_3 = 0.503, \quad c_4 = 0.015. \tag{2.110}$$

To illustrate the point at issue we can consider either one of these. At the point (2.110) the mean fitness is 1.0417 and thus the genetic load (as defined by 2.108) is 0.0233. Suppose now marginal fitness values are found from (2.78) and the load calculated according to (2.108) using these marginal values and the marginal genotypic frequencies. The loads so calculated are 0.0212 for the A locus and 0.0210 for the B locus. The sum of these is almost twice that of the true load (for $R = 0$ it would be exactly twice). Evidently for general fitness schemes some caution is necessary in extending single-locus segregational load arguments to many loci, a point we consider in more detail later.

It is interesting to note that it is possible to devise frequency-dependent selection schemes for which there is no segregational load at a stable equilibrium. Thus in the fitness scheme

A_1A_1	A_1A_2	A_2A_2	
$1 + a(1-2x)$	1	$1 - a(1-2x)$	(2.111)

where x is the frequency of A_1 and a is a small parameter, the point $x = 0.5$ is a point of stable equilibrium, and at this point all genotypes have equal fitness and there is no genetic load.

We turn now to the substitutional load (sometimes called the "evolutionary load" or the "cost of natural selection"). Let fitnesses be of the form (1.24b) with, for convenience, $h = 0.5$. When the frequency of A_1 is x the mean fitness is $1 + sx$ and l, using the approximation (2.109), is $s(1 - x)$. The substitutional load L is defined as the sum of this quantity during the process when x increases from a small value x_1 (at time t_1) to a value x_2 close to unity (at time t_2). Thus

$$L = \Sigma s(1 - x)$$

$$\approx \int_{t_1}^{t_2} s(1 - x)dt$$

$$= 2 \int_{x_1}^{x_2} x^{-1}dx \quad \text{from (1.26)}$$

$$= 2 \log (x_2/x_1).$$

This differs only trivially from $-2 \log x_1$. Unfortunately the value chosen for x_1 will depend to a large extent on the view one takes of the most likely form of genetic evolution (see the discussion in Sect. 1.6). A value often chosen for evolutionary load arguments is $x_1 = 0.0001$ and this gives $L = 18.4$. When $h \neq 0.5$ the load usually exceeds this value, and for operational purposes the "representative value" $L = 30$ is generally used.

What does this calculation really mean? Suppose all selection is through viability differences and the number of reproducing adults in each generation remains constant at N. A considerable proportion of the depletion in population numbers between birth and the age of reproduction is non-genetic. Taking only the genetic component and supposing there is no depletion through genetic deaths of the optimal genotype A_1A_1, a straightforward calculation shows that when the frequency of A_1 is x there must be $N(1 + s)/(1 + sx)$ individuals at birth so that after differential viabilities operate there are N individuals at the age of maturity. Thus the average individual is required to leave approximately $1 + s(1 - x)$ offspring after non-genetic deaths are taken account of, so that there will be $Ns(1 - x)$ "genetic deaths" in each generation associated with the evolutionary process. Summed over the entire process this gives NL individuals in all or an average of NL/T each generation if the substitutional process takes T generations.

Consider now a sequence of loci, all with the same fitness parameters we are considering, at which substitutions start regularly n generations apart. At any one time there will be T/n substitutions in progress and thus a total of $(NL/T)(T/n) = NL/n$ "selective deaths" per generation. Suppose now one sets, with Haldane (1957), an upper limit of $0.1 N$ to this number. Then using the representative value $L = 30$ one calculates a lower limit $n = 300$, so that successive substitutions can start no more frequently than once every 300 generations or else the number of selective deaths will be too large for the population to carry.

It may be argued that if selection is through fertility differences the above reasoning is inappropriate. Nevertheless a parallel argument is possible if one considers the offspring contribution required of the most fit genotype. This is approximately

$1 + s(1 - x)$ for each individual of this genotype (again after "non-genetic" deaths), and the same argument again carries through.

What is the offspring requirement of the most fit individual (whether selection is through viability of fertility or both)? On average he is required to produce $1 + L/T$ offspring for each locus substituting or, using a simple multiplicative model of fitness with linkage equilibrium always holding between loci,

$$(1 + L/T)^{T/n} \approx \exp{(L/n)} \approx \exp{(30/n)} \tag{2.112}$$

offspring in all. (Note that this is approximately 1.1 offspring when $n = 300$ in conformity with the previous calculation.) Clearly this value increases very rapidly as n decreases (i.e., if substitutions at different loci start in comparatively rapid succession). Since, at least in mammals, it has been suggested (Kimura, 1968) that n is quite small, a school of thought has emerged whose members doubt that all, or even most, gene substitutions have had a selective basis. We assess this argument at greater length in Sect. 9.2.

A similar remark applies for the segregational load. Suppose k loci are kept polymorphic by heterozygote selective advantage and that the marginal fitnesses at any locus are $1 - s$, $1 + s$, $1 - s$. The equilibrium mean fitness is unity and, assuming a multiplicative model of selection and no linkage disequilibrium the fitness of the most fit genotype is $(1 + s)^k$ or approximately $\exp{(sk)}$. Apart from non-genetic deaths such individuals are required to produce $\exp{(sk)}$ offspring to maintain stable population size. Since k may be quite large the genetic load argument once more throws doubt on the possibility that observed polymorphisms are largely due to selective agents of this kind. These arguments do not stand up well to a closer scrutiny, as we observe later when considering the "neutral theory" to which they contributed.

2.11 Finite Markov Chains

Some of the arguments presented later in this book use the theory of finite Markov chains and in this section a brief and informal introduction the theory of these is presented.

Consider a discrete random variable X which at time points $0, 1, 2, 3, \ldots$ takes one or other of the values $0, 1, 2, \ldots, M$. We shall say that X, or the system, is in state E_i if X takes the value i. Suppose at any time t that X is in state E_i. Then if the probability p_{ij} that at time $t + 1$ X is in state E_j is independent of t and of the states occupied by X at times $t - 1, t - 2, \ldots$, the variable X is said to be Markovian and its probability laws follow those of a finite Markov chain. If the initial probability (at $t = 0$) that X is in E_i is a_i then the probability that X is in the states $E_i, E_j, E_k, E_l, E_m \ldots$ at times $0, 1, 2, 3, 4$ $\ldots$ is $a_i p_{ij} p_{jk} p_{kl} p_{lm} \ldots$.

Complications to Markov chain theory arise if periodicities occur (for example, if X can return to E_i only at the time points $t_1, 2t_1, 3t_1, \ldots$ for $t_1 > 1$). Further minor complications arise if the states $E_0, E_1, \ldots, E_M$ can be broken down into non-communicating subsets. To avoid unnecessary complications (which never in any event arise in genetical applications) we suppose that no periodicities exist and that, apart from the possibility of a small number of absorbing states (E_i is absorbing if $p_{ii} = 1$) no breakdown into non-communicating subsets occurs.

It is convenient to collect the p_{ij} into a matrix $P = \{p_{ij}\}$, so that

$$P = \begin{pmatrix} p_{00} & p_{01} & \cdots & p_{0M} \\ p_{10} & p_{11} & \cdots & p_{1M} \\ \vdots & & & \\ p_{M0} & p_{M1} & \cdots & p_{MM} \end{pmatrix}. \tag{2.113}$$

The probability $p_{ij}^{(2)}$ that X is in E_j at time $t + 2$, given it is in E_i at time t, is evidently

$$p_{ij}^{(2)} = \sum_k p_{ik} p_{kj}$$

and since the right-hand side is the $(i, j)^{\text{th}}$ element in the matrix P^2, and if $P^{(2)} = \{p_{ij}^{(2)}\}$, then

$$P^{(t)} = P^t \tag{2.114}$$

for $t = 2$. More generally (2.114) is true for any positive integer t. In all cases we consider, P^t can be written in the spectral form

$$P^t = \lambda_0^t r_0 l_0 + \lambda_1^t r_1 l_1' + \ldots + \lambda_M^t r_M l_M' \tag{2.115}$$

where $\lambda_0, \lambda_1, \ldots, \lambda_M$ ($|\lambda_0| \geq |\lambda_1| \geq \ldots \geq |\lambda_M|$) are the eigenvalues of P and $l_0, \ldots, l_M$ and $r_0, \ldots, r_M$, normalized so that

$$l_i' r_i = \sum_{j=0}^M l_{ij} r_{ij} = 1 \tag{2.116}$$

are the corresponding left and right eigenvectors, respectively. Suppose E_0 and E_M are absorbing states and that no other states are absorbing. Then $\lambda_0 = \lambda_1 = 1$ and if $|\lambda_2| > |\lambda_3|$ and $i, j = 1, 2, \ldots, M - 1$,

$$p_{ij}^{(t)} = r_{2i} l_{2j} \lambda_2^t + o(\lambda_2^t) \tag{2.117}$$

for large t. Thus the leading non-unit eigenvalue λ_2 plays an important role in determining the rate at which absorption into either E_0 or E_M occurs.

Let π_i be the probability that eventually E_M (rather than E_0) is entered, given initially that X is in E_i. By considering values of X at consecutive time points it is seen that the π_i satisfy

$$\pi_i = \sum_{j=0}^M p_{ij} \pi_j, \quad \pi_0 = 0, \quad \pi_M = 1. \tag{2.118}$$

Note that for the genetic model (1.43) (with $M = 2N$) the solution of (2.118) was $\pi_i = i/M$. The mean times $\bar{t}_i$ until absorption into E_0 or E_M occurs, given that X is in E_i, similarly satisfy

$$\overline{t_i} = \sum_{j=0}^{M} p_{ij}\overline{t_j} + 1, \quad \overline{t_0} = \overline{t_M} = 0. \tag{2.119}$$

Starting with X in E_i the mean time $\overline{t_{ij}}$ that X is in E_j before absorption satisfies

$$\overline{t_{ij}} = \sum_{k=0}^{M} p_{ik}\overline{t_{kj}} + \delta_{ij}, \quad \overline{t_{0j}} = \overline{t_{Mj}} = 0, \tag{2.120}$$

where $\delta_{ij} = 1$ for $i = j$ and $\delta_{ij} = 0$ otherwise. Note also that

$$\overline{t_{ij}} = \sum_{n=0}^{\infty} p_{ij}^{(n)}, \quad \overline{t_i} = \sum_{j=1}^{M-1} \overline{t_{ij}}. \tag{2.121}$$

An expression can also be found for the variance σ_i^2 of the time before absorption, given initially X in E_i, namely

$$\sigma_i^2 = 2 \sum_{j=1}^{M-1} \overline{t_{ij}}\overline{t_j} - \overline{t_i} - (\overline{t_i})^2. \tag{2.122}$$

It is possible to derive the general form of the distribution of the time that X is in E_j if initially in E_i. Suppose that, starting in E_i, the probability that X ever enters E_j is α_{ij} and that, once in E_j, the probability that X ever returns to E_j is r_j. Then the probability that E_j is occupied exactly n times before absorption takes place at E_0 or E_M is

$$\left. \begin{array}{ll} 1 - \alpha_{ij} & \text{for } n = 0 \\ \alpha_{ij}(r_j)^{n-1}(1 - r_j) & \text{for } n \geqslant 1. \end{array} \right\} \tag{2.123}$$

This is clearly a modified geometric distribution. The mean is thus

$$\overline{t_{ij}} = \alpha_{ij}(1 - r_j) \sum_{n=1}^{\infty} nr_j^{n-1}$$

$$= \alpha_{ij}/(1 - r_j) \tag{2.124}$$

and the variance is

$$\sigma_{ij}^2 = \alpha_{ij}(1 - r_j) \sum_{n=1}^{\infty} n^2 (r_j)^{n-1} - \overline{t_{ij}}^2$$

$$= \overline{t_{ij}} \{1 - \overline{t_{ij}} + 2r_j/(1 - r_j)\}. \tag{2.125}$$

It is possible to find an expression for r_j and hence to calculate (2.125) but we do not enter into details here.

Consider now only those cases for which E_M is the absorbing state eventually entered. Writing X_t for the value of X at time t,

$$p_{ij}^* = \text{Prob } \{X_{t+1} \text{ in } E_j | X_t \text{ in } E_i, E_M \text{ eventually entered}\}$$

$$= \text{Prob } \{X_{t+1} \text{ in } E_j \text{ and } E_M \text{ eventually entered } | X_t \text{ in } E_i\} \div$$

$$\quad \text{Prob } \{E_M \text{ eventually entered } | X_t \text{ in } E_i\}$$

$$= p_{ij} \pi_j/\pi_i, \quad (i, j = 1, 2, ..., M), \tag{2.126}$$

using conditional probability arguments and the Markovian nature of X. Let $\widetilde{P}$ be the matrix derived from P by omitting the first row and first column and let

$$
V = \begin{pmatrix}
\pi_1 & & & & \\
& \pi_2 & & & \\
& & \cdot & & 0 \\
& & & \cdot & \\
& & & & \cdot \\
0 & & & & \pi_M
\end{pmatrix} . \tag{2.127}
$$

Then if $P* = \{p_{ij}^*\}$, Eq. (2.126) shows that

$$
P* = V^{-1}\widetilde{P}V . \tag{2.128}
$$

Standard theory shows that the eigenvalues of $P*$ are identical to those of P (with one unit eigenvalue omitted) and that if $l'(r)$ is any left (right) eigenvector of $\widetilde{P}$, then the corresponding left and right eigenvector of $P*$ are $l'V$, $V^{-1}r$. Further, if $P*^{(n)}$ is the matrix of conditional n step transition probabilities,

$$
P*^{(n)} = (P*)^n = V^{-1}\widetilde{P}^n V
$$

so that

$$
p_{ij}^{*(n)} = p_{ij}^{(n)} \, \pi_j/\pi_i , \tag{2.129}
$$

a conclusion that can be reached directly as with (2.126). If $\overline{t}_{ij}^*$ is the conditional mean time spent in E_j, given initially X in E_i, then

$$
\overline{t}_{ij}^* = \sum_{n=0}^{\infty} p_{ij}^{*(n)}
$$

$$
= (\pi_j/\pi_i) \sum_{n=0}^{\infty} p_{ij}^{(n)} \tag{2.130}
$$

$$
= \overline{t}_{ij}\pi_j/\pi_i .
$$

If there is only one absorbing state interest centers solely on properties of the time until the state is entered. Taking E_0 as the only absorbing state and E_i as the initial state, the mean time t_i until absorption satisfies (2.119) with the single boundary condition $\overline{t}_0 = 0$, and the mean number of visits to E_j satisfies (2.120) with the single condition $\overline{t}_{0j} = 0$.

If there are no absorbing states P will have a single unit eigenvalue and all other eigenvalues will be strictly less than unity in absolute value. Equation (2.115) then shows that

$$
\lim_{t\to\infty} P^t = r_0 l_0' \tag{2.131}
$$

and since r_0 is of the form $(1, 1, 1, \ldots 1)'$,

$$\lim_{t \to \infty} p_{ij}^{(t)} = l_{0j} \text{ for all } i. \tag{2.132}$$

Using a slightly different notation we may summarize this by saying

$$\lim_{t \to \infty} p_{ij}^{(t)} = \alpha_j \tag{2.133}$$

where $\alpha' = (\alpha_0, \alpha_1, \ldots, \alpha_M)$ is the unique solution of the two equations

$$\alpha' = \alpha'P, \quad \sum_{j=0}^{M} \alpha_j = 1. \tag{2.134}$$

The vector α is called the stationary distribution of the process and in genetical applications exists only if fixation of any allele is impossible (e.g. if all alleles mutate at positive rates).

If the matrix P is a continuant (i.e. $p_{ij} = 0$ if $|i - j| > 1$) explicit formulas can be found for most of these quantities. Write $p_{i,i+1} = \lambda_i$ and $p_{i,i-1} = \mu_i$ in conformity with standard notation in this case. If E_0 and E_M are both absorbing states the probability π_i in (2.118) becomes explicitly

$$\pi_i = \sum_{k=0}^{i-1} \rho_k \Big/ \sum_{k=0}^{M-1} \rho_k \tag{2.135}$$

where

$$\rho_0 = 1, \quad \rho_k = \frac{\mu_1 \mu_2 \mu_3 \ldots \mu_k}{\lambda_1 \lambda_2 \ldots \lambda_k}.$$

Further

$$\bar{t}_{ij} = (1 - \pi_i) \left(\sum_{k=0}^{j-1} \rho_k \right) / \rho_{j-1} \lambda_j \quad (j = 1, \ldots, i), \tag{2.136}$$

$$\bar{t}_{ij} = \pi_i \left(\sum_{k=j}^{M-1} \rho_k \right) / \rho_j \lambda_j \quad (j = i+1, ;.., M-1).$$

Equations (2.121) and (2.130) then yield $\bar{t}_i$, $\bar{t}_{ij}^*$ and $\bar{t}_i^*$ immediately. When there is only one absorbing state Eq. (2.121) still holds, but now $\bar{t}_{ij}$ is defined by

$$\bar{t}_{ij} = \begin{cases} \mu_j^{-1} \left\{ 1 + \dfrac{\lambda_{j-1}}{\mu_{j-1}} + \dfrac{\lambda_{j-1}\lambda_{j-2}}{\mu_{j-1}\mu_{j-2}} + \ldots + \dfrac{\lambda_{j-1}\lambda_{j-2}\ldots\lambda_1}{\mu_{j-1}\mu_{j-2}\ldots\mu_1} \right\} & (j = 1, 2, \ldots, i) \\[3mm] \bar{t}_{ii} \left(\dfrac{\lambda_i\lambda_{i+1}\ldots\lambda_{j-1}}{\mu_{i+1}\mu_{i+2}\ldots\mu_j} \right) & (j = i+1, \ldots M) \end{cases} \tag{2.137}$$

if E_0 is the absorbing state and by

$$
\bar{t}_{ij} = \begin{cases}
\dfrac{\mu_{j+1}\mu_{j+2}\cdots\mu_i}{\lambda_j\lambda_{j+1}\cdots\lambda_{i-1}}\,\bar{t}_{ii} & (j = 0, 1, \ldots, i-1) \\[2ex]
\lambda_i^{-1}\left\{1 + \dfrac{\mu_{i+1}}{\lambda_{i+1}} + \dfrac{\mu_{i+1}\mu_{i+2}}{\lambda_{i+1}\lambda_{i+2}} + \ldots + \dfrac{\mu_{i+1}\mu_{i+2}\cdots\mu_{M-1}}{\lambda_{i+1}\lambda_{i+2}\cdots\lambda_{M-1}}\right\} & \\
& (j = i, i+1, \ldots, M-1)
\end{cases}
\tag{2.138}
$$

if E_M is the absorbing state. In this case of course there can be no further concept of a conditional mean absorption time. Finally when there are no absorbing states the stationary distribution α is defined by

$$
\alpha_i = \alpha_0\,(\lambda_0\lambda_1 \ldots \lambda_{i-1})/(\mu_1\mu_2 \ldots \mu_i)
\tag{2.139}
$$

where α_0 is chosen so that $\Sigma\alpha_i = 1$.

A variety of other results is possible for continuant Markov chain models, perhaps the most accessible summary being given in Kemeny and Snell (1960). We will draw on the formulae given above on a number of occasions throughout this book.

We conclude our discussion of finite Markov chains by introducing the concept of time reversibility. Consider a Markov chain admitting a stationary distribution $\{\alpha_0, \alpha_1, \ldots, \alpha_M\}$. Then we define the process to be reversible if, at stationarity,

$$
\text{Prob } \{X_t, X_{t+1}, \ldots, X_{t+n}\} = \text{Prob } \{X_t, X_{t-1}, \ldots, X_{t-n}\}
\tag{2.140}
$$

for every t and n. A necessary and sufficient condition for this is that the stationary state has been reached and that the equation

$$
\alpha_i p_{ij} = \alpha_j p_{ji}
\tag{2.141}
$$

hold for all i, j. Certain classes of Markov chains are always reversible. For example, if the transition matrix is a continuant, Eq. (2.139) and (2.140) jointly easily show that the Markov chain at stationarity is reversible. Certain other chains, in particular several having genetical relevance, are reversible: we will consider these later when showing the uses to which the concept of reversibility can be put.

3. Discrete Stochastic Models

3.1 Introduction

In the last section of the previous chapter some elementary finite Markov chain theory was introduced. In this chapter we apply this theory to various Markov chain models which arise in genetics. We will find that the complexities of these models are such that not all questions of genetical interest can in practice be answered by using this theory, and in the following two chapters we will introduce diffusion theory to arrive at a more complete, although approximate, description of their properties.

3.2 Wright-Fisher Model: Two Alleles

In Chap. 1 we were led to the Wright-Fisher model (1.43) as a reasonable representation of the stochastic behavior of gene frequencies in an idealized finite population. Our first aim is to discuss some of the properties of this model in the light of the theory of Sect. 2.11.

We have already noted that in the model (1.43) the number X of A_1 genes is a Markovian random variable with two absorbing states, $X = 0$ and $X = 2N$. Further, the probability that eventually $X = 2N$, given initially $X = i$, is simply $i/2N$. We now ask whether the theory of Sect. 2.11 gives us further information on the behavior of X before an absorbing state is reached. The most interesting quantities are the mean time $\bar{t}_i$ until absorption, given initially $X = i$, and the mean number of times $\bar{t}_{ij}$ that X takes the value j before absorption. While in principle these expressions can be found from (2.119) and (2.120), in practice solution of these equations seems extremely difficult, and simple expressions for these quantities have not yet been found. It is, on the other hand, possible to find a simple approximation for $\bar{t}_i$ by the following line of argument.

In Eq. (2.119) put $M = 2N$, $i/M = x$, $j/M = x + \delta x$, and $\bar{t}_i = \bar{t}(x)$. We suppose $\bar{t}(x)$ is a twice differentiable function of a continuous variable x. Then (2.119) can be written

$$\bar{t}(x) = \Sigma \, \mathrm{prob} \, \{x \to x + \delta x\} \bar{t}(x + \delta x) + 1 \tag{3.1}$$

$$\doteq \mathrm{E} \, \{\bar{t}(x + \delta x)\} + 1 \tag{3.2}$$

$$\approx \bar{t}(x) + \mathrm{E}(\delta x)\bar{t}'(x) + \tfrac{1}{2} \mathrm{E}(\delta x)^2 \bar{t}''(x) + 1, \tag{3.3}$$

where all expectations are conditional on x and in (3.3) only the first three terms in an infinite Taylor series have been retained. Since from (1.43)

$$E(\delta x) = 0, \qquad E(\delta x)^2 = (2N)^{-1} x(1 - x),$$

Eq. (3.3) gives

$$x(1 - x)\bar{t}''(x) \approx -4N. \tag{3.4}$$

The solution of this equation, subject to the boundary conditions $\bar{t}(0) = \bar{t}(1) = 0$, is

$$\bar{t}(p) \approx -4N\{p \log p + (1 - p) \log (1 - p)\}, \tag{3.5}$$

where p is the initial frequency of A_1. We will see later that this is the so-called diffusion approximation to the mean absorption time although we have here not made any reference to diffusion processes. Note that for $p = (2N)^{-1}$ (the value appropriate if A_1 is a unique new mutation in an otherwise pure A_2A_2 population) Eq. (3.5) reduces to

$$\bar{t}\{(2N)^{-1}\} \approx 2 + 2 \log 2N \text{ generations,} \tag{3.6}$$

while when $p = \frac{1}{2}$,

$$\bar{t}\{\tfrac{1}{2}\} \approx 2.8N \text{ generations.} \tag{3.7}$$

This very long mean time, for equal initial frequencies, is of course intimately connected with the fact that the leading non-unit eigenvalue of (1.43) is very close to unity.

Suppose now the condition is made that A_1 eventually fixes. The possible values for X are 1, 2, 3, ..., $2N$ and (2.126) shows that the conditional transition probability p_{ij}^* is

$$
\begin{aligned}
p_{ij}^* &= \binom{2N}{j} \left(\frac{i}{2N}\right)^j \left(\frac{2N - i}{2N}\right)^{2N - j} \frac{j}{i} \\
&= \binom{2N - 1}{j - 1} \left(\frac{i}{2N}\right)^{j-1} \left(\frac{2N - i}{2N}\right)^{2N - j}.
\end{aligned}
\tag{3.8}
$$

An intuitive explanation for the form of p_{ij}^* is that, under the condition that A_1 fixes, at least one A_1 gene must be produced in each generation. Then p_{ij}^* is the probability that the remaining $2N - 1$ trials produce exactly $j - 1$ A_1 genes. An argument parallel to that leading to (3.4) gives

$$x(1 - x)\{\bar{t}^*(x)\}'' + 2(1 - x)\{\bar{t}^*(x)\}' = -4N \tag{3.9}$$

for the conditional mean time $\bar{t}^*(x)$ to fixation, given a current frequency of x. The solution of (3.9), subject to $\bar{t}^*(1) = 0$ and the requirement

$$\lim_{x \to 0} \ \bar{t}^*(x) \text{ is finite,} \tag{3.10}$$

and assuming initially $x = p$, is

$$\bar{t}^*(p) = -4Np^{-1} (1-p) \log (1-p). \tag{3.11}$$

(We discuss the requirement (3.10) in more detail later.) Note that

$$\bar{t}^* \{(2N)^{-1}\} \approx 4N \text{ generations,} \tag{3.12}$$

$$\bar{t}^* \ \{\tfrac{1}{2}\} \approx 2.8N \text{ generations,} \tag{3.13}$$

$$\bar{t}^* \{1 - (2N)^{-1}\} \approx 2 \log 2N \text{ generations.} \tag{3.14}$$

Equation (3.13) is to be expected from (3.7), by symmetry, since the conditioning is here irrelevant. On the other hand, (3.12) and (3.14) provide new information and show that, while when the initial frequency of A_1 is $(2N)^{-1}$ it is very unlikely that fixation of A_1 will occur, in the small fraction of cases when it does an extremely long fixation time may be expected. Further conclusions will be given later when we consider the diffusion approximation to (1.43).

As noted in Chap. 1, the initial analysis of the model (1.43) by Fisher and Wright paid particular attention to the leading eigenvalue of the transition matrix, regarded as a measure of the rate at which one or other allele is lost from the population. Although, as we see in a moment, the eigenvalues are of less use than expressions like (3.5) and (3.11) for this purpose, they are nevertheless of some interest, as indeed are the corresponding eigenvectors, and we now consider them in some detail.

The eigenvalues of the transition matrix defined by (1.43) were first derived by Feller (1951), but we will compute them here as particular cases of a very general theory given by Cannings (1974). To do this we introduce the concept of an exchangeable process. We consider a population of fixed size $2N$, reproducing (or potentially reproducing) at time points $t = 0, 1, 2, 3, \ldots$. The stochastic rule determining the population structure at time $t + 1$ is quite general provided that any subset of genes at time t has the same distribution of "descendant" genes at time $t + 1$ as any other subset of the same size. In some models a gene present at time t can also be present at time $t + 1$ and is then counted as one of its own descendants. Thus, if the ith gene leaves y_i descendant genes we require only that $y_1 + \ldots + y_{2N} = 2N$ and that the distribution of $(y_i, y_j, \ldots, y_k)$ be independent of $i, j, \ldots k$. The model (1.43) clearly obeys these assumptions since $(y_1, y_2, \ldots, y_{2N})$ have a symmetric multinomial distribution. Let the genes be divided into two allelic classes A_1 and A_2 and let X_t be the number of A_1 genes at time t. Then we have

Theorem 3.1 (Cannings, 1974). If

$$p_{ij} = \text{Prob } \{X_{t+1} = j | X_t = i\}, \quad i, j = 0, 1, 2, \ldots, 2N,$$

the eigenvalues of the matrix $\{p_{ij}\}$ are

$$\lambda_0 = 1, \quad \lambda_j = E\,(y_1 y_2 \ldots y_j), \quad j = 1, 2, \ldots, 2N. \tag{3.15}$$

Since we use this theorem, or generalizations of it, several times below we reproduce here a proof of it, following Cannings (1974).

Proof. Let $P = \{p_{ij}\}$. Suppose a non-singular matrix Z and an upper triangular matrix A can be found such that $PZ = ZA$. Since this equation implies $P = ZAZ^{-1}$, the eigenvalues of P are identical to those of A which, because of the special nature of A, are its diagonal elements. Consider now the (non-singular) matrix Z defined by

$$Z = \begin{pmatrix} 1 & 0 & 0 & 0 & \ldots & 0 \\ 1 & 1 & 1^2 & 1^3 & \ldots & 1^{2N} \\ 1 & 2 & 2^2 & 2^3 & \ldots & 2^{2N} \\ 1 & 3 & 3^2 & 3^3 & \ldots & 3^{2N} \\ \vdots & & & & & \\ 1 & 2N & (2N)^2 & (2N)^3 & \ldots & (2N)^{2N} \end{pmatrix}.$$

With this definition of Z the (i,j)th element of PZ is

$$\sum_k p_{ik} k^j$$

which can be written

$$E\,[\,\{X(t+1)\}^j | X(t) = i\,].$$

Similarly the (i,j)th element of ZA is of the form

$$\sum_{k=0}^{j} a_{kj} i^k$$

which may be written as

$$a_{jj} i^{[j]} + \text{terms in } i^{j-1}, \quad i^{j-2}, \ldots, i^1, i^0.$$

Here $i^{[j]} = i(i-1)(i-2) \ldots (i-j+1)$. It follows from this that if we can write

$$E\,[\,\{X(t+1)\}^j | X(t) = i\,] = a_{jj} i^{[j]} + \text{terms in } i^{j-1}, i^{j-2}, \ldots \tag{3.16}$$

then the a_{jj} ($j = 0, 1, 2, \ldots, 2N$) are the eigenvalues of P. Now in the exchangeable model we have

$$E\,[\,\{X(t+1)\}^j | X(t) = i\,] = E\,\{y_1 + y_2 + \ldots + y_i\}^j$$
$$= i\,E\,\{y_1^j\} + \ldots + i^{[j]}\,E\,(y_1 y_2 \ldots y_j),$$

and it follows that a representation of the form (3.16) is indeed possible for exchangeable models, with

$$a_{jj} = E\,(y_1 y_2 \ldots y_j),\quad j = 0, 1, 2, \ldots, 2N.$$

This completes the proof of the theorem. Cannings also asserts that, except in the trivial case $x_j \equiv 1$, the eigenvalues obey the inequalities

$$1 = \lambda_0 = \lambda_1 > \lambda_2 > \lambda_3 > \ldots > \lambda_k = \lambda_{k+1} = \ldots = \lambda_{2N} = 0$$

for some K. However, Gladstien (1978) notes that this is not quite true and all that can be asserted is

$$1 = \lambda_0 = \lambda_1 > \lambda_2 > \lambda_3 > \ldots > \lambda_k = \lambda_{k+1} = \ldots = \lambda_{2N}.$$

In the model (1.43) any set $y_1, y_2, \ldots, y_j$ has a multinomial distribution with index $2N$ and common parameter $(2N)^{-1}$, so that

$$\lambda_j = \Sigma \ldots \Sigma y_1 y_2 \ldots y_j \,\frac{(2N)!}{y_1! y_2! \ldots y_j!\,(2N - y_1 - \ldots - y_j)!}\left(\frac{1}{2N}\right)^{2N}$$

$$= (2N)(2N-1)\ldots(2N-j+1)/(2N)^j,\quad j = 1, 2, \ldots, 2N. \tag{3.17}$$

The eigenvalues of the matrix (1.43) are thus $\lambda_0 = 1$ together with the set (3.17). Note that $\lambda_1 = 1$ and $\lambda_2 = 1 - (2N)^{-1}$, confirming the values found earlier by other methods.

Although considerable attention has been paid to the leading non-unit eigenvalue λ_2 and to a lesser extent to the complete set (3.17), it is possible to argue that these eigenvalues are of limited usefulness. First, Eq. (2.128) shows that the eigenvalues in the conditional process, where eventual fixation of a specified allele is assumed, are the same as those in the unconditional process. On the other hand, the mean fixation time values are quite different in the two cases, as (3.5) and (3.11) show, and thus are not adequately described by knowledge of the eigenvalues. Second, we shall show later that, at least in the model (1.43), by the time that the term defined by the leading non-unit eigenvalue in the spectral expansion (2.115) dominates the remaining terms it is very likely that loss or fixation of A_1 will already have occurred.

Suppose now that A_1 mutates to A_2 at rate u but that there is no mutation from A_2 to A_1. It is then reasonable to replace the model (1.43) by

$$p_{ij} = \binom{2N}{j}(\psi_i)^j(1 - \psi_i)^{2N-j} \tag{3.18}$$

where $\psi_i = i(1 - u)/2N$. Here interest centers on properties of the time until A_1 is lost, either using eigenvalues or mean time properties. For the moment we consider mean time properties and note that an argument parallel to that leading to (3.4) shows that, to a first approximation, the mean time $\overline{t}\,(x)$, given a current frequency x, satisfies

$$- 4Nux\overline{t}\,'(x) + x(1 - x)\overline{t}\,''(x) = - 4N. \tag{3.19}$$

The solution of this equation, subject to $\overline{t}\,(0) = 0$ and

$\lim_{x \to 1} \bar{t}(x)$ is finite,

(a condition we discuss more fully later), and assuming initially that $x = p$, is

$$\bar{t}(p) = \int_0^1 t(x, p) \, dx \text{ generations,}$$

where

$$\begin{rcases} t(x, p) = 4Nx^{-1}(1 - \theta)^{-1} \{(1 - x)^{\theta - 1} - 1\}, \quad 0 \leqslant x \leqslant p, \\ t(x, p) = 4Nx^{-1}(1 - \theta)^{-1} (1 - x)^{\theta - 1} \{1 - (1 - p)^{1 - \theta}\}, \quad p \leqslant x \leqslant 1, \end{rcases} \quad (3.20)$$

and $\theta = 4Nu$. These equations are more informative than they initially appear since, as we see later, $t(x, p)\delta x$ affords an excellent approximation to the mean number of generations for which the frequency of A_1 takes a value in $(x, x + \delta x)$ before reaching zero.

Suppose next that A_2 also mutates to A_1 at rate v. It is now reasonable to define ψ_i in (3.18) by

$$\psi_i = \{i(1 - u) + (2N - i)v\}/2N. \tag{3.21}$$

There now exists a stationary distribution $\alpha' = (\alpha_0, \alpha_1, \ldots, \alpha_{2N})$ for the number of A_1 genes, given in principle by (2.134). The exact form of this distribution is complex, and we consider later an approximation to it. On the other hand, certain properties of this distribution can be extracted from (3.18) and (3.21). The stationary distribution satisfies the equation $\alpha' = \alpha' P$, where P is defined by (3.18) and (3.21), so that if ξ is a vector with ith element $i(i = 0, 1, 2, \ldots, 2N)$ and μ is the mean of the stationary distribution,

$$\mu = \alpha' \xi = \alpha' P \xi.$$

The ith $(i = 0, 1, 2, \ldots, 2N)$ component of $P\xi$ is

$$\Sigma_j \binom{2N}{j} \psi_i^j (1 - \psi_i)^{2N - j}$$

and from the standard formula for the mean of the binomial distribution, this is $2N \psi_i$ or

$$i(1 - u) + (2N - i)v.$$

Thus,

$$\begin{aligned} \alpha' P\xi &= \Sigma \{i(1 - u) + (2N - i)v\}\alpha_i \\ &= \mu(1 - u) + v(2N - \mu). \end{aligned}$$

It follows that

$$\mu = (1 - u)\mu + v(2N - \mu)$$

or

$$\mu = 2Nv/(u + v). \tag{3.22}$$

Similar arguments show that the variance σ^2 of the stationary distribution is

$$\sigma^2 = 4N^2uv/\{(u + v)^2(4Nu + 4Nv + 1)\} + \text{small order terms.} \tag{3.23}$$

Further moments can also be found, but we do not pursue the details. The above values are sufficient to answer a question of some interest in population genetics, namely "what is the probability that two genes drawn together at random are of the same allelic type?" If the frequency of A_1 is x this probability (ignoring terms of order N^{-1}) is $x^2 + (1 - x)^2$. The required value is the average of this over the stationary distribution, namely

$$E\{x^2 + (1 - x)^2\} = 1 - 2\,E(x) + 2E(x^2).$$

If $u = v, 4Nu = \theta$, (3.22) and (3.23) show that this is

$$\text{Pr \{two genes of same allelic type\}} \approx (1 + \theta)/(1 + 2\theta).$$

This probability can be arrived at in another way, which we now consider since it is useful for purposes of generalization. Let the required probability be F and note that this is the same in two consecutive stationary generations. Two genes drawn at random in any generation will have a common parent gene with probability $(2N)^{-1}$, or different parent genes with probability $1 - (2N)^{-1}$, which will be of the same allelic type with probability F. The probability that neither of the genes drawn is a mutant, or that both are, is $u^2 + (1 - u)^2$, while the probability that precisely one is a mutant is $2u(1 - u)$. It follows that

$$F = \{u^2 + (1 - u)^2\}(2N)^{-1}\{1 + F(2N - 1)\}$$
$$+ 2u(1 - u)(1 - F)(2N - 1)(2N)^{-1}.$$

Thus exactly

$$F = \frac{1 + 2u(1 - u)(2N - 2)}{1 + 4u(1 - u)(2N - 1)}$$

and approximately

$$F = (1 + \theta)/(1 + 2\theta) \tag{3.24}$$

in agreement with the previous formula. Later we consider a third approach, which again yields the same answer.

To find the eigenvalues of the matrix defined by (3.18) and (3.21) we use a second theorem due to Cannings (1974). Suppose that if mutation does not exist the conditions for Theorem 3.1 hold. Now assume that A_1 mutates to A_2 at rate u, with reverse mutation at rate v. Write $x_i = y_i + z_i$, where $y_i = 1$ or 0 depending on whether or not the ith gene at time t continues to exist at time $t + 1$. Thus, $y_i = 0$ in the model (3.18), but we are considering now more general conditions than those specified by this equation. The variable z_i is the number of offspring genes from the ith gene at time t. If this gene is of type A_1, define z_{i1} as the (random) number of its A_1 (i.e. non-mutated) offspring: z_{i1} has a distribution which depends on z_i. Similarly if the ith gene is of type A_2 let z_{i2} be the random number of its A_1 (i.e. mutant) offspring.

Theorem 3.2 (Cannings, 1974). The eigenvalues of the matrix P describing the stochastic behavior of the number of A_1 genes are

$$\lambda_0 = 1, \quad \lambda_j = \Sigma \Pr(z_1, \ldots, z_j) \left\{ E \prod_{i=1}^{j} (y_i + z_{i1} - z_{i2} \mid z_1, \ldots, z_j) \right\},$$

$$(j = 1, 2, \ldots, 2N). \tag{3.25}$$

The proof of this theorem is omitted here. In the model defined by (3.18) and (3.21), $y_i \equiv 0$ and $z_1 \ldots z_j$ have a multinomial distribution with index $2N$ and common parameter $(2N)^{-1}$. Further, given z_i, z_{i1} and z_{i2} have binomial distributions with respective parameters $1 - u$ and v. Thus,

$$E(z_{i1} - z_{i2} \mid z_i) = (1 - u - v)z_i$$

and

$$\lambda_j = \Sigma \Pr(z_1, \ldots, z_j)(1 - u - v)^j z_1 \ldots z_j$$

$$= (1 - u - v)^j E(z_1 \ldots z_j) \tag{3.26}$$

$$= (1 - u - v)^j \{2N(2N - 1) \ldots (2N - j + 1)/(2N)^j\} \quad j = 1, 2, \ldots, 2N.$$

The conclusion of (3.17) has been used in reaching this formula. Note that the leading non-unit eigenvalue λ_1 is $1 - u - v$ and is thus independent of N. This is extremely close to unity and suggests a very slow rate of approach to stationarity in this model. Note also that the eigenvalues (3.26) apply also in the one-way mutation model where we simply put $v = 0$. We return to Theorem 3.2 later when considering models other than (3.18).

Suppose now that selection exists and that the genotypes A_1A_1, A_1A_2, and A_2A_2 have fitnesses given by (1.24a). In view of (1.23) a reasonable stochastic model is found by assuming that the transition matrix for the number of A_1 individuals is (3.18), where now

$$\psi_i = \bar{w}^{-1} [\{w_{11}x^2 + w_{12}x(1 - x)\}(1 - u) + \{w_{12}x(1 - x) + w_{22}(1 - x)^2\}v],$$

$$\tag{3.27}$$

where $x = i/2N$ and $\bar{w}$ is defined by (1.37). The qualitative properties of this model are clear: when $u = v = 0$, one or other absorbing state, $X = 0, X = 2N$, is eventually reached. When $u > 0, v = 0, A_1$ is eventually lost from the population while when $u, v > 0$ there will exist a stationary distribution for the number of A_1 genes. Essentially no quantitative results concerning this behavior are known and the best that can be done is to consider approximations. We will do this (in Chap. 5) by using diffusion theory and for the moment foreshadow this approach by deriving an approximate formula for absorption probabilities when $u = v = 0$.

Suppose $w_{11} = 1 + s$, $w_{12} = 1 + sh$ and $w_{11} = 1$, where s is of order N^{-1}. Put $\alpha = 2Ns$ and, in Eq. (2.118), write $i = 2Nx, j = 2N(x + \delta x)$. Then this equation may be written

$$\pi(x) = \Sigma \Pr (x \rightarrow x + \delta x)\pi(x + \delta x)$$
$$\approx \Sigma \Pr (x \rightarrow x + \delta x) \{\pi(x) + \delta x\pi'(x) + \tfrac{1}{2} (\delta x)^2 \pi''(x)\}$$
$$= \pi(x) + E(\delta x)\pi'(x) + \tfrac{1}{2} E(\delta x)^2 \pi''(x).$$

Under the assumptions we have made

$$E(\delta x) = (2N)^{-1} \alpha x(1 - x)\{x + h(1 - 2x)\} + 0(N^{-2}),$$
$$E(\delta x)^2 = (2N)^{-1}x(1 - x) + 0 \, (N^{-2}).$$

Thus to the order of approximation we use

$$2\alpha\{x + h(1 - 2x)\}\pi'(x) + \pi''(x) = 0.$$

The solution of this equation, subject to $\pi(0) = 0, \pi(1) = 1$, is

$$\pi(x) = \int_0^x \psi(y)dy / \int_0^1 \psi(y)dy, \tag{3.28}$$

where

$$\psi(y) = \exp [-\alpha y\{2h + y(1 - 2h)\}].$$

In the particular case $h = \tfrac{1}{2}$ (so that the heterozygote is intermediate in fitness between the two homozygotes), this reduces to

$$\pi(x) = \{1 - \exp (-\alpha x)\}/\{1 - \exp (-\alpha)\}. \tag{3.29}$$

It is of some interest to use this approximate formula to get some idea of the effect of the selective differences on the probability of fixation of A_1. Suppose for example that $N = 10^5, s = 10^{-4}$, and $x = 0.5$. Then $\alpha = 20$ and, from (3.29), $\pi(0.5) = 0.999955$. Note that for $s = 0$ we would have $\pi(0.5) = 0.5$. Evidently the rather small selective advantage 0.0001, which is certainly too small to be observed in laboratory experiments, is nevertheless large enough in evolutionary terms to have a significant effect on the fixation probability. Clearly this occurs because, while selection can have but a minor effect in any generation, the number of generations until fixation occurs is so very large that the cumulative selective effect is considerable. We consider this problem at greater length later when more general models are considered and when a more powerful theory is available to handle them.

3.3 Moran Models: Two Alleles

The conclusions reached so far depend on the assumption that the appropriate model to describe the stochastic behavior of the number of A_1 genes is one or other form of (3.18). Different conclusions are reached for models other than these, and we consider now a model due to Moran (1958) for which this is so. Moran's model has the additional advantage of allowing explicit expressions for many quantities of evolutionary interest, although it applies strictly only for haploid populations.

Consider then a haploid population in which, at time points $t = 1, 2, 3 \ldots$, an individual is chosen at random to reproduce. After reproduction has occurred, an individual is chosen to die (possibly the reproducing individual but not the new offspring individual). The model can be generalized by allowing mutation and selection, the latter being introduced by weighting the probability that an individual of a specific genotype is chosen to give birth (or to die).

We consider first the simplest case where there is no selection or mutation. Suppose the population consists of $2N$ haploid individuals (we use this notation to allow direct comparison with the diploid case), each of whom is either A_1 or A_2. Suppose at time t the number of A_1 individuals is i. Then at time $t + 1$ there will be $i - 1$ A_1 individuals if an A_2 is chosen to give birth and an A_1 to die. The probability of this, under our assumptions, is clearly

$$p_{i, i-1} = i(2N - i)/(2N)^2.$$ (3.30a)

Similar reasoning shows that

$$p_{i, i+1} = i(2N - i)/(2N)^2,$$ (3.30b)

$$p_{i, i} = \{i^2 + (2N - i)^2\}/(2N)^2 = 1 - p_{i, i-1} - p_{i, i+1}.$$ (3.30c)

The matrix defined by the transition probabilities (3.30) is a continuant so that much of the theory of Sect. 2.11 can be applied to it. In the notation of that section,

$$\lambda_i = \mu_i = i(2N - i)/(2N)^2, \quad \rho_i = 1, \quad i = 0, 1, 2, \ldots, 2N.$$ (3.31)

It follows that the probabilitiy π_i of fixation of A_1, given currently $i A_1$ individuals, is

$$\pi_i = i/2N,$$ (3.32)

and that

$$\bar{t}_{ij} = 2N(2N - i)/(2N - j), \quad j = 1, 2, \ldots, i,$$
$$\bar{t}_{ij} = 2Ni/j, \quad j = i + 1, \ldots, 2N - 1. \tag{3.33}$$

Thus immediately

$$\bar{t}_i = 2N(2N - i) \sum_{j=1}^{i} (2N - j)^{-1} + 2N \sum_{j=i+1}^{2N-1} j^{-1}, \tag{3.34}$$

$$\bar{t}_{ij}^* = 2N(2N - i)j/\{i(2N - j)\}, \quad j = 1, 2, \ldots, i,$$
$$\bar{t}_{ij}^* = 2N, \quad j = i + 1, \ldots, 2N - 1, \tag{3.35}$$

$$\bar{t}_i^* = 2N(2N - i)i^{-1} \sum_{j=1}^{i} j(2N - j)^{-1} + 2N(2N - i - 1). \tag{3.36}$$

An interesting example of these formulae is the case $i = 1$, corresponding to a unique A_1 mutant in an otherwise purely A_2 population. Here $\bar{t}_{1j}^* = 2N$ for all j so that, given the mutant is eventually fixed, the number of A_1 genes takes, on average, each of the values 1, 2, ..., $2N - 1$ a total of $2N$ times. The conditional mean fixation time is clearly $2N(2N - 1)$ time units: the variance of the conditional absorption time can also be written down but we do not do so here.

The eigenvalues of the matrix (3.30) can be found by using Theorem 3.1. Take any collection of j genes and note that the probability that one of these is chosen to reproduce is $j/2N$, with the same probability that one is chosen to die. For this model a gene can be (and indeed usually is) one of its own "descendants". Using the notation of Theorem 3.1, the product $y_1 y_2 \ldots y_j$ can take only three values:

0   if one of these genes is chosen to die and the gene so chosen is not chosen to reproduce,
2   if one of the genes is chosen to reproduce and none is chosen to die,
1   otherwise.

Thus $\lambda_0 = 1$ and

$$\lambda_j = E(y_1 y_2 \ldots y_j)$$
$$= 0\{j(2N - 1)/(2N)^2\} + 2j(2N - j)/(2N)^2 + 1 - j(4N - 1)/(2N)^2$$
$$= 1 - j(j - 1)/(2N)^2, \quad j = 1, 2, \ldots, 2N. \tag{3.37}$$

Various expressions for the corresponding eigenvectors, first by Watterson (1961) using Tchebycheff polynomials and most recently by Gladstien (1977), have been given. We are particularly interested in the largest non-unit eigenvalue and its associated eigenvectors. The required eigenvalue is

$$\lambda_2 = 1 - 2/(2N)^2, \tag{3.38}$$

an elementary calculations show that the corresponding right eigenvector $\mathbf{r}$ and left eigenvector $\mathbf{l}'$ are

$$\mathbf{r} = (0, 1(2N-1), 2(2N-2), \ldots, i(2N-i), \ldots, 1(2N-1), 0)' \tag{3.39}$$

$$\mathbf{l}' = (-\tfrac{1}{2}(2N-1), 1, 1, 1, \ldots, 1, -\tfrac{1}{2}(2N-1)).$$

Thus the asymptotic distribution of the number X_t of A_1 genes for large t, given $X_t \neq 0, 2N$, is uniform over the values $\{1, 2, 3, \ldots, (2N-1)\}$. Note also that the fact that λ_2 is very close to unity agrees with the very large mean absorption times (3.34) for intermediate values of i.

If mutation from A_1 to A_2 is allowed (at rate u), with no reverse mutation, A_1 must eventually become lost, and interest centers on the time required for this to occur. The model (3.30) is now amended to

$$p_{i, i-1} = \{i(2N-i) + ui^2\}/(2N)^2 = \mu_i$$

$$p_{i, i+1} = i(2N-i)(1-u)/(2N)^2 = \lambda_i \tag{3.39}$$

$$p_{i, i} = 1 - p_{i, i-1} - p_{i, i+1}.$$

Equation (2.137) can now be used to find $\bar{t}_{ij}$ and thus $\bar{t}_i$. We do not present explicit expressions since it will be more useful (see (3.45) below) to proceed via approximations. If mutation from A_2 to A_1 (at rate v) is also allowed, (3.30) becomes

$$p_{i, i-1} = \{i(2N-i)(1-v) + ui^2\}/(2N)^2 = \mu_i$$

$$p_{i, i+1} = \{i(2N-i)(1-u) + v(2N-i)^2\}/(2N)^2 = \lambda_i \tag{3.40}$$

$$p_{i, i} = 1 - p_{i, i-1} - p_{i, i+1}.$$

The stationary distribution $\boldsymbol{\alpha}$ for the number of A_1 genes is found from (2.139) to be

$$\alpha_j = \frac{\alpha_0 (2N)! \, \Gamma\{j+A\} \, \Gamma\{B-j\}}{j!(2N-j)! \, \Gamma\{A\} \Gamma\{B\}} \tag{3.41}$$

where $\Gamma\{\cdot\}$ is the well known gamma function, $A = 2Nv/(1-u-v)$, $B = 2N(1-v)/(1-u-v)$, $\alpha_0 = \Gamma\{B\} \, \Gamma\{A+C\}/[\Gamma\{D\}\Gamma\{C\}]$ and $C = 2Nu/(1-u-v)$, $D = 2N/(1-u-v)$. Although these expressions are exact they are most unwieldy, and we consider in a moment a simple approximation for α_j.

Note that the Markov chain defined by (3.40), having a stationary distribution and a continuant transition matrix, is automatically reversible (see the closing remarks in Chap. 2). This is not necessarily true for other genetical models; for example it can be shown that the chain defined jointly by (3.18) and (3.21) (i.e. the Wright-Fisher model) is in general not reversible. What does reversibility mean in genetical terms? All the theory we have considered so far is *prospective*, i.e. given the current state of a Markov chain, probability statements are made about its future behavior. Recent developments in population genetics theory often concern the *retrospective* behavior:

the present state is observed, and questions are asked about the evolution leading to this state. For reversible processes these two aspects have many properties in common, and information about the prospective behavior normally yields almost immediately useful information about the retrospective behavior. We shall see later how the identity of prospective and retrospective probabilities can be used to advantage in discussing various evolutionary questions.

The eigenvalues of (3.40) can be found by applying Theorem 3.2. Here $y_i = 1$ unless the ith gene has been chosen to die, in which case $y_i = 0$. Similarly z_i, z_{i1} and z_{i2} are zero unless the ith gene has been chosen to reproduce. Exploiting symmetry arguments we deduce from (3.25) that $\lambda_0 = 1$ and

$$\lambda_j = (2N)^{-1}(2N - j) \left\{ E \prod_{i=1}^{j} y_i \right\} + (2N)^{-1} j \left\{ E(y_1 + 1 - u - v) \prod_{i=2}^{j} y_i \right\}$$

$$= 1 - j(u + v)/(2N) - j(j - 1)(1 - u - v)/(2N)^2, \quad j = 1, 2, \ldots, 2N. \tag{3.42}$$

These eigenvalues apply also in the case $v = 0$. Note that the leading non-unit eigenvalue is $1 - (u + v) / (2N)$, and since $2N$ time units in the process we consider may be thought to correspond to one generation in the Wright-Fisher model, this agrees closely with the value $1 - u - v$ found in (3.26) in that model.

We now obtain approximations for several of the above quantities. It is evident, from (3.34), that

$$\bar{t}(p) \approx - (2N)^2 \{p \log p + (1 - p) \log (1 - p)\}, \tag{3.43}$$

where $p = i/2N$. Note the similarity between this formula and (3.5). A factor of $2N$ may be allowed in comparing the two to convert from birth-death events to generations. There remains a further factor of 2 to explain, and we show in the next section why this factor exists.

Consider next the expression (3.41). Put $x = j/(2N)$, $u = \alpha/(2N)$, $v = \beta/(2N)$ and let j and $2N$ increase indefinitely with x, α and β fixed. Using the Stirling approximation $\Gamma\{y + a\}/\Gamma\{y\} \sim y^a$ for large y, moderate a, (3.41) reduces to

$$2N\alpha_j \sim \frac{\Gamma\{\alpha + \beta\}}{\Gamma\{\alpha\}\Gamma\{\beta\}} x^{\beta-1}(1 - x)^{\alpha-1}, \tag{3.44}$$

at least for values of x not extremely close to 0 or 1. Clearly this approximate expression is far simpler than the exact value (3.41). In a similar way the values for $\bar{t}_{ij}$, calculated from (2.137) and (3.39), become

$$(2N)^{-1}\bar{t}_{ij} \approx x^{-1}\{(1 - x)^{\theta/2-1} - 1\}\{1 - \tfrac{1}{2}\theta\}^{-1}, \quad j \leqslant i$$

$$(2N)^{-1}\bar{t}_{ij} \approx x^{-1}(1 - x)^{\theta/2-1}\{1 - (1 - p)^{1-\theta/2}\}\{1 - \tfrac{1}{2}\theta\}^{-1}, \quad j \geqslant i \tag{3.45}$$

where $p = i/(2N)$, $x = j/(2N)$ and $\theta = 4Nu$. It follows that

$$\bar{t_i} \approx (2N)^2 \int_0^p x^{-1} \{(1-x)^{\theta/2-1} - 1\} dx +$$

$$+ \int_p^1 x^{-1}(1-x)^{\theta/2-1} \{1 - (1-p)^{1-\theta/2}\} dx](1 - \tfrac{1}{2}\theta)^{-1} \qquad (3.46)$$

birth-death events. In the particular case $p = (2N)^{-1}$ this is, to a close approximation,

$$\bar{t_i} \approx 2N [1 + \int_{(2N)^{-1}}^p x^{-1} (1-x)^{\theta/2-1} dx] \qquad (3.47)$$

birth-death events. When $\theta = 2$ the form of $\bar{t_i}$ may be found from (3.46) by application of L'Hospital's rule.

Selection can be incorporated in this model by assuming differential birth rates or differential death rates. The two approaches give similar results so we consider here only the case where death rates differ, so that if at any time there are i A_1 genes in the population the probability that the next individual chosen to die is A_1 is

$$\mu_1 i / \{\mu_1 i + \mu_2 (2N - i)\}. \qquad (3.48)$$

If $\mu_1 = \mu_2$ there is no selection while if $\mu_1 < \mu_2$ the allele A_1 has a selective advantage over A_2. It follows that the transition matrix for the number of A_1 individuals has elements

$$p_{i,i-1} = \mu_1 i(2N - i)/[2N\{\mu_1 i + \mu_2(2N - i)\}],$$

$$p_{i,i+1} = \mu_2 i(2N - i)/[2N\{\mu_1 i + \mu_2(2N - i)\}], \qquad (3.49)$$

$$p_{i,i} = 1 - p_{i,i-1} - p_{i,i+1}.$$

The matrix defined by (3.49) is a continuant and the theory of Sect. 2.11 applies. In the notation of that section

$$\rho_0 = 1, \quad \rho_k = (\mu_1/\mu_2)^k$$

and the probability π_i of eventual fixation of A_1, given an initial number of i A_1 individuals, is

$$\pi_i = \{1 - (\mu_1/\mu_2)^i\}/\{1 - (\mu_1/\mu_2)^{2N}\}. \qquad (3.50)$$

If now $\mu_1/\mu_2 = 1 - \tfrac{1}{2}s$, where s is small and positive, A_1 has a slight selective advantage over A_2 and (3.50) can be approximated by

$$\pi(x) \approx \{1 - \exp - \tfrac{1}{2}\alpha x)\}/\{1 - \exp(-\tfrac{1}{2}\alpha)\}, \qquad (3.51)$$

where $x = i/2N$ and $\alpha = 2Ns$. This formula differs from (3.29) by a factor of 2 in the exponents: this is not because the selective differences differ by a factor of 2 (since indeed they do not), but from a more deep-rooted difference between the two models which we examine in the next section.

It is possible to use (3.49) in conjunction with the continuant formulae of Sect. 2.11 to get expressions for mean absorption times, conditional mean absorption times, and so on. We do not do this here since the formulae become very unwieldy and uninformative, and since also we later consider simple approximations for these quantities. It may be remarked finally that no formula is known for the eigenvalues of the matrix defined by (3.49).

3.4 Other "Exchangeable" Models

The Wright-Fisher model and the Moran model, both considered in some detail above, are but two of a wide variety of models that could be introduced to describe the stochastic behavior of genetic populations. We consider now a general class of models which includes these two as particular cases and which will show why there are slight differences in some of the formulae applying for them. We consider only the selectively neutral case since it is here perhaps that the effects of stochastic fluctuations are most interesting. More specifically we consider only the exchangeable processes defined before Theorem 3.1 and a particular case of these, namely conditional branching processes.

Theorem 3.1 shows that, for exchangeable processes, the leading non-unit eigenvalue is $\lambda_2 = E(y_1 y_2)$, where (as defined before Theorem 3.1) y_i is the number of descendent genes of the ith gene in the population. Now by symmetry

$$2N \text{ var } (y_i) + 2N(2N - 1) \text{ covar } (y_i, y_j) = 0,$$

or

$$\text{covar } (y_i, y_j) = - \sigma_g^2/(2N - 1), \tag{3.52}$$

where $\sigma_g^2 = \text{var } (y_i)$. Immediately then

$$\begin{aligned}
\lambda_2 &= E(y_1 y_2) \\
&= \text{covar } (y_1, y_2) + E(y_1) \, E(y_2) \\
&= 1 - \sigma_g^2/(2N - 1).
\end{aligned} \tag{3.53}$$

To confirm this formula, in the Wright-Fisher model y_i has a binomial distribution with index $2N$ and parameter $(2N)^{-1}$, so that

$$\lambda_2 = 1 - \{1 - (2N)^{-1}\}/(2N - 1) = 1 - (2N)^{-1}$$

agreeing with (3.17), and a parallel calculation can be made for the Moran model.

Exchangeable models have further properties in common. It is clear by symmetry that the probability of eventual fixation of any allele in such a model must be its initial frequency. Further, suppose that there are $X(t) \, A_1$ genes in such a model at time t.

Suppose $X(t) = i$ and relabel genes so that the first i genes are A_1. Then

$$\text{var } \{X(t+1)|X(t)\} = \text{var } (y_1 + \ldots + y_i)$$
$$= i\sigma_g^2 + i(i-1) \text{ covar } (y_1, y_2)$$
$$= i(2N-i)\sigma_g^2/(2N-1) \qquad (3.54)$$

from (3.52). Writing $x(t) = X(t)/2N$ it is clear that

$$\text{var } \{x(t+1)|x(t)\} = x(t)\{1 - x(t)\}\sigma_g^2/(2N-1). \qquad (3.55)$$

Arguments parallel to those leading to (3.5) now show that if two alleles A_1 and A_2 are allowed in the population, the mean time until fixation of one or other allele is

$$\bar{t}(p) \approx -(4N-2)\{p \log p + (1-p) \log (1-p)\}/\sigma_g^2, \qquad (3.56)$$

where p is the initial frequency of A_1. This formula explains the factor of 2 discussed after equation (3.43). In the Wright-Fisher model $\sigma_g^2 \approx 1$ while in the Moran model $\sigma_g^2 \approx 2/(2N)$. Setting aside the factor $2N$ as explained by the conversion from generations to birth-death events it is clear that the crucial factor is the difference in the variance in offspring distribution. It is also this factor which leads to the difference between (3.29) and (3.51) and between other similar pairs of formulae.

A particular case of exchangeable processes arises with conditional branching process models. Here we suppose that each gene produces (independently) k offspring with probability $f_k(k = 0, 1, 2, 3, \ldots)$. Put $f(s) = \Sigma f_i s^i$: then the generating function of the distribution of the total number of offspring genes is $[f(s)]^{2N}$. We now make the condition that the total number of such offspring is $2N$. If at time t there were $i A_1$ genes, the probability p_{ij} that at time $t+1$ there will be $j A_1$ genes is then

$$p_{ij} = \frac{\text{coeff } t^i s^{2N} \text{ in } [f(ts)]^i [f(s)]^{2N-i}}{\text{coeff } s^{2N} \text{ in } [f(s)]^{2N}}. \qquad (3.57)$$

Transition probabilities of this form were introduced by Moran and Watterson (1959) who used them to find explicit expressions for the leading non-unit eigenvalue in dioecious populations with various family structures. Extensions to this theory were given by Feldman (1966).

Karlin and McGregor (1965) have analysed such models in detail. They show in particular that the eigenvalues of the matrix $\{p_{ij}\}$ are

$$\lambda_0 = \lambda_1 = 1, \quad \lambda_k = \frac{\text{coeff } s^{2N-k} \text{ in } [f(s)]^{2N-k}[f'(s)]^k}{\text{coeff } s^{2N} \text{ in } [f(s)]^{2N}}, \qquad (3.58)$$

$$k = 2, 3, \ldots, 2N.$$

Note that these must agree with the values found from (3.16) since a conditional branching process is an exchangeable model. We check that this agreement holds for the eigenvalue λ_2. It is clear from (3.57) that

$$\sum_j p_{ij}t^j = \frac{\text{coeff } s^{2N} \text{ in } [f(ts)]^i[f(s)]^{2N-i}}{\text{coeff } s^{2N} \text{ in } [f(s)]^{2N}}.$$

Differentiating twice with respect to t and putting $t = 1$,

$$\sum_j j(j-1)p_{ij} = \lambda_2 i(i-1) + \eta_2 i, \qquad (3.59)$$

where λ_2 is defined by (3.58) and η_2 is some constant independent of i and j. Note that $\sum_j p_{ij} = i$ by symmetry and that $\sum_j j(j-1)p_{1j} = \sigma_g^2$, where σ_g^2 is defined after (3.52). Thus putting $i = 1$ in (3.59) we get $\eta_2 = \sigma_g^2$ and then putting $i = 2$,

$$\sum_j j(j-1)p_{2j} = 2\lambda_2 + 2\sigma_g^2$$

so that

$$\sum_j j^2 p_{2j} = 2\lambda_2 + 2\sigma_g^2 + 2. \qquad (3.60)$$

But the left-hand side in (3.60) is $E(y_1 + y_2)^2$, where y_i is the random number of offspring genes left by parental gene i. It follows that

$$2 + 2\sigma_g^2 + 2E(y_1 y_2) = 2\lambda_2 + 2\sigma_g^2 + 2$$

or

$$\lambda_2 = E(y_1 y_2),$$

as required. Parallel calculations can be made for the remaining eigenvalues, but we do not pursue the details here. Note finally that there are exchangeable models which cannot be cast in the form of a conditional branching process model (the Moran model is a case in point), and thus the former are more general than the latter: however, the extensive Karlin and McGregor theory for conditional branching processes often allows a very rapid examination of their properties.

3.5. K-Allele Models

The above models can easily be extended to allow K different alleles at the locus in question, where K is an arbitrary positive number. In this case the population configuration at any time can be described by a vector $(X_1, X_2, ..., X_K)$, where X_i is the number of genes of allelic type A_i. (Note that if we assume, as is usual, that $X_1 + X_2 +$

... + $X_K = 2N$, only $K - 1$ elements in the above vector are independent.) The most interesting cases of such models arise when there is no mutation and an exchangeable model determines the evolution of the population. In this case any allele A_i can be treated on its own (all other alleles being classed simply as non-A_i) and much of the theory of the preceding sections can be applied. One problem for which the preceding theory is inadequate is to find the mean time until loss of the first allele lost, the mean time until loss of the second allele lost, and so on.

We consider in particular the natural generalization of the model (1.43), namely

Pr $\{Y_i$ genes of allele A_i at time $t + 1|X_i$ genes of allele

$$i \text{ at time } t, \quad i = 1, 2, ..., K\}$$

$$= \frac{(2N!)}{Y_1! Y_2! ... Y_K!} \; \psi_1^{Y_1} \psi_2^{Y_2} ... \psi_K^{Y_K}. \tag{3.61}$$

Here ψ_i could be a complicated function involving mutation and selection parameters as well as $X_1 ... X_K$. We consider here only the simple case where such factors do not arise so that $\psi_i = X_i/(2N)$. In this case the model (3.61) is exchangeable and the comments made above concerning such models apply. The eigenvalues of the matrix defined by (3.61) are precisely the values (3.17) where now λ_j has multiplicity $(K + j - 2)!/\{(K - 2)!j!\}, (j = 2, 3, ..., 2N)$. The eigenvalue $\lambda_0 = \lambda_1 = 1$ has total multiplicity K. These eigenvalues have the interesting interpretation that

$$\text{Pr \{at least } j \text{ allelic types remain present at time } t\} \sim \text{const } \lambda_j^t. \tag{3.62}$$

Expressions for the mean times between losses of alleles are given explicitly later [cf. (5.105) and (5.106)]: the eigenvalue expression (3.62) does not give useful information for these mean times.

When mutation exists between all alleles there will exist a multidimensional stationary distribution of allelic numbers. The means, variances and covariances in this distribution can be found by procedures analogous to those leading to (3.22) and (3.23). We consider in detail only the case where mutation is symmetric: here the probability that any gene mutates is assumed to be u, and given that a gene of allelic type A_i has mutated, the probability that the new mutant is A_j is $(K - 1)^{-1}, (j \neq i)$. Clearly, in the stationary distribution, the mean number of A_i alleles is $2N/K$. On the other hand, it sometimes occurs that this is not a likely value for the number of A_i genes, and we see this best by finding the probability F that two genes taken at random from the population are of the same allelic type. Generalizing the argument that led to (3.24) we find, ignoring terms of order u^2,

$$F = [(2N)^{-1} + \{1 - (2N)^{-1}\}F](1 - 2u) + \{1 - (2N)^{-1}\}(1 - F)\{2u/(K - 1)\}$$

or, if $\theta = 4Nu$,

$$F \approx (K - 1 + \theta)/(K - 1 + K\theta). \tag{3.63}$$

Note that this agrees with (3.24) for $K = 2$ and, letting $K \to \infty$,

$$F \approx (1 + \theta)^{-1}. \tag{3.64}$$

This formula is of particular value, particularly for the theory of Chap. 9. Our immediate interest in it is to consider the case where θ is small. Then $F \approx 1$, and this implies that it is very likely that one or other allele appears with high frequency with the remaining alleles having negligible frequency. This situation is quite different from that predicted by considering the mean vector only.

The eigenvalues of the matrix defined by the symmetric mutation model are the values (3.17) if λ_i is multiplied by $\{1 - uK(K - 1)^{-1}\}^i$. The multiplicity of λ_i is $(i + K - 2)!/\{i!(K - 2)!\}$.

3.6. Infinitely Many Alleles

The model of the previous section can be extended to the case where an infinite number of alleles can arise at the locus in question. It is then assumed that a gene mutates with probability u and that the new mutant is of an entirely novel allelic type not currently or previously existing in the population. (We discuss in Chap. 8 why such a model might be considered.) To be specific, we assume that if in generation t, there are X_i genes of allelic type A_i ($i = 1, 2, 3, \ldots$), then the probability that in generation $t + 1$ there will be Y_i genes of allelic type A_i, together with Y_0 new mutant genes (all necessarily of novel allelic type) is

$$\Pr \{Y_0, Y_1, Y_2, \ldots |X_1, X_2, \ldots\} = \frac{(2N)!}{\Pi \, Y_i!} \, \Pi \, \pi_i^{Y_i}, \tag{3.65}$$

where $\pi_0 = u$ and $\pi_i = X_i(1 - u)/(2N), i = 1, 2, 3, \ldots$.

This model differs fundamentally from the K-allele model of the previous section in that, since each allele will sooner or later be lost from the population, there can exist no non-trivial stationary distribution for the frequency of any allele. Nevertheless we are interested in stationary behaviour, and it is thus important to consider what concepts of stationarity exist for this model. To do this we consider delabelled configurations of the form $\{a, b, c, \ldots\}$, where such a configuration implies that there exist a genes of one allelic type, b genes of another allelic type, and so on. The specific allelic types involved are not of interest. The possible configurations can be written down as $\{2N\}$, $\{2N - 1, 1\}$, $\{2N - 2, 2\}$, $\{2N - 2, 1, 1\}$, $\ldots$, $\{1, 1, 1, \ldots, 1\}$ in dictionary order: the number of such configurations is $p(2N)$, the number of partitions of $2N$ into positive integers. For small values of N values of $p(2N)$ are given by Abramowitz and Stegun (1965), Table 24.5) who provide also asymptotic values for large N. It is clear that (3.65) implies certain transition probabilities from one configuration to another. Although these probabilities are extremely complex and the Markov chain of configurations has an extremely large number of states, nevertheless standard theory show that there exists a stationary distribution (of configurations), some of the characteristics of which we now explore.

Consider first the probability that two genes drawn at random are of the same allelic type. For this to occur neither gene can be a mutant and further both must be descended from the same parent gene (probability $(2N)^{-1}$) or different parent genes which were of the same allelic type. Writing $F_2^{(t)}$ for the desired probability in generation t, we get

$$F_2^{(t+1)} = (1-u)^2[(2N)^{-1} + \{1-(2N)^{-1}\}F_2^{(t)}].\tag{3.66}$$

At equilibrium, $F_2^{(t+1)} = F_2^{(t)} = F_2$ and thus

$$F_2 = \{1 - 2N + 2N(1-u)^{-2}\}^{-1} \sim (1+\theta)^{-1},\tag{3.67}$$

where $\theta = 4Nu$. Consider next the probability $F_3^{(t+1)}$ that three genes drawn at random in generation $t+1$ are of the same allelic type. These three genes will all be descendants of the same gene in generation t, (probability $(2N)^{-2}$), of two genes (probability $3(2N-1)(2N)^{-2}$) or of three different genes (probability $(2N-1)(2N-2)(2N)^{-2}$). Further, none of the genes can be a mutant, and it follows that

$$F_3^{(t+1)} = (1-u)^3(2N)^{-2}[1 + 3(2N-1)F_2^{(t)} + (2N-1)(2N-2)F_3^{(t)}].\tag{3.68}$$

At equilibrium $F_3^{(t+1)} = F_3^{(t)} = F_3$ and simple rearrangement in (3.68) yields

$$F_3 \approx 2(2+\theta)^{-1}F_2 \approx 2!/[(1+\theta)(2+\theta)].\tag{3.69}$$

Continuing in this way we find

$$F_i^{(t+1)} = (1-u)^i[(2N-1)(2N-2)\ldots(2N-i+1)(2N)^{1-i}F_i^{(t)} + \text{terms in}$$
$$F_{i-1}^{(t)}, \ldots, F_2^{(t)}]\tag{3.70}$$

and that for small values of i,

$$F_i \approx (i-1)!/[(1+\theta)(2+\theta)\ldots(i-1+\theta)].\tag{3.71}$$

Note that we can also interpret F_i as the probability that a sample of i genes contains only one allelic type, or, in other words, that the sample configuration is $\{i\}$. This conclusion may be used to find the probability of the sample configuration $\{i-1, 1\}$. The probability that, in a sample of i genes, the first $i-1$ genes are of one allelic type while the last gene is of a new allelic type is $F_{i-1} - F_i$. The probability we require is, for $i \geqslant 3$, just i times this, or

$$\Pr\ \{i-1, 1\} = i\{F_{i-1} - F_i\} \approx i(i-2)!\theta/[(1+\theta)(2+\theta)\ldots(i-1+\theta)].\tag{3.72}$$

For $i = 2$ the required probability is

$$\Pr\ \{1, 1\} \approx \theta/(1+\theta).\tag{3.73}$$

The probabilities of other configurations can be found in a similar way and we illustrate this by considering the probability $F_{2,2}^{(t+1)}$ that, of four genes drawn at random in generation $t+1$, two are of one allelic type and two of another. Clearly none of the genes can be a mutant, and furthermore they will be descended from four different parent genes (of configuration $\{2, 2\}$), from three different parent genes (of configuration $\{2, 1\}$, the singleton being transmitted twice), or from two different parent genes, both transmitted twice. Considering the probabilities of the various events, we find

$$F_{2,2}^{(t+1)} = (1-u)^4 (2N)^{-3} [(2N-1)(2N-2)(2N-3)F_{2,2}^{(t)} + 2(2N-1)(2N-2)$$
$$F_{2,1}^{(t)} + 3(2N-1)F_{1,1}^{(t)}]. \tag{3.74}$$

Retaining only higher-order terms and letting $t \to \infty$,

$$F_{2,2} \approx (3+\theta)^{-1} F_{2,1} = 3/[(1+\theta)(2+\theta)(3+\theta)]. \tag{3.75}$$

Continuing in this way we find (Karlin and McGregor, 1972), that for sample size $r(r \gg N)$,

$$F_{r_1, r_2, \ldots, r_k} = \frac{r! \theta^k}{1^{\alpha_1} 2^{\alpha_2} \ldots r^{\alpha_r} \alpha_1! \alpha_2! \ldots \alpha_r! S_r(\theta)}. \tag{3.76}$$

Here $r_1 + r_2 + \ldots + r_k = r$, α_i indicates the number of values in the set $(r_1, r_2, \ldots, r_k)$ equal to i, $S_r(\theta)$ is defined as $\theta(\theta+1)(\theta+2) \ldots (\theta+r-1)$, and the left-hand side may be interpreted as the probability that a sample of r genes contains k different allelic types, r_1 genes of one unlabelled allelic type, r_2 of another, and so on.

By suitable summation in (3.76) the probability distribution of the random variable k may be found as

$$\text{Pr } \{k \text{ allelic types in sample}\} = |S_r^k| \theta^k / S_r(\theta) \tag{3.77}$$

where $|S_r^k|$ is the coefficient of θ^k in $S_r(\theta)$ and is hence a Stirling number of the first kind (see Abramowitz and Stegun, 1965). Combining (3.76) and (3.77),

$$\text{Pr } \{r_1, r_2, \ldots, r_k | k\} = \frac{r!}{|S_r^k| 1^{\alpha_1} 2^{\alpha_2} \ldots r^{\alpha_r} \alpha_1! \alpha_2! \ldots \alpha_r!}. \tag{3.78}$$

These formulae will be of considerable interest to us in the sequel. For the moment we leave them, noting only that, from (3.77),

$$\text{E}(k) = \frac{\theta}{\theta} + \frac{\theta}{\theta+1} + \frac{\theta}{\theta+2} + \ldots + \frac{\theta}{\theta+r-1}. \tag{3.79}$$

The mode of derivation of the above formulae allows us to make a further observation. Equation (3.66) can be rewritten in the form

$$F_2^{(t+1)} - F_2^{(\infty)} = (1-u)^2 \{1 - (2N)^{-1}\} \{F_2^{(t)} - F_2^{(\infty)}\} \tag{3.80}$$

and this implies that $(1 - u)^2 \{1 - (2N)^{-1}\}$ is an eigenvalue of the configuration process transition matrix. A similar argument using (3.68) shows that a second eigenvalue is $(1 - u)^3 \{1 - (2N)^{-1}\}\{1 - 2 (2N)^{-1}\}$. Equations (3.70) and (3.74) suggest that $(1 - u)^4 \{1 - (2N)^{-1}\}\{1 - 2 (2N)^{-1}\}\{1 - 3 (2N)^{-1}\}$ is an eigenvalue of multiplicity 2. It is indeed found that

$$\lambda_i = (1 - u)^i \{1 - (2N)^{-1}\}\{1 - 2(2N)^{-1}\} \dots \{1 - (i - 1)(2N)^{-1}\} \qquad (3.81)$$

is an eigenvalue of the configuration process matrix and that its multiplicity is $p(i) - p(i - 1)$. This provides a complete listing of all the eigenvalues. For details see Ewens and Kirby (1975).

We consider next the number of alleles existing in the population at any time. Any specific allele A_m will be introduced with frequency $(2N)^{-1}$ into the population and after a random number of generations will leave the population, never to return. The frequency of A_m is a Markovian random variable with transition matrix (3.18) (with ψ_i defined immediately below). There will exist a mean time T until A_m leaves the population. The mean number of new alleles to be formed each generation is $2Nu$, and the mean number to be lost each generation through mutation and random drift is $\bar{K}/T$, where $\bar{K}$ is the mean number of alleles existing in each generation. It follows, by balancing the number of alleles gained each generation with the number lost that, at stationarity,

$$\bar{K} = 2Nu\, T.$$

An approximation to T is found by putting $p = (2N)^{-1}$ in (3.20) and this gives, to a close approximation,

$$\bar{K} \approx \theta[1 + \int_{(2N)^{-1}}^{1} x^{-1}(1 - x)^{\theta - 1}dx]. \qquad (3.82)$$

A more detailed approximation is possible. If $\bar{K}(x_1, x_2)$ is the mean number of alleles present in the population with frequency in (x_1, x_2) $((2N)^{-1} \leqslant x_1 < x_2 \leqslant 1)$, then

$$\bar{K}(x_1, x_2) \approx \theta \int_{x_1}^{x_2} x^{-1} (1 - x)^{\theta - 1}dx. \qquad (3.83)$$

This equation can be used to confirm (3.79). An allele whose population frequency is x is observed in a sample of size r with probability $1 - (1 - x)^r$. From this and (3.83) it follows that the mean number of different alleles observed in a sample of size r is approximately

$$\theta \int_{0}^{1} \{1 - (1 - x)^r\}x^{-1}(1 - x)^{\theta - 1}dx$$

and the value of this expression is precisely (3.79). The function $\theta x^{-1}(1 - x)^{\theta - 1}$ is called the "frequency spectrum" of the process considered and can be used (as we have just seen) to arrive rapidly at results reached more laboriously by discrete distribution methods. Thus for example

Pr {only one allele observed in a sample of i genes}

$$\approx \theta \int_0^1 x^i \{x^{-1}(1-x)^{\theta-1}\}dx$$

$$= (i-1)!/[(1+\theta)(2+\theta) \dots (i-1+\theta)]$$

and this is precisely (3.71). More complex formulae such as (3.76) can be rederived using multivariate frequency spectra, but we do not pursue the details. We shall later find a number of uses for frequency spectra, all arising through their definitions in equations of the form (3.83).

Analogous, and indeed more explicit, formulae can be found for similar infinite allele processes. Thus in the Moran model considered in Sect. 3.3 one can assume that mutation occurs, each new mutant being of an entirely novel allelic type. The possible configurations of the process are the same as those given above, but here an explicit probability can be given for each configuration. To do this it is convenient to change notation and let $\beta_j (j = 1, 2, \dots, 2N)$ be the number of allelic types with exactly j representative genes. Hence $\Sigma j\beta_j = 2N$, and it can be shown (see Trajstman, 1974) that the stationary distribution of the process is

$$\Pr\{\beta_1, \dots, \beta_{2N}\} = \left(\begin{matrix} \gamma + 2N - 1 \\ 2N \end{matrix}\right)^{-1} \prod_{j=1}^{2N} (\gamma/j)^{\beta_j}/\beta_j!, \tag{3.84}$$

where $\gamma = 2Nu/(1-u)$ and u is the mutation rate. It can further be shown (Kelly, 1976) that this process is reversible. In other words, if we write down the possible states $E_1, E_2, \dots, E_{p(2N)}$ of the process and the transition probabilities between them, Eq. (2.140) holds true when we interpret α_i as the stationary probability of the ith configuration. Thus the prospective and retrospective behaviors of the process are identical, a fact we shall take advantage of later in discussing certain properties of this model.

Various generalizations of the selectively neutral model are possible. One generalization supposes that all heterozygotes have the same fitness $1 + s$ $(s > 0)$ and all homozygotes fitness 1. An extreme example arises for self-sterility alleles where homozygotes cannot appear: here we put $s = \infty$. For such models the simple symmetry arguments, which lead to (3.82), no longer apply and a more complex analysis is necessary. We consider this further in Chap. 5. A second generalization supposes that alleles fall into two classes such that individuals having two "favored" alleles have fitness $1 + 2s$, those having only one favored allele have fitness $1 + s$, and those having no favored allele, a fitness 1. This model also is considered in more detail in Chap. 5.

3.7 Frequency-Dependent Selection

There is no requirement that the fitness values w_{ij} in the model defined by (3.18) and (3.27) should be fixed constants (although so far we have assumed that they are), and the analysis we have carried out, in particular the derivation of (3.28), continues to be valid when the w_{ij}, (or s and h in (3.28)) are functions of x. This can lead to some interesting consequences. Thus for small t the fitness scheme

$$w_{11} = 1 + t(1 - \tfrac{1}{2}x), \quad w_{12} = 1, \quad w_{22} = 1 + \tfrac{1}{2} tx,$$

is equivalent to (1.24b) if we put $h = \tfrac{1}{2}x/(1-x)$. Using this value in (3.28) gives $\pi(x)$ $= x$. Thus survival probabilities for this frequency-dependent fitness scheme are the same as those obtaining when there are no fitness differentials. We defer further consideration of these remarks until later.

3.8 Two Loci

We consider in this section two-locus Markov chain analogs of the one-locus model (1.43). While a good deal of progress on the problems considered in this section (and other problems) is possible using diffusion theory (Ohta and Kimura, 1969a, b; Littler, 1973), we defer consideration of these analyses to a later section (Sect. 6.6).

Two Markov chain models have been introduced as reasonable extensions of (1.43). The first is the "random union of zygotes" model of Kimura (1963), Watterson (1970), Serant and Villard (1972) and Littler (1973) and the second is the "random union of gametes" model of Karlin and McGregor (1968) and Hill and Robertson (1966, 1968). For convenience we call these here the RUZ and RUG models, respectively. A general theory of Weir and Cockerham (1974a, b) yields many results for both models. We will follow as far as possible the notation of Sect. 2.9 in discussing these models.

The RUZ model is defined as follows. Suppose a population of fixed size N contains, in generation t, $X_{ij}(t)$ individuals whose genotype is made up of gametic types i and j $(i \leqslant j)$ (for a definition of a gamete of type i, see Sect. 2.9). Let α_{ijk} be the probability that a gamete produced by such an individual is of type k. (When $i,j = 1$, 4 or 2, 3 these probabilities will involve the recombination fraction R.) Then

$$c_k(t) = \sum_{i \leqslant j} \sum X_{ij}(t)\alpha_{ijk}N^{-1}$$

is the probability that a gamete chosen at random forming generation $t + 1$ is of type k. It is now assumed that the N individuals in generation $t + 1$ have their gametes determined by $2N$ independent trials with the probability of a gamete of type k on any trial being $c_k(t)$.

The values $X_{ij}(t + 1)$ are thus determined from the $X_{ij}(t)$ only through the quantities $c_k(t)$, and it follows that the quadruple (c_1, c_2, c_3, c_4) forms a Markov chain. The transition matrix of this chain can be found from that of the X_{ij} and is perhaps best written down in terms of the joint moment-generation function

$$E[e^{\Sigma\theta jcj(t+1)}|c_i(t)] = \left[\sum_{i=1}^{4} \sum_{j=1}^{4} c_i(t)c_j(t)e^{k\overset{4}{\underset{=1}{\Sigma}}\alpha_{ijk}\theta\,k/N} \right]^N \tag{3.85}$$

This equation was given by Watterson (1970) and requires the biologically reasonable definition $\alpha_{ijk} = \alpha_{jik}$.

Before examining the consequences of (3.85) we define the RUG model. Here we ignore the zygote stage in our formulation of the model and simply assume, following (2.64), that if generation t produces $n_i(t)$ gametes of type i ($i = 1, ..., 4$), the probability that generation $t + 1$ produces $n_i(t + 1)$ gametes of type $i(i = 1, ..., 4)$ is

$$p(\mathbf{n}(t) \to \mathbf{n}(t+1)) = \frac{(2N)!}{\prod\limits_{i=1}^{4} n_i(t+1)!} \prod_{i=1}^{4} \psi_i^{n_i(t+1)} \tag{3.86}$$

where

$$\psi_i = c_i(t) + \eta_i R\{c_1(t)c_4(t) - c_2(t)c_3(t)\}, \qquad c_i(t) = n_i(t)/(2N),$$

and the η_i have been defined in (2.64). The models defined by (3.85) and (3.86) are not equivalent in general, but are so in the limiting case $R = 0$.

The qualitative behaviors of both models are identical. All four gametes will segregate in the population (with the possibility that one gamete is temporarily absent not excluded) until, after a random time the distribution of which is determined by R and N and the initial gamete frequencies, one or other allele is lost from the population. From this time on segregation continues at one locus only and the model (1.43) applies. Eventually one or other allele at this locus is lost, and the population consists entirely of one of the four gametic types. Clearly questions of particular interest concern the time that all four gametes co-exist, the gamete eventually fixed and, because linkage disequilibrium is of major interest in two-locus systems, the transient behavior of the coefficient of linkage disequilibrium, defined in generation t by

$$D(t) = c_1(t)c_4(t) - c_2(t)c_3(t). \tag{3.87}$$

We attack these questions quantitatively by discussing, for both models, (i) the eigenvalue μ for which, for large t,

$$\text{Prob (segregation continues at both loci)} \sim c\mu^t, \tag{3.88}$$

(ii) the probability of ultimate fixation of gamete type i and (iii) the mean $E\{D(t)\}$ and mean square $E\{D(t)\}^2$ of the coefficient of linkage disequilibrium at time t.

To find the eigenvalue μ defined by (3.87) we follow the approach of Watterson (1970, 1972). We consider the variable $D(t)$, defined by (3.87), and $S(t)$ and $Z(t)$, defined by

$$S(t) = c_1(t)c_4(t) + c_2(t)c_3(t),$$

$$Z(t) = \{c_1(t) + c_2(t)\}\{c_1(t) + c_3(t)\}\{c_2(t) + c_4(t)\}\{c_3(t) + c_4(t)\}.$$

It is then found that, conditional on $c_1(t), c_2(t), c_3(t), c_4(t)$,

$$E\{S(t+1)\} = a_{11}S(t) + a_{12}\{D(t)\}^2 + a_{13}Z(t),$$

$$E\{D(t+1)\}^2 = a_{21}S(t) + a_{22}\{D(t)\}^2 + a_{23}Z(t),$$

$$E\{Z(t+1)\} = a_{31}S(t) + a_{23}\{D(t)\}^2 + a_{33}Z(t).$$

Here the a_{ij} are constants whose values depend only on N and R. It is clear that, given the initial values $c_i(0)$,

$$E \begin{pmatrix} S(t) \\ \{D(t)\}^2 \\ Z(t) \end{pmatrix} = A^t \begin{pmatrix} S(0) \\ \{D(0)\}^2 \\ Z(0) \end{pmatrix}, \qquad (3.89)$$

where $A = \{a_{ij}\}$. Since $S(t)$ is positive if and only if all four alleles continue to segregate in generation t, a generalization of the method of Appendix B shows that the leading non-unit eigenvalue of the transition matrix implied by (3.89) is the leading eigenvalue of A. This may be calculated either algebraically or numerically. Watterson shows that the largest eigenvalue μ can be written in the form

$$\mu = 1 - (2N)^{-1} - y(2N)^{-1},$$

where y is the solution of a certain cubic equation (see Watterson, 1970, 1972b; Littler, 1973). The most useful discussion of the properties of μ is through numerical examples, but some limiting cases are of special interest. Thus $R = 0$ implies $y = 0$ and hence

$$\mu = 1 - (2N)^{-1}.$$

This is to be expected since for $R = 0$ the model is equivalent to a one-locus model with four alleles for which this eigenvalue has already been established. Perhaps of more importance for consideration of properties of two-locus systems is to consider the behavior of μ for R moderate and N large. Here Watterson (1970, 1972) show that

$$\mu = \{1 - (2N)^{-1}\}^2 + 0(N^{-3}).$$

Indeed μ is very close to the value $\{1 - (2N)^{-1}\}^2$ for R not extremely small and N greater than about 50. Some numerical values (taken from Littler, 1973) are given in Table 3.1. The reason for this behaviour is obvious enough. When N and R are not both small the two loci behave almost independently, so that the probability that segregation continues at both loci is close to the square of the probability that segregation continues at any one locus. As $R \to 0$ the two segregation behaviors become more dependent.

A parallel evaluation of μ for the RUG model has been made by Littler (1973). Here Littler sets up equations of the form

Table 3.1. Values of the eigenvalue μ (see text for definition) in both RUG and RUZ models

				R	
		0.01	0.10	0.20	0.50
	10	0.941	0.910	0.905	0.903
N	25	0.972	0.961	0.961	0.960
	50	0.984	0.980	0.980	0.980

$$E\{[D(t+1)]^2 | c_i(t)\} = b_{11}\{D(t)\}^2 + b_{12}I(t) + b_{13}Z(t),$$

$$E[I(t+1)|c_i(t)] = b_{21}\{D(t)\}^2 + b_{22}I(t) + b_{23}Z(t),$$

$$E[Z(t+1)|c_i(t)] = b_{31}\{D(t)\}^2 + b_{23}I(t) + b_{33}Z(t),$$

where

$$I(t) = \{c_1(t)c_4(t) - c_2(t)c_3(t)\}\{1 - 2c_1(t) - 2c_2(t)\}\{1 - 2c_1(t) - 2c_3(t)\},$$

$Z(t)$ is as defined above and the b_{ij} are constants depending only on N and R. This gives, conditional on the values $c_i(0)$,

$$E\begin{pmatrix} \{D(t)\}^2 \\ I(t) \\ Z(t) \end{pmatrix} = B^t \begin{pmatrix} \{D(0)\}^2 \\ I(0) \\ Z(0) \end{pmatrix} \tag{3.90}$$

where $B = \{b_{ij}\}$. This is analogous to (3.89), and the leading eigenvalue μ for this model is the leading eigenvalue of B.

As for the RUZ model, μ decreases from $1 - (2N)^{-1}$ (at $R = 0$), being always greater than $\{1 - (2N)^{-1}\}^2$ for $R \leqslant 0.5$ (Karlin and McGregor, 1968). For the combinations of N and R values listed in Table 3.1, the numerical values given apply (to the order of accuracy shown) for the RUG model also, although the formulae for the two eigenvalues are different, and for values of N and R not listed the agreement between the two is not quite so close. Nevertheless, the general discussion given for the values of μ in the RUZ model also apply here for the RUG model.

We now turn to probabilities of fixation for the various gametes. These were found by Kimura (1963) for the RUZ model and by Karlin and McGregor (1968) for the RUG model. Suppose we can find functions $\phi_i \{c_1(t), c_2(t), c_3(t), c_4(t)\}$, which we abbreviate to $\phi_i(t)$, having the properties

$$E\{\phi_i(t+1)|c_1(t), c_2(t), c_3(t), c_4(t)\} = \phi_i(t), \quad i = 1, \ldots, 4, \tag{3.91a}$$

$$\phi_i(\infty) = 1 \text{ if gamete } i \text{ eventually fixes,} \tag{3.91b}$$

$$= 0 \text{ otherwise.}$$

Then by iteration in (3.89),

$$E\{\phi_i(\infty)|c_1(0), c_2(0), c_3(0), c_4(0)\} = \phi_i(0).$$

But the left-hand side is just the probability that gamete i fixes: we conclude that if we can find functions $\phi_i(t)$ satisfying (3.91), gamete fixation probabilities are given by the values of $\phi_i(0)$. Functions $\phi_i(t)$ satisfying (3.91) always exist and can usually be found after some trial and error.

One complication arises with this procedure. The RUZ model concerns zygotes rather than gametes and the initial composition of the population then relates to zygote frequencies rather than gamete frequencies. Nevertheless the essence of the above procedure still applies. For the RUZ model it is found that functions $\phi_i(t)$ satisfying (3.91) are

$$\phi_i(t) = c_i(t) + \frac{\eta_i 2NR\, D(t)}{2NR + 1}$$

while for the RUG model,

$$\phi_i(t) = c_i(t) + \frac{\eta_i 2NR\, D(t)}{2NR + 1 - R}.$$

It follows immediately for the RUG model that the probability of fixation of gametes of type i is

$$c_i(0) + \frac{\eta_i 2NR\, D(0)}{2NR + 1 - R}. \qquad (3.92)$$

Matters are slightly more complicated for the RUZ model since we must give probabilities in terms of initial zygotic frequencies. It is found (see Watterson, 1970) that the required value is

$$c_i^* + \frac{\eta_i 2NR\, D^* + \Delta(2N)^{-1}}{2NR + 1},$$

where c_i^* is the frequency of the gamete i among the zygotes of the initial generation, $D^* = c_1^* c_4^* - c_2^* c_3^*$ and $\Delta = \{X_{14}(0) - X_{23}(0)\}(2N)^{-1}$.

Note in both cases that if R is fixed and moderate and N is large, the fixation probability of gamete i is approximately $c_i + \eta_i D$, where c_i stands for c_i^* in the RUZ model and $c_i(0)$ in the RUG model. But this is just the initial value of the product of the frequencies of the two alleles making up the ith gamete so that, to a close approximation, the probability that any gamete fixes is simply the product of the probabilities that the two corresponding alleles fix. This arises because the segregation processes at the two loci are effectively independent. When R is small this is no longer true, and the association between loci must be taken into account when computing fixation probabilities.

We turn finally to the behavior of the linkage disequilibrium function $D(t)$. A considerable part of two-locus theory relates to this quantity so it is of some value to discuss its behavior in the two models under consideration.

Consider first the model (3.85). It is easy to show for this model that

$$E\{D(t+1)|D(t)\} = \{1 - (2N)^{-1} - R\}D(t), \qquad (3.93)$$

and hence the mean value of $D(t)$ decreases to zero geometrically fast with t at rate $1 - (2N)^{-1} - R$. Unless R is close to zero this is quite a rapid rate. A parallel remark applies for the model (3.86) where we find

$$E\{D(t+1)|D(t)\} = \{1 - (2N)^{-1}\}\{1 - R\}D(t).$$

The convergence to zero of $E\{D(t)\}$ is only slightly slower for this model.

More detailed information about the behavior of $D(t)$ will depend on knowledge of the variance of $D(t)$ or equivalently, since the above expressions easily yield the mean of $D(t)$, the expected mean square $E\{D(t)\}^2$. Suppose the initial value of D is zero. Since eventually $D = 0$ (once fixation of one or other allele has occurred), we might expect that in this case the variance of D will increase from zero as t increases and, after achieving a maximum for intermediate values of t, decrease again to zero. If initially D is non-zero we might perhaps expect the variance of D monotonically to decrease to zero.

Although the behavior of the variance of D is by no means simple, these expectations are in essence confirmed for the RUG model by Hill and Robertson (1968) and for both models by Littler (1973). Note that (3.89) and (3.90), respectively, show that $E\{D(t)\}^2$ can be written in the form

$$E\{D(t)\}^2 = \alpha_1\mu_1^t + \alpha_2\theta_1^t + \alpha_3\theta_2^t,$$

for the RUZ model and

$$E\{D(t)\}^2 = \beta_1\mu_2^t + \beta_2\rho_1^t + \beta_3\rho_2^t,$$

for the RUG model. Here the $\alpha_i, \beta_i, \theta_i$ are constants depending on N and R and μ_1, μ_2 are the leading eigenvalues of A and B, respectively. While for large t the behavior of $E\{D(t)\}^2$ in both cases is determined largely by the maximum eigenvalue μ, the behavior for small t is quite complicated and all eigenvalues (and eigenvectors) are needed to describe it. For the RUZ model calculations of Littler (1973) reveal the behavior shown in Fig. 3.1. When D is initially zero, the variance of D increases to a maximum value of order 0.01 and then decreases. The maximum is reached sooner and is slightly greater for small values of N. For large t, the variance of D is smallest for small values of N. When D is not zero initially, $E\{D(t)\}^2$ decreases with t, and for large t is minimized for small values of N. Note that the eigenvalue μ determining the ultimate behaviour of $E\{D(t)\}^2$

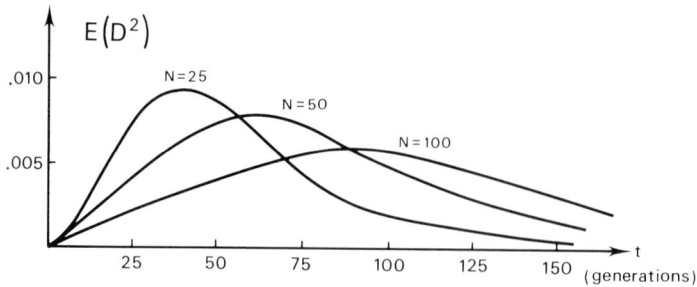

Fig. 3.1. Value of $E(D^2)$ in generation t for various N. All gamete frequencies initially 0.25, R = 0.005. Taken from Littler (1973).

is much closer to unity than the value $1 - (2N)^{-1} - R$ implied by (3.93) for the ultimate behavior of $E\{D(t)\}$. Because of this it has sometimes been asserted that observed values of D can differ significantly form the mean. In view of the above results this conclusion cannot really be drawn, and it is clear that the eigenvalues on their own do not give a complete picture of the true behavior of $D(t)$. In the RUG model, $E\{D(t)\}^2$ can be found by a procedure similar to that used in the RUZ model. Numerical examples have been given by Hill and Robertson (1968). Thus if all gametes have initial frequency 0.25 and NR = 1, var $\{D(t)\}$ reaches a maximum of 0.006 at about N generations, while when NR = 4 a maximum of 0.003 is reached after about N generations. In general the larger NR, the smaller the maximum and the sooner it is reached. For large populations and unlinked loci, var $\{D(t)\}$ is always extremely small, indicating that, by random effects only, it is unlikely that D will ever assume a large value.

All the above has assumed that there is no mutation, so that eventual fixation of one or other gamete is certain. If mutation exists at positive rate from A_i to A_j $(i \neq j)$ and B_i to B_j $(i \neq j)$ there will exist a stationary distribution of gamete frequencies and in particular a stationary distribution of D. Since we are interested in the extent of likely variation of D from zero, we consider now the stationary mean value of D^2 in this mutation case, following the analysis of Ohta and Kimura (1969b). Suppose that the mutation rates are u_1 (from A_1 to A_2), v_1 (from A_2 to A_1), u_2 (from B_1 to B_2) and v_2 (from B_2 to B_1). We consider the three quantities $Z(t), I(t)$ and $\{D(t)\}^2$ introduced above. Ohta and Kimura set up a recurrence relation similar to that given above for the case without mutation, where the coefficients b_{ij} now include mutation terms. By letting $t \to \infty$, non-degenerate limits are found for the expectations of these quantities and in particular for $E\{D(\infty)\}^2$. We defer giving an explicit formula here since a slightly simpler formula is available later (Sect. 6.6) using diffusion theory.

3.9 Effective Population Size

We return now to one-locus models. While the model (1.43) is no more plausible than several others as a description of biological reality, it has (perhaps for historical reasons) assumed a central place in population genetics theory. We have already noted three properties of this model:

(i) maximum non-unit eigenvalue = $1 - (2N)^{-1}$,
(ii) probability that two genes taken at random are descendants of the same parent gene = $(2N)^{-1}$,
(iii) var $\{x(t + 1)|x(t)\} = x(t)\{1 - x(t)\}/(2N)$,
where $x(t)$ is the fraction of A_1 genes in generation t.

In view of these properties it is perhaps natural, if the model (1.43) is used as a standard, to define the effective population size in diploid models other than (1.43) in the following way:

$$N_e^{(e)} = \text{eigenvalue effective population size} = \tfrac{1}{2}(1 - \lambda_{max})^{-1}, \qquad (3.94)$$

$$N_e^{(i)} = \text{inbreeding effective population size} = (2\pi_2)^{-1}, \qquad (3.95)$$

$$N_e^{(v)} = \text{variance effective population size} = \frac{x(t)\{1 - x(t)\}}{2\,\text{var}\,\{x(t + 1)|x(t)\}}. \qquad (3.96)$$

Here λ_{max} is the largest non-unit eigenvalue of the transition matrix of the model considered and π_2 is the probability that two genes taken at random in any generation are descendents of the same parent gene. Our aim is to compute these effective population sizes for two classes of models that generalize (1.43). The first such class is that of general "exchangeable" models (see Sect. 3.4), and the second comprises models that incorporate complicating features such as two sexes, geographical subdivision, fluctuating population sizes, etc.

We consider first general exchangeable models (as defined in Sect. 3.4) and limit attention for the moment to those models where generations do not overlap. Equations (3.53) and (3.94) show immediately for these models that the eigenvalue effective population size $N_e^{(e)}$ is given by

$$N_e^{(e)} = (N - \tfrac{1}{2})/\sigma_g^2, \qquad (3.97)$$

where σ_g^2 is the variance in the number of offspring genes from any given gene. Equations (3.55) and (3.96) show that the variance effective population size $N_e^{(v)}$ is given by

$$N_e^{(v)} = (N - \tfrac{1}{2})/\sigma_g^2. \qquad (3.98)$$

A value for $N_e^{(i)}$ can be found in the following way. Suppose that the ith gene in generation t leaves m_i offspring genes in generation $t + 1$, $(\Sigma m_i = 2N)$. Then the probability, given $m_1, ..., m_{2N}$, that two genes drawn at random in generation $t + 1$ are descendents of the same gene is

$$\sum_{i=1}^{2N} m_i(m_i - 1)/\{2N(2N - 1)\}. \qquad (3.99)$$

The probability π_2 in (3.95) is the expectation of this quantity. Now m_i has mean unity and variance σ_g^2, so that, on taking expectations, $\pi_2 = \sigma_g^2/(2N - 1)$. It follows that

$$N_e^{(i)} = (N - \tfrac{1}{2})/\sigma_g^2 \tag{3.100}$$

and hence, for general exchangeable models, that all three effective population sizes are equal.

An inbreeding effective population number is sometimes defined where attention is focussed on the diploid nature of the organisms in the population. This number will be denoted $N_e^{(id)}$ and is defined as the reciprocal of the probability that two genes taken at random in generation $t + 1$ are descended from the same individual in generation t. This is tantamount, for exchangeable models, to selecting two genes at random in generation t and asking whether two genes drawn at random in generation $t + 1$ are both descended from one or other or both of these. In the notation of (3.99) the probability of this event can be written as the expected value of

$$\sum_{i=1}^{N} (m_i + m_{N+i})(m_i + m_{N+i} - 1)/\{2N(2N-1)\}. \tag{3.101}$$

It is not hard to see that this leads to

$$N_e^{(id)} = \frac{4N - 2}{\sigma_d^2 + 2}, \tag{3.102}$$

where σ_d^2 is the variance of the number of offspring genes from each (diploid) individual. This will agree with (3.100) if, and only if,

$$\sigma_g^2 = 1 - (2N)^{-1}. \tag{3.103}$$

(Note that for exchangeable models $\sigma_d^2 = 2\sigma_g^2 (2N - 2)/(2N - 1)$.) The only frequently occurring model for which (3.103) holds is the model (1.43) itself, so that the practice of regarding (3.100) and (3.102) as equivalent formulas is not justified except for this model.

The above definitions are not appropriate for models (such as 3.30) where generations overlap. Writing N_e for any one of the effective population sizes defined in (3.94)–(3.96), it seems reasonable for such models to define the effective population size as $N_e k/(2N)$, where k is the number of individuals to die each time unit. Since $k = 2N$ for models where generations do not overlap, this leaves (3.94)–(3.96) unchanged for such models. For the model (3.30), where $k = 1$, it yields

$$N_e^{(e)} = N_e^{(i)} = N_e^{(v)} = \tfrac{1}{2}N. \tag{3.104}$$

We turn next to the second class of models where a definition of effective population size is useful, namely those which attempt to incorporate biological complexity more than does (1.43). In all these models the fundamental feature of (1.43), namely the binomial sampling of genes to form one generation from another, is in essence maintained. (It is possible to consider the effective population size for a wider class of such models, for example for general exchangeable models incorporating the biological complexity required, but we do not enter systematically into such a generalization here.) The first model considered assumes the existence of two sexes. Suppose in any generation there are N_1 diploid males and N_2 diploid females, with $N_1 + N_2 = N$. The model assumes that the genetic make-up of each individual in the daughter generation is found by drawing one gene at random (and with replacement) from the male pool of genes and similarly from the female pool. If $X_1(t)$ represents the number of A_1 genes among males in generation t and $X_2(t)$ the corresponding number among females, then $X_1(t + 1)$ can be represented in the form

$$X_1(t + 1) = i(t + 1) + j(t + 1),$$ (3.105)

where $i(t + 1)$ has a binomial distribution with index N_1 and parameter $X_1(t)/(2N_1)$ and $j(t + 1)$ has a binomial distribution with index N_1 and parameter $X_2(t)/(2N_2)$. A similar remark applies to $X_2(t + 1)$, where now the index is N_2 rather than N_1. Evidently the pair $\{X_1(t), X_2(t)\}$ is Markovian and there will exist a transition matrix whose leading non-unit eigenvalue we require to calculate $N_e^{(e)}$.

To do this we use the theory of Appendix 2. It is necessary to find some function $Y(X_1, X_2)$ which is zero in the absorbing states of the system, positive otherwise, and for which

$$E[Y\{X_1(t + 1), X_2(t + 1)\}|X_1(t), X_2(t)] = \lambda Y(t)$$ (3.106)

for some constant λ. Such a function always exists, but some trial and error is usually necessary to find it. In the present case it is found, after much labor, that a suitable function is

$$Y(X_1, X_2) = \tfrac{1}{2} C\{X_1(2N_1 - X_1)(2N_1)^{-2} + X_2(2N_2 - X_2)(2N_2)^{-2}\}$$
$$+ \{1 - (X_1 - N_1)(X_2 - N_2)N_1^{-1} N_2^{-1}\},$$ (3.107)

where

$$C = \tfrac{1}{2}\{1 + (1 - 2N_1^{-1} - 2N_2^{-1})^{1/2}\}.$$

With this definition the eigenvalue λ becomes

$$\lambda = \tfrac{1}{2} [1 - (4N_1)^{-1} - (4N_2)^{-1} + \{1 + N^2(4N_1N_2)^{-2}\}^{1/2}],$$ (3.108)

or approximately

$$\lambda \approx 1 - (N_1 + N_2)(8N_1N_2)^{-1}.$$ (3.109)

From this it follows that, to a close approximation,

$$N_e^{(e)} = 4N_1N_2N^{-1}.$$ (3.110)

Note that if $N_1 = N_2 (= \frac{1}{2} N)$, then $N_e^{(e)} \approx N$, as we might expect, while if N_1 is very small and N_2 is large, $N_e^{(e)} \approx 4N_1$. This latter value is sometimes of use in certain animal breeding programs.

The inbreeding population size is found much more readily. Two genes taken at random in any generation will have identical parent genes if both are descended from the same "male" gene or both from the same "female" gene. The probability of identical parentage is thus

$$\pi_2 = \frac{1}{2} \frac{N-1}{2N-1} \{(2N_1)^{-1} + (2N_2)^{-1}\},$$

and from this it follows that

$$N_e^{(i)} = (2\pi_2)^{-1} \approx 4N_1 N_2 N^{-1}. \tag{3.111}$$

The variance effective population size cannot be found so readily and indeed strictly it is impossible to use (3.96) to find such a quantity since an equation of this form does not exist. The fraction of A_1 genes is not a Markovian variable and in particular, using the notation of (3.96), the variance of $x(t+1)$ cannot be given in terms of $x(t)$ alone. This indicates a real deficiency in this mode of definition of effective population size. On the other hand, we shall see in the next chapter that sometimes a "quasi-Markovian" variable exists in terms of which a generalized expression for the variance effective population size may be defined. In the present case the weighted fraction of A_1 genes, defined as

$$x(t) = X_1(t)/(4N_1) + X_2(t)/(4N_2)$$

has the required quasi-Markovian properties and

$$\text{var } \{x(t+1)|x(t)\} = x(t)\{1 - x(t)\}N(8N_1 N_2)^{-1} + 0(N_1^{-2}, N_2^{-2}).$$

From this a generalized variance effective population size may be defined, in conjunction with (3.96), as

$$N_e^{(v)} = 4N_1 N_2 N^{-1}. \tag{3.112}$$

Thus note that for this model, $N_e^{(e)} \approx N_e^{(i)} \approx N_e^{(v)}$ although strict equality does not hold for any of these relations.

We return now to the case of a monoecious population and consider complications due to geographical structure. A simplified model for this situation which, despite its obvious biological unreality, is most useful in revealing the effect of such subdivision, has been given by Moran (1962). It is supposed that the total population, of size $N(H+1)$, is subdivided into $H+1$ subpopulations each of size N, and that in each generation K genes chosen at random migrate from subpopulation i to subpopulation j for all i, j ($i \neq j$). Suppose that in subpopulation i there are $X_i(t)$ A_1 genes in generation t. There is no single Markovian variable describing the behavior of the total population, but the quantities $X_i(t)$ are jointly Markovian and to find $N_e^{(e)}$ it is necessary to find some function $Y\{X_1(t) ..., X_{H+1}(t)\}$ obeying the requirements of Appendix B. Again after some trial and error it is found that a suitable function Y is defined by

$$Y(t) = [A - D + \{(A - D)^2 + 4BC\}^{1/2}] \sum_i X_i(t)\{2N - X_i(t)\}$$

$$+ 2B \sum_{i \neq j} \sum X_i(t)\{2N - X_j(t)\}. \tag{3.113}$$

Here

$$A = (4N^2 + H^2K^2 + K^2H - 2N - 4NKH)/4N^2,$$
$$B = (4KN - K^2H - K^2)/(4N^2),$$
$$C = (4HKN - K^2H^2 - K^2H)/(4N^2),$$
$$D = (4N^2 + HK^2 + K^2 - 4HK)/(4N^2).$$

Further, with this definition of $Y(t)$, the eigenvalue λ satisfying

$$E\{Y(t + 1)|X_1(t), ..., X_{H+1}(t)\} = \lambda Y(t)$$

becomes

$$\lambda = \tfrac{1}{2}[A + D + \{(A - D)^2 + 4BC\}^{1/2}]. \tag{3.114}$$

If small order terms are ignored, this yields eventually

$$N_e^{(e)} \approx N(H + 1)\{1 + \{2K(H + 1)\}^{-1}\} \tag{3.115}$$

for large H and K. This equation is in fact accurate to within 10% even for $H = K = 1$, and it thus reveals that population subdivision leads to a slight increase in the eigenvaue effective population size compared to the value $N(H + 1)$ obtaining with no subdivision.

The inbreeding effective population size $N_e^{(i)}$ can be found most efficiently by noting that it is independent of K, since the act of migration is irrelevant to the computation of its numerical value. Thus immediately from (3.100)

$$N_e^{(i)} = \{N(H + 1) - \tfrac{1}{2}\}/\{1 - (2N)^{-1}\}, \tag{3.116}$$

since each gene produces a number of offspring according to a binomial distribution with index $2N$ and parameter $(2N)^{-1}$. This value clearly differs only trivially from the true population size $N(H + 1)$ although, for small H and K, it differs slightly from $N_e^{(e)}$).

The computation of $N_e^{(v)}$ is beset with substantial difficulties since there exists no Markovian variable for the model nor indeed, unless migration rates are of a large order of magnitude, is there even a "quasi-Markovian" variable. Because of this no satisfactory value for $N_e^{(v)}$ has yet been put forward.

We consider finally a population whose size assumes cyclically the sequence of values $N_1, N_2, N_3, ..., N_k, N_1, N_2, ...$. There is no unique value of $N_e^{(e)}, N_e^{(i)}$ or $N_e^{(v)}$ in this case, and it is convenient to extend our previous definition to cover k consecutive generations of the process. If the population size in generation $t + k$ is N_i, it is easy to see that if $X(t)$ is the number of A_1 genes in generation t, and in each generation reproduction occurs according to the model (1.43),

$$E[X(t + k)\{2N_i - X(t + k)\}|X(t)] = X(t)\{2N_i - X(t)\} \prod_{i=1}^{k} \{1 - (2N_i)^{-1}\}.$$

Defining now $N_e^{(e)}$ by the equation

$$\{1 - (2N_e^{(e)})^{-1}\}^k = \prod_{i=1}^{k} \{1 - (2N_i)^{-1}\},$$

it is clear from Appendix 2 that if k is small and the N_i large,

$$N_e^{(e)} \approx k \{N_1^{-1} + ... + N_k^{-1}\}^{-1}. \tag{3.117}$$

Thus the eigenvalue effective population size is effectively the harmonic mean of the various population sizes taken during the k-generation cycle. A parallel formula holds for $N_e^{(i)}$ although here it is easier to work through the probability $Q(t + k)$ that two genes in generation $t + k$ do *not* have the same ancestor in generation t. Clearly

$$Q(t + k) = \{1 - (2N_{i-1})^{-1}\}Q(t + k - 1)$$

and iteration over k generations gives

$$Q(t + k) = \prod_{i=1}^{k} \{1 - (2N_i)^{-1}\}Q(t).$$

Elementary calculations now show that $N_e^{(i)}$ is also essentially equal to the harmonic mean of the population size. Again, if $x(t)$ is the fraction of A_1 genes in generation t,

$$\text{var } \{x(t + k)|x(t)\} = \tfrac{1}{2} k \{N_1^{-1} + N_2^{-1} + ... + N_k^{-1}\}x(t)\{1 - x(t)\} + 0(N_i^{-2}).$$

This shows that, to a suitable approximation, $N_e^{(v)}$ is also the harmonic mean of the population size.

This conclusion has been generalized by Karlin (1968). Karlin assumed that in any generation the population size takes one or other of the values $N_1, N_2, N_3, ..., N_m$ according to Markov chain rules, so that there exists a probability q_{ij} of a transition from a population of size N_i to a population of size N_j. The cyclic case just considered arises if $q_{i,i+1} = 1$ for $i = 1, 2, ..., m - 1$, $q_{m1} = 1$. The leading non-unit eigenvalue, and hence $N_e^{(e)}$, depends on the transition matrix $\{q_{ij}\}$ as well as the particular form for

$f(s)$ assumed in (3.57). Explicit values are hard to achieve in general but in all cases for which expressions can be found, $N_e^{(e)}$ is close to the weighted harmonic mean of the possible population sizes. Thus if $\{f_k\}$ is assumed to be a Poisson distribution, leading to a generalized binomial transition probability extending (1.43), and if $q_{ij} = q_j$ for all i, the leading non-unit eigenvalue is

$$1 - \Sigma\, q_j (2N_j)^{-1},$$

so that

$$N_e^{(e)} = [\Sigma\, q_j N_j^{-1}]^{-1}.$$

We conclude with some general observations. First, we have only considered one source of complexity at a time in computing effective population sizes. Thus it would, for example, be possible to extend the definition of $N_e^{(i)}$ for the dioecious model to general exchangeable models (to arrive at the formula

$$N_e^{(i)} = 4N^2 \{N_1 \sigma_m^2 + N_2 \sigma_f^2 + N^2 N_1^{-1} - 2N\}^{-1},$$

where σ_m^2 is the variance in the number of offspring genes from a "male" gene, and σ_f^2 is similarly defined for "female" genes). Again, we could consider subdivided populations admitting two sexes. There is little doubt that the effective population sizes found in such cases is found by a natural composition of the values given above.

Second, many papers and several textbooks use what appears to be a definition of $N_e^{(i)}$ defined by the outcome of a given experiment or by a given field observation. Thus, for example, in the (diploid) formula (3.102) the symbol σ_d^2 is sometimes replaced by V, where V is defined by

$$V = \Sigma\, (n_i - 2)^2 / N,$$

where n_i is the (random) number of genes produced by the ith individual. While such a definition might be of use in a retrospective analysis of a given experiment, it is not allowable for theoretical purposes since V is a random variable and thus can take quite different values for two different populations that have identical properties and thus must have the same value of $N_e^{(v)}$.

Third, all three effective population sizes suffer some defects. Thus, $N_e^{(e)}$ and $N_e^{(v)}$ are defined assuming two alleles at the locus in question. $N_e^{(i)}$ is not defined in terms of allelic type and is thus possibly superior to $N_e^{(e)}$ and $N_e^{(v)}$ as a pure measure of population structure although as we note in a moment each expression has its special interpretation and usefulness and further $N_e^{(i)}$ is not of much value in characterizing various properties of the geographically structured case.

Fourth, while the three effective sizes are often nearly equal in the examples considered above, they can in other cases differ substantially. This occurs particularly in populations with non-constant size. An extreme example arises when a single heterozygote in generation t gives rise to a very large number of offspring in generation $t + 1$. Here both $N_e^{(v)}$ and $N_e^{(e)}$ are very large, but $N_e^{(i)}$ is unity. Thus $N_e^{(v)}$ and $N_e^{(e)}$ tend to be defined in terms of the future evolution of the population, whereas $N_e^{(i)}$ is concerned more with its past. In some circumstances one effective size is of most-interest and for other circumstances, another.

Fifth, we have not defined an effective population size for continuous-time models. There is no reason to believe that these differ significantly from those given above. Some specific formulae are given by Felsenstein (1971), Hill (1972) and Kimura and Crow (1972). Nor have we considered the complications that can occur when fertility is inherited (see, for example, Nei, 1966), or in a variety of other situations.

Finally, we recall what is perhaps our main motive in defining an effective population size, namely to conside whether various complex population structures can lead to a significantly increased importance for random drift compared to its importance in the model (1.43). Our conclusions show that this occur when there is extremely large variance in offspring number, when the population size is cyclic and the smallest size the population assumes during the cycle is very small and when, in a dioecious population, the number of breeding individuals in one sex is very small. In all other cases, particularly in the case of geographically subdivided populations, there appears to be little significant scope for random drift beyond that applying for the model (1.43). We discuss the consequences of some of these observations later in this book.

3.10 Age Properties and Retrospective Behavior

At the end of Chap. 2 it was mentioned that much interest currently attaches to retrospective, as contrasted to prospective, properties of stochastic processes in genetics. The concept of reversibility was introduced, and in Sect. 3.6 it was remarked that the infinite allele Moran process is reversible. In this section we show briefly how the concept of reversibility can be used to derive certain retrospective properties and then consider some more general formulae. The whole topic of retrospective behavior was opened up by Kimura and Ohta (1973) and Maruyama (1974), but we rely here on the more recent work of Levikson (1977), Sawyer (1977), Kelly (1976, 1977), and Watterson (1976a, 1977b).

To be specific we consider the infinite allele Moran model the stationary distribution of which is given by (3.84). (We have not given the full details of this model: these are provided for example by Watterson, 1976a.) We consider a given allele A_1 which arises by (a unique) mutation and after a random time leaves the population. The behavior of the number of A_1 genes is governed by (3.39), and certain properties of the process can be found by inserting the values given in this equation into (2.137). We suppose that the process has continued for a sufficiently long time so that stationarity has been reached and is then observed at a given time, at which it is noted that the number of A_1 genes is i. We consider properties of the age t of the A_1 allele, i.e. of the (random) time in the past that it arose by mutation.

We consider in particular the mean number of times that the number of A_1 genes has taken the value j from the time it first arose by mutation until the time at which it is observed that the number is i. By reversibility arguments (see Watterson, 1976a), this is identical to the mean number of times the value j will be taken in the future, given a current value of i. This is just $\bar{t}_{ij}$, defined jointly by (2.137) and (3.39). The mean age of A_1 is simply the sum of these values over all j, so that

$$E \text{ (age } A_1 | i \text{ genes currently observed)} = \sum_{j=1}^{2N} \bar{t}_{ij}, \qquad (3.118)$$

where we have replaced M in (2.137) by $2N$. It is clear that a diffusion approximation is available for this value: we discuss this later.

Levikson (1977a) gives a formula more general than the above in that it applies to processes that are not necessarily reversible (and thus applies to the Wright-Fisher infinite alleles model). In the more abstract notation of (2.121) this is

$$E \text{ (age } A_1 | i \text{ genes currently observed)} = \sum_{j=1}^{2N} \bar{t}_{1j} \bar{t}_{ji} / \bar{t}_{1i}. \qquad (3.119)$$

Again this has the more detailed interpretation that the mean time μ_{ij} that the number of A_1 genes has been j before the time at which it is observed that this number is i is

$$\mu_{ij} = \bar{t}_{1j} \bar{t}_{ji} / \bar{t}_{1i}. \qquad (3.120)$$

It is therefore clear that for the Moran model,

$$\bar{t}_{ij} = \bar{t}_{1j} \bar{t}_{ji} / \bar{t}_{1i} \qquad (3.121)$$

where $\bar{t}_{ij}$ is defined jointly by (2.137) and (3.39). Indeed it is easy to see by substitution that (3.121) holds for any model leading to (2.137) and thus shows that the concept of reversibility can be extended at least to certain classes of absorbing Markov chains, specifically to those where there exists a single absorbing state E_0, an initial state E_1 and a continuant transition matrix. This extension can in fact best be effected by defining a new process for which $p_{01} = 1$ and then using the stationarity reversibility theory. We take this up in more detail later.

A different approach to age distribution questions and time reversal problems for the Wright-Fisher model (3.85) is taken by Sawyer (1977). Choose some gene in generation $T(T \gg 1)$ and trace its ancestry back in time. In each generation $t < T$ this gene will have exactly one ancestor which, because of mutation, is perhaps of different allelic form from that of the chosen gene. Let $R_t(1 \leq R_t \leq 2N)$ be the total number of genes in generation $t(t \leq T)$ that are of the same allelic type as this ancestor gene in generation t. Recalling that each daughter gene in the model (3.65) is a mutant of a unique allelic type with probability u, we find

$$p_{ij} = \text{Prob } \{R_{t+1} = j | R_t = i\} = u\delta_{ij} + (1 - u)q_{ij}, \qquad (3.122)$$

where

$$q_{ij} = \binom{2N-1}{j-1} \{(1-u)i/(2N)\}^{j-1} \{1 - (1-u)i/(2N)\}^{2N-j} \tag{3.123}$$

and δ_{ij} is the Kronecker delta. Evidently R_t is a Markovian variable and it is easy to approximate its evolution by a diffusion process. From this age distribution properties of the gene originally chosen can be obtained, and we examine these in more detail in Chap. 5.

Returning to the Moran model, we note in conclusion that various further time reversal properties are known: these are listed by Watterson and Guess (1977) and by Kelly (1976); several of them are unexpected. For example, the age of the oldest allele in the population in equilibrium is independent of its current frequency and indeed of the entire population allelic configuration. Explicit formulae are given for the probability that a sample of m genes contains the oldest allele in the population and for the probability that an allele observed j times in a sample of m genes is the oldest in the population. The formula for the latter shows that the probability is independent of the actual configuration of allele frequencies in the sample. Some of these results will be of interest to use later when considering the neutral theory (in Chap. 9).

4. Diffusion Theory

4.1 Introduction

In the previous chapter we encountered some difficulty in deriving explicit formulae for several quantities of evolutionary interest, particularly when the population behavior was described by the Wright-Fisher model (1.43) or one of its generalizations. Even for models such as (3.30) where explicit formulae can often be found, the effects of the genetic parameters are sometimes obscured by the complexities of the expressions which arise. For both these reasons it would be most useful to us if we could approximate these quantities by reasonably accurate expressions which are comparatively simple and display explicitly the effects of the various genetic parameters. Fortunately there exists a general approach which very often does all this for us, namely in approximating the discrete process by a continuous diffusion process.

A substantial and mathematically deep theory of diffusion processes exists (see, for example, Ito and McKean, 1965; Freedman, 1971; Mandl, 1968): we make some reference to this theory in Sect. 4.7. Our approach to diffusion processes does not proceed through this theory, being often rather intuitive and avoiding theoretical niceties. This is in part because for us the fundamental process is always a discrete one (usually a finite Markov chain), for which some of these niceties are irrelevant, and in part because the mathematical depth of formal diffusion theory is inappropriate to the level of this book. We shall in particular assume without question that a unique diffusion process having certain properties we require does exist.

Diffusion theory has a long and honorable place in population genetics theory, going back to Fisher (1922). In this chapter we consider the elements of the theory divorced from specific genetical applications, while in Chap. 5 the theory now developed will be applied to a variety of genetical models.

4.2 The Forward and Backward Kolmogorov Equations

We consider a discrete Markov chain with state space $\{0, 1, 2, \ldots, M\}$, transition matrix $P = \{p_{ij}\}$ and initial value k: this notation will be used throughout this chapter. For convenience write $p_{ki}^{(t)}$ as $f(i; k, t)$, so that

$$f(j; k, t+1) = \sum_i f(i; k, t) p_{ij}. \tag{4.1}$$

We rescale the space axis by a factor M^{-1} and consider the new variables

$$p = kM^{-1}, \quad x = iM^{-1}, \quad z = jM^{-1} = x + \delta x. \tag{4.2}$$

In all applications of interest to us, $E(\delta x | x) = 0(M^{-\gamma})$ and $\text{Var}(\delta x | x) = 0(M^{-\gamma})$, where $\gamma = 1$ or 2; now change the time scale so that possible changes in the random variable can occur at time points $\delta t, 2\delta t, 3\delta t, \ldots$, where $\delta t = M^{-\gamma}$. The rescaled process is of course essentially identical to the original process and in particular is still a discrete process. Nevertheless we feel that as $M \to \infty$ the process converges in some way to a continuous-time continuous-space diffusion process, and our aim is to identify this diffusion process and to discover some of its properties.

Suppose that in the discrete process the moments of the change δx, given the current value x at time t, satisfy the equations

$$E(\delta x) = a(x)\delta t + o(\delta t), \tag{4.3}$$

$$\text{Var}(\delta x) = b(x)\delta t + o(\delta t), \tag{4.4}$$

$$E(|\delta x|^3) = o(\delta t). \tag{4.5}$$

Here $a(x)$ and $b(x)$ are assumed to be functions of x but not of t. We write (4.1) in the form

$$f(z; p, t + \delta t) = \int f(x; p, t) f(z; x, \delta t) dx,$$

where here and below all integrals have terminals 0 and 1. Let $Q(x)$ be a function having two finite derivatives and for which $Q(0) = Q(1) = Q'(0) = Q'(1) = 0$. Multiplying both sides of this equation by $Q(z)$ and integrating, we get

$$\int Q(z) f(z; p, t + \delta t) dz = \int \int Q(z) f(x; p, t) f(z; x, \delta t) dx dz. \tag{4.6}$$

Now

$$Q(z) = Q(x) + \delta x\, Q'(x) + \tfrac{1}{2} (\delta x)^2\, Q''(x) + 0(|\delta x|^3),$$

and if we insert this expression on the right-hand side of (4.6), we obtain

$$\int \int Q(x) f(x; p, t) f(z; x, \delta t) dx dz + \int \int Q'(x) f(x; p, t) \{\delta x f(z; x, \delta t)\} dx dz$$
$$+ \tfrac{1}{2} \int \int Q''(x) f(x; p, t) \{(\delta x)^2 f(z; x, \delta t)\} dx dz + o(\delta t).$$

Carrying out the integrations with respect to z, we get

$$\int Q(x) f(x; p, t) dx + \delta t \int Q'(x) a(x) f(x; p, t) dx + \tfrac{1}{2} \delta t \int Q''(x) b(x) f(x; p, t) dx + o(\delta t).$$

The second term in this expression can be integrated by parts to give

$$[Q(x)a(x)f(x;p,t)]_0^1 - \int Q(x)\frac{\partial}{\partial x}\{a(x)f(x;p,t)\}dx.$$

The first term here vanishes because of our assumption $Q(0) = Q(1) = 0$. Similarly the third term can be integrated by parts twice to give

$$\tfrac{1}{2}\int Q(x)\frac{\partial^2}{\partial x^2}\{b(x)f(x;p,t)\}dx.$$

Putting all these expressions in (4.6) and rearranging terms, we get

$$\int Q(x)\{f(x;p,t+\delta t) - f(x;p,t)\}dx = \delta t \int Q(x)H(x)dx + o(\delta t),$$

where

$$H(x) = -\frac{\partial}{\partial x}\{a(x)f(x;p,t)\} + \tfrac{1}{2}\frac{\partial^2}{\partial x^2}\{b(x)f(x;p,t)\}.$$

Dividing by δt and letting $\delta t \to 0$ we get

$$\int Q(x)\frac{\delta}{\delta t}\{f(x;p,t)\}dx = \int Q(x)H(x)dx.$$

Since this is to be true for all functions $Q(x)$ satisfying the above conditions we must have

$$\frac{\partial}{\partial t}\{f(x;p,t)\} = H(x).$$

Dropping the dependence on p, we arrive at the following partial differential equation satisfied by $f(x,t)$:

$$\frac{\partial f(x;t)}{\partial t} = -\frac{\partial}{\partial x}\{a(x)f(x;t)\} + \tfrac{1}{2}\frac{\partial^2}{\partial x^2}\{b(x)f(x;t)\}. \tag{4.7}$$

Since the process $\delta t \to 0$ is equivalent to $M \to \infty$, we now assume that there exists a diffusion process on $[0, 1]$ also satisfying (4.3)–(4.5) and which possesses a density function $f(x;t)$ which satisfies (4.7). We expect this process to approximate the original discrete process in the sense that, for $0 < z < y < 1$,

$$\int_z^y f(x;t)dx \tag{4.8}$$

provides a good approximation to the probability that the original unscaled discrete random variable is between Mz and My at time $M^\gamma t$. Equation (4.7) is called the forward Kolmogorov (Fokker-Planck or diffusion) equation and is of fundamental importance in the theory of population genetics, and in the next chapter we will see how $a(x)$ and $b(x)$ can be determined for various important genetical models.

In the derivation of (4.7) no mention has been made of the initial value p of the diffusion variable. The function $f(x; t)$ should be written more fully $f(x; p, t)$, but since the value of p does not appear elsewhere in (4.7) it is often suppressed, entering in the solution of (4.7) through an obvious boundary condition. There is, however, a second equation that makes fundamental use of the value of p. If we consider instead of the time points $(0, t, t + \delta t)$ the new time points $(0, \delta t, t + \delta t)$ we arrive at the equation

$$f(x; p, t + \delta t) = \int g(\delta p; p) f(x; p + \delta p, t) d(\delta p). \tag{4.9}$$

Here δp is the change in the value of the random variable in the time interval $(0, \delta t)$ and $g(\delta p; p)$, its probability density function. Expanding the integrand as above and retaining leading terms we arrive at the equation

$$\frac{\partial f(x; p, t)}{\partial t} = a(p) \frac{\partial f(x; p, t)}{\partial p} + \tfrac{1}{2} b(p) \frac{\partial^2 f(x; p, t)}{\partial p^2}. \tag{4.10}$$

This is the backward Kolmogorov equation, which for several purposes is more useful than the forward equation (4.7).

Some care must be exercised in the interpretation of (4.10). The density function $f(x; p, t)$ clearly depends on p and all that is claimed is that, as a function of p, it satisfies (4.10). The statement sometimes made that (4.10) implies a time reversal and that p is a random variable with x fixed is incorrect: the random variable in Eq. (4.10) is the current gene frequency x.

An explicit solution of (4.7), or of (4.10), can sometimes be achieved, as we see in the next chapter. The solution is usually of the eigenfunction expansion form

$$f(x; p, t) = \sum_{i=1}^{\infty} g_i(x, p) \exp(-\lambda_i t) \tag{4.11}$$

where the λ_i ($0 \leqslant \lambda_1 < \lambda_2 < \lambda_3 \ldots$) are eigenvalue constants and the $g_i(x, p)$, the associated eigenfunctions. This form of solution is clearly analogous to (2.115), a parallel we examine in more detail in particular cases. Remarkably, a considerable amount of information concerning the diffusion process (4.7) can be found without computing the explicit solution (4.11), as we now see.

4.3. Fixation Probabilities

In this and the next three sections we assume without question the existence of a diffusion process on $[0, 1]$ satisfying (4.3)–(4.5) and admitting a density function satisfying (4.7) and (4.10).

An equation parallel to (4.10) can be found by replacing $f(x; p, t)$ by $F(x; p, t)$ throughout, where

$$F(x; p, t) = \int_0^x f(y; p, t) dy,$$ (4.12)

so that

$$\frac{\partial F(x; p, t)}{\partial t} = a(p) \frac{\partial F(x; p, t)}{\partial p} + \frac{1}{2} b(p) \frac{\partial^2 F(x; p, t)}{\partial p^2}.$$ (4.13)

Suppose now that both $x = 0$ and $x = 1$ are absorbing states of the diffusion process. From this we can arrive at the equation

$$\frac{\partial P_0(p; t)}{\partial t} = a(p) \frac{\partial P_0(p; t)}{\partial p} + \frac{1}{2} b(p) \frac{\partial^2 P_0(p; t)}{\partial p^2},$$ (4.14)

where $P_0(p; t)$ is the probability that absorption has occurred at $x = 0$ at or before time t. The same equation holds for the probability $P_1(p; t)$ that absorption has occurred at $x = 1$ at or before time t: although $P_0(p; t)$ and $P_1(p; t)$ obey the same equation their values differ due to different boundary conditions. By letting $t \to \infty$, the probability $P_0(p)$ that absorption ever occurs at $x = 0$ satisfies the equation

$$0 = a(p) \frac{dP_0(p)}{dp} + \frac{1}{2} b(p) \frac{d^2 P_0(p)}{dp^2},$$ (4.15)

and since $P_0(p)$ clearly satisfies the boundary conditions $P_0(0) = 1, P_0(1) = 0$, it is possible to solve (4.15) explicitly to get

$$P_0(p) = \int_p^1 \psi(y) dy / \int_0^1 \psi(y) dy,$$ (4.16)

where

$$\psi(y) = \exp\left[-2 \int^y \{a(z)/b(z)\} dz \right].$$ (4.17)

Similarly the probability $P_1(p)$ that absorption eventually occurs at $x = 1$ is found to be

$$P_1(p) = \int_0^p \psi(y) dy / \int_0^1 \psi(y) dy.$$ (4.18)

Note that we already found these formulae as approximations to the values in a finite Markov chain (see Eqs. (3.28) and (3.29), where a different notation is used) without reference to diffusion processes. Although we have carried out a scaling of the time axis in passing from the original Markov chain to the diffusion process, there is no need to rescale the values (4.16) and (4.18) when using them as approximations in the Markov chain. This is no longer true for questions concerning the time until absorption, as we now see.

4.4. Absorption Time Properties

We start by assuming that both $x = 0$ and $x = 1$ are absorbing barriers and consider the mean time until one or other boundary is reached. Equation (4.14) and the corresponding equation for $x = 1$ show that if $\phi(t;p)$ is the density function of the time t until absorption occurs, then $\phi(t;p)$ satisfies the equation

$$\frac{\partial \phi(t;p)}{\partial t} = a(p) \frac{\partial \phi(t;p)}{\partial p} + \tfrac{1}{2} b(p) \frac{\partial^2 \phi(t;p)}{\partial p^2} \,. \tag{4.19}$$

Then

$$-1 = -\int_0^\infty \phi(t;p)dt$$

$$= -[t \, \phi \, (t;p)]_0^\infty + \int_0^\infty t \frac{\partial \phi}{\partial t} dt$$

$$= 0 + \int_0^\infty t \{a(p)\frac{\partial \phi}{\partial p} + \tfrac{1}{2} b(p) \frac{\partial^2 \phi}{\partial p^2} \} \, dt$$

so that

$$-1 = a(p) \frac{d\bar{t}\,(p)}{dp} + \tfrac{1}{2} b(p) \frac{d^2 \bar{t}(p)}{dp^2}\,, \tag{4.20}$$

providing an interchange in the order of integration and differentiation is justified and that $t\phi(t;p) \to 0$ as $t \to \infty$. Here

$$\bar{t}\,(p) = \int_0^\infty t\phi(t;p)dt \tag{4.21}$$

is the mean time until one or other absorbing boundary is reached and clearly satisfies the boundary conditions $\bar{t}(0) = \bar{t}(1) = 0$. The solution of (4.20), subject to these boundary conditions, is best expressed in the form

$$\bar{t}\,(p) = \int_0^1 t(x;p)dx\,, \tag{4.22}$$

where

$$t(x;p) = 2P_0(p)[b(x)\psi(x)]^{-1} \int_0^x \psi(y)dy, \quad 0 \leqslant x \leqslant p, \tag{4.23}$$

$$t(x;p) = 2P_1(p)[b(x)\psi(x)]^{-1} \int_x^1 \psi(y)dy, \quad p \leqslant x \leqslant 1. \tag{4.24}$$

For the original Markov chain we approximate the mean absorption time by

$$M^\alpha \bar{t}(p). \tag{4.25}$$

The representation (4.22) is not fortuitous. We shall see later (Sect. 4.7) that the function $t(x, p)$ has the interpretation that

$$\int_{x_1}^{x_2} t(x;p)dx \tag{4.26}$$

is the mean time in the diffusion process that the random variable spends in the interval (x_1, x_2) before absorption. Correspondingly, we approximate the mean number of times in the Markov chain that the discrete random variable takes the value $j(=Mx)$ before absorption by

$$\bar{t}_{k,j} \approx M^{\alpha-1}t(x;p). \tag{4.27}$$

The representation (4.26) allows further conclusions to be drawn. Let $g(x)$ be any well-behaved function of x and consider the integral $I_g(p)$ of this function over the time until absorption occurs. This integral is a random variable (since its value will depend on the actual path traced out by the diffusion variable) and its mean value from (4.26) is clearly

$$E[I_g(p)] = \int_0^1 g(x)t(x;p)dx. \tag{4.28}$$

In a similar way, if $\Sigma g(x)$ is the sum of the function $g(x)$ in the discrete process,

$$E[\Sigma g(x)] \approx M^\alpha \int_0^1 g(x)t(x;p)dx. \tag{4.29}$$

There is an alternative way of deriving (4.28) akin to the derivation of the backward Eq. (4.10). We note that the integral of $g(p)$ over $(0, \delta t)$ is approximately $g(p)\delta t$, so that, writing $E[I_g(p)]$ as $\mu(p)$ for convenience,

$$\mu(p) \approx g(p)\delta t + E[\int_0^1 g(x)t(x;p+\delta p)dx]$$

$$\approx g(p)\delta t + \mu(p) + a(p)\delta t \frac{d\mu}{dp} + \tfrac{1}{2} b(p)\delta t \frac{d^2\mu}{dp^2} + o(\delta t).$$

Dividing by δt and letting $\delta t \to 0$, we get

$$a(p)\frac{d\mu(p)}{dp} + \tfrac{1}{2} b(p)\frac{d^2\mu(p)}{dp^2} = -g(p). \tag{4.30}$$

This equation generalizes (4.20) and may be solved, subject to the boundary conditions $\mu(0) = \mu(1) = 0$, to rederive (4.28).

It is possible to derive higher moments of the absorption time (and more generally of $I_g(p)$). For the absorption time we have

$$-2\overline{t}(p) = -2\int_0^\infty t\phi(t;p)dt$$

$$= -[t^2\phi(t;p)]_0^\infty + \int_0^\infty t^2 \frac{\partial\phi}{\partial t} dt$$

$$= \int_0^\infty \{a(p)\frac{\partial t^2 \phi(t;p)}{\partial p} + \tfrac{1}{2} b(p)\frac{\partial^2 t^2 \phi(t;p)}{\partial p^2}\}dt$$

$$= a(p)\frac{dS(p)}{dp} + \tfrac{1}{2} b(p)\frac{d^2 S(p)}{dp^2}, \tag{4.31}$$

formally interchanging the order of integration and differentiation, where $S(p)$ is the second moment of the absorption time. Equation (4.31) can be solved (subject to $S(0) = S(1) = 0$), for $S(p)$ and hence a formula for the variance of the absorption time can be found. Clearly this procedure can be generalized to find any moment, but the formulae become complicated and we present here only an expression for the variance $\sigma^2(p)$. This is

$$\sigma^2(p) = 4[P_1(p)\int_p^1 \psi(x) \int_0^x \xi(y)dydx - P_0(p)\int_0^p \psi(x) \int_0^x \xi(y)dydx] - [\overline{t}(p)]^2 \tag{4.32}$$

where $\psi(x)$ has been defined in (4.17) and

$$\xi(x) = [b(x)\psi(x)]^{-1}\overline{t}(x). \tag{4.33}$$

It is also possible to find higher moments of the random variable $I_g(p)$. This has been done by Nagylaki (1974a), and we here only outline the method. Denoting the nth moment of this variable by $\mu^{(n)}(p)$, the successive moments satisfy the recurrence relation

$$-(n+1)f(p)\mu^{(n)}(p) = a(p)\frac{d\mu^{(n+1)}(p)}{dp} + \tfrac{1}{2}b(p)\frac{d^2\mu^{(n+1)}(p)}{dp^2}, \qquad (4.34)$$

and the boundary conditions $\mu^{(n)}(0) = \mu^{(n)}(1) = 0$. For $n = 1$ this generalizes (4.31), and higher moments may be found from (4.34) by iteration. In particular by choosing $g(x) = 1, (0 < x < 1), g(x) = 0$, otherwise, higher moments of the absorption time can be found from (4.34).

We now use the diffusion process to find an approximation for the distribution of the sojourn time in any state of the Markov chain. Equation (2.123) shows that the distribution of the sojourn time depends on two parameters (there denoted by α_{ij} and r_j), and Eq. (2.124) shows how these parameters are related to the mean of the sojourn time. Suppose first we wish to approximate the distribution of the sojourn time at j, where $j > k$, k being the initial value. The parameter α_{kj} is the probability that state $\{j\}$ is ever reached and equation (4.18) is readily adapted to give

$$\alpha_{kj} \approx \int_0^p \psi(y)dy / \int_0^x \psi(y)dy, \qquad (4.35)$$

where $k = pM, j = xM$, and the drift and diffusion parameters $a(y)$ and $b(y)$, needed to calculate $\psi(y)$, are associated with the diffusion process approximating the Markov chain. We approximate r_j by using Eqs. (2.124) and (4.27) to find

$$\alpha_{kj}/(1 - r_j) \approx M^{\alpha-1}\bar{t}(x;p). \qquad (4.36)$$

Combining (4.35) and (4.36), the sojourn time distribution is given by (2.123) where α_{kj} is approximated by (4.35) and r_j by

$$r_j \approx 1 - \tfrac{1}{2}M^{1-\alpha}[\{b(x)\psi(x)\}/\{\int_0^x \psi(y)dy \int_0^1 \psi(y)dy\}]. \qquad (4.37)$$

When $j < k$ Eq. (2.124) continues to hold, but here we approximate α_{kj} by

$$\alpha_{kj} \approx \int_p^1 \psi(x)dx / \int_x^1 \psi(x)dx. \qquad (4.38)$$

The approximation (4.37) remains unchanged.

All of the above formulae require modification when there is only one absorbing state. We do not go into details here and merely note the conclusions. If 0 is the only absorbing state (4.22) continues to hold, but $t(x;p)$ must be redefined as

$$t(x;p) = 2[b(x)\psi(x)]^{-1}\int_0^x \psi(y)dy, \quad 0 \leq x \leq p, \qquad (4.39)$$

$$t(x;p) = 2[b(x)\psi(x)]^{-1}\int_0^p \psi(y)dy, \quad p \leq x \leq 1. \qquad (4.40)$$

Similarly when 1 is the only absorbing state we have

$$t(x;p) = 2[b(x)\psi(x)]^{-1} \int_p^1 \psi(y)dy, \quad 0 \leqslant x \leqslant p, \tag{4.41}$$

$$t(x;p) = 2[b(x)\psi(x)]^{-1} \int_x^1 \psi(y)dy, \quad p \leqslant x \leqslant 1. \tag{4.42}$$

In both cases Eqs. (4.25)–(4.30) hold, except that revised boundary conditions are needed for (4.30) to produce the solution (4.28), where now $t(x;p)$ is given either by (4.39) and (4.40) or by (4.41) and (4.42).

4.5 The Stationary Distribution

We have assumed above that in the Markov chain we are interested in there has existed at least one absorbing state. In several cases of interest there are no absorbing states, and there exists a stationary distribution $\{\alpha_j\}$ satisfying (2.133) and given implicitly as the solution to (2.134). Since an explicit expression for this distribution has not been found in many examples of genetic interest, we aim in this section to approximate this distribution by finding the stationary distribution of the approximating diffusion process. It will turn out that this leads to a very simple form for this approximating distribution in which the effects of the genetical parameters are clearly displayed.

Our starting point is the forward Kolmogorov Eq. (4.7). If we integrate throughout formally with respect to x, there results eventually

$$\frac{\partial}{\partial t}[1 - F(x;t)] = a(x)f(x;t) - \frac{1}{2}\frac{\partial\{b(x)f(x;t)\}}{\partial x}. \tag{4.43}$$

Here $F(x;t)$ is the distribution function

$$F(x;t) = \int_0^x f(y;t)dy. \tag{4.44}$$

This formal derivation suggests that the right-hand side in (4.43) is the rate of flow of probability (from left to right) across the point x at time t. This interpretation can be verified and we thus call the right-hand side in (4.43) the probability flux of the diffusion process. If a stationary distribution $f(x)$ exists this probability flux will be zero if $f(x; t)$ is replaced by $f(x)$, so that the stationary distribution satisfies the equation

$$- a(x)f(x) + \frac{1}{2}\frac{d\{b(x)f(x)\}}{dx} = 0. \tag{4.45}$$

Elementary integration shows that the solution of this equation is

$$f(x) = \text{const } [b(x)]^{-1} \exp \{2 \int^x a(y)/b(y)dy\}, \tag{4.46}$$

where the constant is allocated so that

$$\int_0^1 f(x)dx = 1. \tag{4.47}$$

So far as the original Markov chain is concerned, our interpretation is that the diffusion approximation to the stationary probability that the random variable lies in $[Mx_1, Mx_2]$ is given by

$$\Pr \{Mx_1 \leqslant X \leqslant Mx_2\} \approx \int_{x_1}^{x_2} f(x)dx. \tag{4.48}$$

This approximation turns out to be satisfactory except when $x_1 \approx 0$ or $x_2 \approx 1$, in which case special arguments (which we will consider later) are needed.

4.6 Conditional Processes

In this section we consider diffusion processes where 0 and 1 are both absorbing barriers. It is often of interest to single out those diffusions for which a nominated absorbing barrier is eventually reached, and we do this by the theory of conditional processes. For definiteness we assume the barrier in question is $x = 1$ although we will give some formulae applying when it is $x = 0$.

Since there can be no non-trivial stationary distribution for such processes and since also there is no interest in fixation probabilities, interest centers almost entirely on properties of the time until fixation. Regarding the diffusion as an approximation to a Markov chain it is clear from (2.130) that the sojourn time density (4.23) and (4.24) should be replaced by

$$t^*(x;p) = t(x;p)P_1(x)/P_1(p). \tag{4.49}$$

This gives

$$t^*(x;p) = 2P_0(p)P_1(x)[P_1(p)b(x)\psi(x)]^{-1} \int_0^x \psi(y)dy, \quad 0 \leqslant x \leqslant p, \tag{4.50}$$

$$t^*(x;p) = 2P_1(x)[b(x)\psi(x)]^{-1} \int_x^1 \psi(y)dy, \quad p \leqslant x \leqslant 1. \tag{4.51}$$

We consistently use the asterisk notation (*) to denote functions computed conditional on eventual absorption at $x = 1$ and the double asterisk notation (**) when eventual absorption is at $x = 0$. Thus conditional on eventual absorption at $x = 0$, the sojourn time density, by arguments parallel to those just given, is

$$t^{**}(x;p) = 2P_0(x)[b(x)\psi(x)]^{-1} \int_0^x \psi(y)dy, \quad 0 \leqslant x \leqslant p, \tag{4.52}$$

$$t^{**}(x;p) = 2P_0(x)P_1(p)[P_0(p)b(x)\psi(x)]^{-1} \int_x^1 \psi(y)dy, \quad p \leqslant x \leqslant 1. \tag{4.53}$$

Note that (2.129) suggests an even stronger result than these, namely that the conditional density functions $f^*(x;p, t)$ and $f^{**}(x;p, t)$ of the diffusion variable at time t satisfy

$$f^*(x;p, t) = f(x;p, t)P_1(x)/P_1(p), \tag{4.54}$$

$$f^{**}(x;p, t) = f(x;p, t)P_0(x)/P_0(p). \tag{4.55}$$

It is clear that Eqs. (4.50)–(4.53) can be used immediately to find the conditional mean times before absorption. These were originally found by Kimura and Ohta (1969) by a method other than that just outlined. We now indicate a third way in which these conditional mean times can be derived, namely by finding the conditional process analogs to the Kolmogorov Eqs. (4.7) and (4.10), and to do this we must find the conditional process drift and diffusion coefficients analogous to those defined by (4.3) and (4.4). Let A be the event that absorption eventually occurs at $x = 1$, and $p^*(x \to x + \delta x)$ be the conditional probability density, given A, of a transition from x to $x + \delta x$ in time δt. Then

$$p^*(x \to x + \delta x) = p(x \to x + \delta x \text{ and } A) / \text{Prob} (A)$$
$$= p(x \to x + \delta x)P_1(x + \delta x)/P_1(x)$$
$$\approx p(x \to x + \delta x)[1 + \delta x P_1'(x)/P_1(x)],$$

where we use the dash notation (') to refer to differentiation with respect to x. Hence, in an obvious notation,

$$a^*(x)\delta t = \int (\delta x)p^*(x \to x + \delta x)d(\delta x)$$
$$\approx \int (\delta x)p(x \to x + \delta x)[1 + (\delta x)P_1'(x)/P_1(x)]d(\delta x)$$
$$= \{a(x) + b(x)P_1'(x)/P_1(x)\}\delta t.$$

Thus it follows that

$$a^*(x) = a(x) + b(x)P_1'(x)/P_1(x) \tag{4.56}$$

and similarly

$$b^*(x) = b(x). \tag{4.57}$$

(These arguments can be made more rigorous by suitable handling of small-order terms.) The conditional density $f^*(x;p, t)$ now satisfies the forward equation

$$\frac{\partial f^*(x;p,t)}{\partial t} = -\frac{\partial \{a^*(x)f^*(x;p,t)\}}{\partial x} + \frac{1}{2}\frac{\partial^2 \{b^*(x)f^*(x;p,t)\}}{\partial x^2} \tag{4.58}$$

and the backward equation

$$\frac{\partial f^*(x;p,t)}{\partial t} = a^*(p)\frac{\partial f^*(x;p,t)}{\partial p} + \frac{1}{2}b^*(p)\frac{\partial^2 f^*(x;p,t)}{\partial p^2}. \tag{4.59}$$

Using (4.54), (4.56) and (4.57) it is easy to check that these are consistent with (4.7) and (4.10). The conditional mean absorption time may now be found by using $a^*(x)$ and $b^*(x)$ in (4.41) and (4.42), and the resulting value agrees with that found from (4.50) and (4.51). Note that this final approach is more general in that it uses the defining Eqs. (4.58) and (4.59) and thus can be used to find higher moments of the conditional absorption time. We take this point up later when considering specific applications.

4.7 Diffusion Theory

So far in this chapter we have used diffusion processes as approximations to finite Markov chains and have been rather casual about the mathematical theory of such processes. We have, for example, assumed the existence and uniqueness of certain diffusion processes and the existence of certain density functions when in fact we have not verified that these assumptions are justified. In this section we outline several results from the theory of diffusion processes that assist with this verification and which confirm and extend some of the formulae we have derived informally. This account will be disconnected: for more complete details the reader should consult the books referred to in Sect. 4.1. This section may be omitted at least at a first reading as the mathematical level is deeper than in the rest of the book and most of the results noted will be given explicitly again elsewhere.

A one-dimensional diffusion is a strong Markov process $X(t)$ with temporally homogeneous transition probabilities and continuous sample paths. By "strong Markov" we mean that

$$\text{Prob } \{c < X(t_1 + t_2) < d \,|\, X(t), t \leqslant t_1\} = \text{Prob } \{c < X(t_2) < d\} \tag{4.60}$$

where the probability on the right-hand side is calculated supposing the process starts at time zero at the value $X(t_1)$. Here t_1 is a Markov time, i.e. a random time such that the occurence or otherwise of the event $\{t_1 < t\}$ is determined entirely by the values $X(s), s \leqslant t$.

Associated with such a diffusion is the operator $T_t f(x)$, defined by

$$T_t f(x) = \text{E } [f\{X(t)\} \,|\, X(0) = x]. \tag{4.61}$$

This operator has several important properties: it transforms continuous functions into continuous functions, it is linear $(T(\alpha f + \beta g) = \alpha T(f) + \beta T(g))$, positive $(f \geqslant 0 \rightarrow T(f) \geqslant 0)$, conservative $(T(1) = 1)$ and, introducing the norm $|f| = \max |f(x)|$, contractive $(|T(f)| \leqslant |f|)$. Furthermore, the set of such operators forms a semi-group $(T_t(T_s) = T_{t+s}, T_0 = $ identity operator) and this semi-group is strongly continuous $(|T_t(f) - f| \rightarrow 0$ as $t \rightarrow 0$ if f is continuous). A semi-group possessing all these properties is said to be a Fellerian semi-group, and the powerful analytic theory of such semi-groups may be used to arrive at properties of the diffusion process not perhaps readily obtainable otherwise.

The semi-group of operators T_t possesses a generator A defined for all suitable functions f by

$$Af(x) = \lim_{t \to 0} t^{-1} (T_t f(x) - f(x)). \tag{4.62}$$

Specifically, functions f in the domain of A (i.e. for which Af is defined) are those continuous functions for which

$$\lim_{t \to 0} |t^{-1}(T_t f - f) - Af| = 0. \tag{4.63}$$

If f is in the domain of A, then so also is $T_t f$ and (4.62) yields

$$A(T_t f)(x) = \lim_{\delta \to 0} \delta^{-1}(T_{t+\delta}f(x) - T_t f(x))$$

or, letting $h(t, x) = T_t f(x)$,

$$\frac{\partial}{\partial t} h(t, x) = Ah(t, x).$$

In view of (4.68) below, this is a generalized Kolmogorov backward equation. We return to the generator A and results deriving from it in a moment.

We limit consideration to diffusions on $[0, 1]$. The qualitative behavior of the diffusion on $(0, 1)$ and also at the boundary points is determined by the so-called scale function $p(x)$ and speed measure $m(x)$ of the diffusion process. These are related to A through the equation

$$Af(x) = \frac{d}{dm} \frac{d}{dp} f(x): \tag{4.64}$$

we will use this equation in a moment to derive more explicit formulae for $p(x)$ and $m(x)$.

Interest naturally centers on those diffusions having properties analogous to (4.3)–(4.5). Specifically we shall consider those diffusions for which, if $X(0) = x$,

$$E \{X(t) - x | x\} = a(x)t + o(t, x), \tag{4.65}$$

$$E\ \{(X(t) - x)^2|x\} = b(x)t + o(t, x),\tag{4.66}$$

$$E\ \{|X(t) - x|^3|x\} = o(t, x),$$

where the error term is uniformly small for all x, i.e.,

$$\sup_x t^{-1}|o(t, x)| \to 0 \quad \text{as} \quad t \to 0.\tag{4.67}$$

Note that not all diffusions satisfy (4.65)–(4.67), and we henceforth limit considera-tion to those that do. We now employ the Taylor series expansion

$$f(X(t)) - f(x) = \{X(t) - x\}\frac{df}{dx} + \tfrac{1}{2}\ \{X(t) - x\}^2\frac{d^2f}{dx^2} + 0\ (|X(t) - x|^3).$$

Taking expectations of both sides, letting $t \to 0$, and recalling the definition of $Af(x)$ in (4.62), we arrive at

$$Af(x) = a(x)\frac{df(x)}{dx} + \tfrac{1}{2}\ b(x)\ \frac{d^2f(x)}{dx^2}\ .\tag{4.68}$$

This defines the operator A in terms of the drift and diffusion coefficients $a(x)$ and $b(x)$. Equations (4.64) and (4.68) jointly now show that

$$p(x) = \int_c^x \exp\left(-\ 2\int^y a(z)/b(z)dz\right)dy,\tag{4.69}$$

$$m(x) = 2\int_c^x \{b(y)\}^{-1}\ \exp\left(2\int^y a(z)/b(z)dz\right)dy,\tag{4.70}$$

for some arbitrary constant c. It is of interest to note that, up to a linear transform, $p(x)$ is identical to the fixation probability $P_1(x)$. A diffusion is said to be on its natural scale if $p(x) = x$, which, from (4.69), is equivalent to $a(x) \equiv 0$. For any diffusion not on its natural scale it is possible to find a transformed diffusion (indeed the trans-formation is $x \to p(x)$) that is, and this explains the intimate link between $p(x)$ and $P_1(x)$.

The functions $p(x)$ and $m(x)$ are central to many properties of diffusion processes, and we now show how they can be used to elucidate boundary behavior. Let r be an arbitrary point in $(0, 1)$ and s be one or other boundary point (i.e. $s = 0$ or $s = 1$). Compute the functions

$$u(s) = \int_r^s m(x)dp(x),\tag{4.71}$$

$$v(s) = \int_r^s p(x)dm(x).\tag{4.72}$$

For processes where (4.65)–(4.67) hold the nature of the boundary s is exhibited as follows:

$u(s)$	$v(s)$	boundary type	accessible?	absorbing?	
$< \infty$	$< \infty$	regular	yes	no	
$< \infty$	$= \infty$	exit	yes	yes	(4.73)
$= \infty$	$< \infty$	entrance	no	no	
$= \infty$	$= \infty$	natural	no	yes	

A boundary is accessible if there exists positive probability that it can be reached from a given interior point and is absorbing if the process remains forever at the boundary if it starts there. We later give genetic examples of these various boundaries: note that the terminology of boundary type follows Feller (1954) (see Mandl, 1968), and that different terminologies are possible (e.g. Prohorov and Rozanov, 1969).

One powerful result of diffusion theory is Dynkin's formula, which we give below in two different forms. Let $f(x)$ be any function in the domain of A and τ be any Markov time for which $E_x(\tau) < \infty$, the notation E_x indicating expectation conditional on $X(0) = x$. Then

$$E_x[\int_0^\tau Af(X(t))dt] = E_x[f(X(\tau))] - f(x).$$

(4.74)

This equation has the following consequence. Suppose both boundaries 0 and 1 are accessible and let τ be the (random) time until one or other boundary is reached. Then

$$E_x[\int_0^\tau F(X(t))dt] = \int_0^1 G(x,y)F(y)dm(y),$$

(4.75)

for any bounded function $F(x)$, where

$$G(x,y) = [p(1) - p(0)]^{-1}[p(y) - p(0)][p(1) - p(x)], \quad y \leqslant x,$$
$$G(x,y) = [p(1) - p(0)]^{-1}[p(x) - p(0)][p(1) - p(y)], \quad y \geqslant x.$$

(4.76)

It is not hard to use (4.75) to justify several of the formulae given in previous sections for diffusion with two accessible absorbing barriers. Thus, if we choose $F(x) \equiv 1$, the left-hand side in (4.75) is the mean absorption time and the integrand on the right-hand side is easily seen, using the definitions (4.69) and (4.70), to be identical to the $t(x;p)$ of (4.22), defined in (4.23) and (4.24). By choosing $F(x) = 1$ for $x_1 \leqslant x \leqslant x_2$ and $F(x) = 0$ otherwise we can derive the more detailed result (4.26). We may also properly derive the fixation probability (4.18) using (4.74). Choosing $f(z) = p(z)$,

$$Af(z) = \frac{d}{dm} \frac{d}{dp} p(z)$$

$$= \frac{d}{dm} 1 \tag{4.77}$$

$$= 0.$$

Equation (4.74) now gives, for $f(z) = p(z)$,

$$0 = \pi p(1) + (1 - \pi)p(0) - p(x)$$

or

$$\pi = \{p(x) - p(0)\}/\{p(1) - p(0)\},$$

where π is the probability of eventually reaching the boundary 1. But this is just Eq. (4.18).

We have shown above why the description "scale function" is appropriate for $p(y)$. The terminology "speed measure" may be explained by noting that for a process on its natural scale,

$$\frac{dm(y)}{dy} = 2[b(y)]^{-1}. \tag{4.78}$$

It is a standard result for a diffusion process with $a(x) = 0, b(x) = b$ that the mean time to reach $\pm\delta$ from zero is $b^{-1}\delta^2$ and is thus inversely proportional to the diffusion coefficient b. While in our process $b(x)$ is not constant, it may be regarded as such (to a sufficient approximation) in an interval $y \pm \delta$ and this shows that, to this level of approximation, $dm(y)/dy$ is proportional to the mean time to leave such an interval. This justifies the term "speed measure" although since larger values of $dm(y)/dy$ correspond to larger mean times for leaving the internal $y \pm \delta$, a better name for $m(y)$ would perhaps be "inertia measure".

We may note in passing that these considerations can be used to give an intuitive explanation for the form of the Eq. (4.75). For processes on a natural scale (4.75) becomes, with the choice $F(x) \equiv 1$,

$$E_x(\tau) = \int_0^1 G(x, y)dm(y), \tag{4.79}$$

where

$$G(x, y) = y(1 - x), \quad y \leqslant x, \tag{4.80}$$

$$G(x, y) = x(1 - y), \quad y \geqslant x, \tag{4.81}$$

and $dm(y)/dy$ is given by (4.78). To help understand the form (4.79) consider a symmetric random walk on $(0, n^{-1}, 2n^{-1}, \ldots, 1 - n^{-1}, 1)$ with initial value $x = in^{-1}$. It is a standard result for such a process that the mean number of visits paid to $y \ (= jn^{-1})$ before reaching 0 or 1 is

$$G^*(x, y) = 2ny(1 - x), \quad y \leqslant x, \tag{4.82}$$

$$G^*(x, y) = 2nx(1 - y), \quad y \geqslant x. \tag{4.83}$$

Now define the interval I_j as $((j - 1)n^{-1}, (j + 1)n^{-1})$ and consider the diffusion process on $(0, 1)$. Starting at $x = in^{-1} = z_0$ the diffusion will eventually reach one or other boundary of I_i. Let the time required to do this be t_1, and let z_1, (either $(i - 1)n^{-1}$ or $(i + 1)n^{-1}$) be the value of the diffusion variable when a boundary is reached. Because the process is on a natural scale both boundaries are equally likely to be reached. Now consider a similar interval (either I_{i-1} or I_{i+1}) about the boundary just reached and let t_2 be the time until one or other boundary of this new interval is reached. Define the value of the diffusion variable at this time z_2, and continue in this way until one or other boundary 0 or 1 is attained. The process $z_1, z_2, \ldots$ performs a symmetric random walk, and we can apply (4.82) and (4.83) to it. Let m_j be the mean time to reach a boundary of I_j, starting at jn^{-1}. Then

$$E_x(t_k) = \sum_j \Pr(z_{k-1} = jn^{-1}) m_j. \tag{4.84}$$

Since

$$\sum_k \Pr(z_{k-1} = jn^{-1}) = G^*(x, jn^{-1}),$$

we have, on summing (4.84) over all k,

$$E_x(\tau) = E_x(t_1 + t_2 + \ldots) = \sum_j G^*(x, jn^{-1}) m_j. \tag{4.85}$$

We have noted earlier that

$$m_j \approx \tfrac{1}{2} n^{-2} dm(y)/dy \quad (y = jn^{-1})$$

and since $G^*(x, y) = 2nG(x, y)$ we may approximate (4.85) by

$$E_x(\tau) \approx \sum_j G(x, y) n^{-1} dm(y)/dy$$

$$\approx \int_0^1 G(x, y) dm(y),$$

which is (4.79). This argument can be made more rigorous by proper consideration of small-order terms.

In our informal derivations in previous sections we have assumed without proof that density functions for diffusion processes exist. McKean (1956) has shown that a diffusion does have a density function $g(y; x, t)$ with respect to $dm(y)$: specifically,

$$\Pr[c \leqslant X(t) \leqslant d] = \int_c^d g(y; x, t) dm(y) \tag{4.86}$$

for all x, c and d other than boundary points. The corresponding density with respect to dy is $g(y; x, t) dm(y)/dy$. The density g has the remarkable symmetry property

$$g(y; x, t) = g(x; y, t) \tag{4.87}$$

which can be used, as we see later, to discuss time reversibility properties of the diffusion process. The density $g(y; x, t)$ also satisfies the equation

$$\frac{\partial g(y; x, t)}{\partial t} = \frac{d}{dm} \frac{d}{dp} g(y; x, t). \tag{4.88}$$

If we think of the derivatives on the right-hand side as being with respect to $m(y)$ and $p(y)$, then (4.88) reduces to the forward Kolmogorov Eq. (4.7), while if they are regarded as being with respect to $m(x)$ and $p(x)$, it becomes the backward Eq. (4.10). These two equations are thus more intimately linked than our derivation in Sect. 4.2 suggests. The density $g(y; x, t)$ possesses the properties

$$g(s; x, t) = 0 \tag{4.89}$$

if s (either 0 or 1 or both) is an exit boundary and

$$\frac{d}{dp(y)} g(y; x, t) = 0 \quad \text{at } y = s \tag{4.90}$$

if s is a regular or entrance boundary. Finally, when there are no natural boundaries, $g(y; x, t)$ possesses (Elliott, 1955) the eigenfunction expansion

$$g(y; x, t) = C + \sum_{n=1}^{\infty} e^{\lambda_n t} \phi_n(x) \phi_n(y) \tag{4.91}$$

where $0 > \lambda_1 > \lambda_2 > \ldots$ are distinct eigenvalues. The constant C is zero unless there exists a stationary distribution, when it takes the value

$$C = [\int_0^1 dm(y)]^{-1}. \tag{4.92}$$

Furthermore, in this case,

$$\lim_{t \to \infty} \Pr[c \leqslant X(t) \leqslant d] = C \int_c^d dm(y), \tag{4.93}$$

thus defining the stationary distribution and confirming our more informally derived
(4.46). The eigenfunctions $\phi_n(x)$ satisfy the normalizing property

$$\int_0^1 \{\phi_n(y)\}^2 dm(y) = 1, \tag{4.94}$$

this equation holding whether or not there exists a stationary distribution. The function $\phi_1(x)$ is analogous to the right eigenvector corresponding to the leading-non-unit
eigenvalue of the transition matrix P (see Sect. 4.2), while $\phi_1(y)dm(y)/dy$ is analogous
to the corresponding left-eigenvector.

4.8 Multidimensional Processes

So far we have considered diffusion processes in one dimension only. In a number of
cases in population genetics theory we are, however, required to consider a vector of
random variables rather than a single variable, and this leads to the consideration of
multidimensional diffusion processes. We now informally extend to the multivariate
case some of the derivations given in Sect. 4.2.

Consider first a set of linearly independent, jointly Markovian variables $x_1, \ldots, x_k$
for which, after a suitable rescaling of time and space axes,

$$E(\delta x_i) = a_i(x_1, \ldots, x_k)\delta t + o(\delta t),$$

$$\mathrm{Var}(\delta x_i) = b_i(x_1, \ldots, x_k)\delta t + o(\delta t), \tag{4.95}$$

$$\mathrm{Covar}(\delta x_i, \delta x_j) = c_{ij}(x_1, \ldots, x_k)\delta t + o(\delta t),$$

with higher absolute moments of order $o(\delta t)$. Let $f(x_1, \ldots, x_k; t)$ be the joint density
function of these random variables at time t. Then proceeding as in Sect. 4.2, this density function satisfies the forward equation

$$\frac{\partial}{\partial t} f(x_1, \ldots, x_k; t) = -\sum_i \frac{\partial}{\partial x_i} \{a_i(x_1, \ldots, x_k)f(x_1, \ldots, x_k; t)\}$$

$$+ \frac{1}{2}\sum_i \frac{\partial^2}{\partial x_i^2} \{b_i(x_1, \ldots, x_k)f(x_1, \ldots, x_k; t)\}$$

$$+ \sum_{i<j}\sum \frac{\partial^2}{\partial x_i \partial x_j} \{c_{ij}(x_1, \ldots, x_k)f(x_1, \ldots, x_k; t)\}. \tag{4.96}$$

There will also exist a backward equation of obvious form corresponding to (4.10).
Suppose now a joint stationary density $f = f(x_1, \ldots, x_k)$ exists. Then from (4.96) this
density function will satisfy the equation

$$-\sum_i \frac{\partial}{\partial x_i} \{a_i f\} + \frac{1}{2}\sum_i \frac{\partial^2}{\partial x_i^2} \{b_i f\} + \sum_{i<j}\sum \frac{\partial^2}{\partial x_i \partial x_j} \{c_{ij} f\} = 0. \tag{4.97}$$

Unfortunately the concept of a probability flux in several dimensions is more complex than in one dimension and perhaps as a consequence, no simple explicit formula is known for the stationary distribution generalizing (4.46).

A most important question in multidimensional diffusion processes concerns the possible existence of a "second-order diffusion" or "quasi-Markovian variable". We illustrate this concept by an example. In the two-sex model of Sect. 3.9 the pair (x_t, y_t), where $x_t(y_t)$ is the frequency of A_1 among males (females) in generation t, is jointly Markovian. One suspects that x_t and y_t will seldom differ significantly from each other and that some weighted average of the two would behave in a "quasi-Markovian" manner. Such a possibility was investigated by Moran (1958) and Watterson (1962); we present here an outline of the definitive work of Norman (1975c) on this point, simplified to cover specifically genetical applications.

Consider a population of size N reproducing at time points $n = 0, 1, 2, 3, \ldots$, and suppose there exists at time n a random variable X_n $(0 \leqslant X_n \leqslant 1)$ having the properties

$$E\{X_{n+1} - X_n\} = \tau_N a(X_n) + e_{1,n}^N,$$

$$E\{X_{n+1} - X_n\}^2 = \tau_N b(X_n) + e_{2,n}^N, \qquad (4.98)$$

$$E\{|X_{n+1} - X_n|\}^3 = e_{3,n}^N.$$

Here all expectations are conditional on $X_0, X_1, \ldots, X_n$, $\tau_N > 0$ and $\to 0$ as $N \to \infty$, and the "error" terms $e_{i,n}^N$ are all $o(\tau_N)$ in the sense that, for any finite t,

$$\sum_{n < [t/\tau_N]} E\{|e_{i,n}^N|\} \to 0 \text{ as } N \to \infty. \qquad (4.99)$$

The conditions (4.98) are reminiscent of the conditions (4.3)–(4.5) and (4.65)–(4.67), although we emphasize that X_n is not necessarily a Markovian variable. [In the two-sex model just mentioned, for example, the quantity we use later for X_n (see Sect. 5.2) is not Markovian.] Note also that a function X_n satisfying (4.98) does not necessarily exist, and if it does it is not necessarily unique: there may be several "quasi-Markovian" variables satisfying conditions like (4.98). We expect that under certain reasonable conditions the behaviour of X_n will mimic that of a diffusion variable, and make this expectation more precise by specializing to the genetic case a general theorem of Norman (1975c).

Theorem 4.1. Suppose in Eq. (4.98) that $a(x)$ and $b(x)$ are polynomials with $a(0) \geqslant 0$, $a(1) \leqslant 0$, $b(0) = b(1) = 0$ and $b(x) > 0$, $0 < x < 1$. Define $X(t)$ as a diffusion variable having initial value $X(0) = X_0$ and drift and diffusion coefficients $a(x)$, $b(x)$ respectively. Then for any time points $0 < t_1 < t_2 < \ldots < t_j$, the joint distribution of X_{n_1}, $X_{n_2}, \ldots, X_{nj}$ converges to that of $X(t_1), X(t_2), \ldots, X(t_j)$ as $N \to \infty$, $n_i \to \infty$ and $n_i \tau_N \to t_i$.

We do not prove this remarkable theorem here and note only the simplicity of the conditions for its applicability. In particular, as we see in the next chapter, the two-sex model of Chap. 3 satisfies these conditions, and this will lead to a definition of a variance effective population size for this model.

4.9 Time Reversibility

We now discuss, rather informally, the extent to which a diffusion process is time reversible in the sense outlined for Markov chains in Chap. 2 and 3. We recall the definition (2.140) of time reversibility for Markov chains and note that this implies

$$\alpha_i p_{ij}^{(t)} = \alpha_j p_{ji}^{(t)} \tag{4.100}$$

for any positive integer t. Consider now a diffusion process on $[0, 1]$ possessing a stationary distribution $f(x)$, given by (4.46) or equivalently $dm(x)/dx$ given by (4.70). The diffusion analogue of (4.100) is that

$$f(x)f(y;x,t) = f(y)f(x;y,t) \tag{4.101}$$

where $f(y;x,t)$ is the transition density of the diffusion. We note from (4.86) that

$$f(y;x,t) = g(y;x,t)dm(y)/dy = g(y;x,t)f(y),$$

and the property (4.87) now shows immediately that (4.101) holds. We therefore declare any diffusion process admitting a stationary distribution to be time reversible. We can extend (4.101) to any finite collection of time values and hence show that the complete distribution of the diffusion process is time reversible.

We have noted that for Markov chains certain questions involving time reversibility can be considered even though no stationary distribution exists. Various devices enable us to do the same thing for diffusion processes. Here we note only a useful general device, due again to Norman (1978), which we take up in more detail in the next chapter when considering genetical examples. It is possible that one boundary of the diffusion process is absorbing and thus that no stationary distribution exists. When this is so the reversibility argument cannot be applied directly. On the other hand, it is sometimes possible to alter the diffusion process by inserting a small parameter ϵ such that the original process corresponds to $\epsilon = 0$ and, when $\epsilon > 0$, a stationary distribution does exist. Thus for $\epsilon > 0$ the time reversibility argument holds and if we now let $\epsilon \to 0$ it is sometimes possible to derive meaningful results for the original process by continuity. A specific example of this is given in Sect. 5.9.

4.10 Expectations of Functions of Diffusion Variables

In Eq. (4.28) we found the expected value of the integral of a function $g(x)$ over the entire time taken before absorption has been reached. In some cases it is of interest to find the expected value at a single time point t, viz.

$$h(t) = E_t[g(x)], \tag{4.102}$$

where $f(x; p, t)$ is the density function of the diffusion variable at time t. This expecta-
tion can be used in a variety of ways and has been exploited with particular success by
Ohta and Kimura (1969a, 1971a): see also Kimura and Ohta (1971, pp. 183–190).
Suppose at time $t + \delta t$ the random variable takes the value $x + \delta x$. Then

$$h(t + \delta t) = E[g(x + \delta x)]$$
$$= E_t E_x [g(x + \delta x)] \tag{4.103}$$

where E_x refers to the expectation operator conditional on the observed value x at
time t and E_t refers to expectation with respect to the distribution of x. Now

$$E_x[g(x + \delta x)] \approx g(x) + E(\delta x)g'(x) + \tfrac{1}{2} E(\delta x)^2 g''(x).$$

Inserting these values in (4.103) we get

$$h(t + \delta t) \approx h(t) + E_t [a(x)g'(x) + \tfrac{1}{2} b(x)g''(x)]\delta t$$

and hence

$$\frac{d}{dt} h(t) = E_t [a(x)g'(x) + \tfrac{1}{2} b(x)g''(x)]. \tag{4.104}$$

Note in particular that if the diffusion admits a stationary distribution, the limiting
case $t \to \infty$ in (4.104) yields

$$E[a(x)g'(x) + \tfrac{1}{2} b(x)g''(x)] = 0. \tag{4.105}$$

A generalization of these equations is possible for multidimensional diffusions. If the
diffusion process involves linearly independent variables $x_1, x_2, ..., x_k$, and if $h(t)$ is the
expected value of some function $g(x_1, ..., x_k)$ at time t, then

$$\frac{d}{dt} h(t) = E_t [\Sigma a_i(x_1, ..., x_k) \frac{\partial g}{\partial x_i} + \tfrac{1}{2} \Sigma b_i(x_1, ..., x_k) \frac{\partial^2 g}{\partial x_i^2}$$
$$+ \Sigma \Sigma c_{ij}(x_1, ..., x_k) \frac{\partial^2 g}{\partial x_i \partial x_j}] \tag{4.106}$$

in an obvious notation. If a stationary distribution exists then at stationarity

$$E[\Sigma a_i(x_1, ..., x_k) \frac{\partial g}{\partial x_i} + \tfrac{1}{2} \Sigma b_i(x_1, ..., x_k) \frac{\partial^2 g}{\partial x_i^2} + \Sigma \Sigma c_{ij}(x_1, ..., x_k) \frac{\partial^2 g}{\partial x_i \partial x_j}] = 0.$$
$$\tag{4.107}$$

We give examples later (see in particular Sect. 6.6) of the use of these formulae.

5. Applications of Diffusion Theory

5.1 Introduction

In this chapter we apply some of the diffusion theory considered in the previous chapter to various Markov chain models arising in population genetics to arrive at conclusions of evolutionary interest.

Our first aim is to see how the behavior of a given Markov chain can be mimicked by a diffusion process on $[0, 1]$, and to do this it is convenient to start with the general Wright-Fisher model specified by (3.18) and (3.27). In this model the variable considered is the number j of A_1 genes in a diploid population of fixed size N and thus has state-space $\{0, 1, 2, ..., 2N\}$. To work with a variable whose state space is closer to that of the diffusion process we consider instead the fraction x of A_1 genes in the population, whose state space is $\{0, (2N)^{-1}, ..., 1\}$. It is convenient to adopt the notation (see (1.24))

$$w_{11} = 1 + s, \quad w_{12} = 1 + sh, \quad w_{22} = 1. \tag{5.1}$$

We assume throughout this chapter (except in Sect. 5.10, where more complex expressions are required) the notation x for the frequency of A_1 (and implicitly $1 - x$ for A_2). Further, except in Sects. 5.8 and 5.10, a Wright-Fisher model with the fitness parameters (5.1), and mutation rates $u(A_1 \to A_2)$ and $v(A_2 \to A_1)$, is assumed. We also define throughout the initial value of x to be p.

The diffusion model we concentrate on requires that s, u and v are all $0(N^{-1})$ and we put

$$\alpha = 2Ns, \quad \beta_1 = 2Nu, \quad \beta_2 = 2Nv \tag{5.2}$$

where α, β_1, and β_2 are all $0(1)$. Then standard binomial formulae for the model (3.18) show that

$$E(\delta x | x) = [\alpha x(1 - x) \{x + h(1 - 2x)\} - \beta_1 x + \beta_2(1 - x)](2N)^{-1} + o(N^{-1}),$$
$$\text{Var}(\delta x | x) = x(1 - x)(2N)^{-1} + o(N^{-1}), \tag{5.3}$$
$$E\{|\delta x|^3\} = o(N^{-1}).$$

These moments fit into the format (4.3)–(4.5) provided we choose

$$\delta t = (2N)^{-1}, \tag{5.4}$$

$$b(x) = x(1 - x), \tag{5.5}$$

$$a(x) = \alpha x(1 - x)\{x + h(1 - 2x)\} - \beta_1 x + \beta_2(1 - x). \tag{5.6}$$

The requirement (5.4) is met by assuming that unit time in the diffusion process corresponds to $2N$ generations in the Markov chain: it is important to keep this scaling in mind when considering "time" properties in the diffusion process. We now consider some properties of the diffusion process on [0, 1] with drift and diffusion parameters given respectively by (5.6) and (5.5).

Before proceeding we note that in practical applications the idealized model (3.18) will probably have to be replaced by something more complex, perhaps one or other of the models discussed in Chap. 3 in connection with effective population sizes. At the end of the next section we pursue this point for one particular such complex model. Although the theory is by no means clear it seems likely that all the diffusion results given below will continue to hold, at least to a good approximation, when N is replaced by the variance effective population size $N_e^{(v)}$. Except for the case considered at the end of the next section we make no further explicit mention of this point in this chapter.

The first step in discussing properties of the diffusion process with the drift and diffusion coefficients (5.5) and (5.6) is to compute the scale function and speed measure of the process, defined by (4.69) and (4.70). These become

$$p(x) = \int_c^x y^{-2\beta_2} (1 - y)^{-2\beta_1} \exp \{\alpha(2h - 1)y^2 - 2\alpha hy\}dy, \tag{5.7}$$

$$m(x) = 2 \int_c^x y^{2\beta_2 - 1} (1 - y)^{2\beta_1 - 1} \exp \{-\alpha(2h - 1)y^2 + 2\alpha hy\}dy, \tag{5.8}$$

for an arbitrary constant c. We first use these expressions to consider boundary behavior. Use of (5.7) and (5.8) in (4.71) and (4.72) shows that near $x = 1$, the functions $u(x)$ and $v(x)$ take the form (for $\beta_1 \neq \frac{1}{2}$)

$$u(x) = A + 0(1 - x)^{1 - 2\beta_1}, \quad v(x) = B + 0(1 - x)^{2\beta_1}.$$

Here A and B are constants whose precise values are unimportant. It follows that $v(x)$ is always finite at $x = 1$, but that $u(x)$ is finite only if $\beta_1 < \frac{1}{2}$. From this we conclude that the boundary $x = 1$ is regular (accessible but non-absorbing) if $\beta_1 < \frac{1}{2}$ and entrance (inaccessible and non-absorbing) if $\beta_1 > \frac{1}{2}$. The same conclusion holds for the boundary $x = 0$ replacing β_1 by β_2: note that the values of α and h are irrelevant to these boundary descriptions. The case $\beta_1 = \frac{1}{2}$ is easily hand led seperately.

The intuitive meaning of these conclusions is clear enough. If the mutation rate from A_1 to A_2 and the population size are jointly large enough there is zero probability that the frequency of A_1 can ever achieve the value unity. Of course this conclusion applies for the diffusion process and is not true for the Markov chain (3.18), a fact we shall take up again later when considering the accuracy of diffusion approximations, particularly at a boundary.

If $\beta_1 = 0$ the boundary $x = 1$ is found to be exit (accessible and absorbing), and this again accords with what we expect since if the boundary is reached the absence of mutation from A_1 to A_2 means that the frequency of A_1 remains forever at unity. (The fact that the boundary is accessible is less obvious intuitively: in Sect. 5.8 we will meet (natural) boundaries which, although absorbing, are not accessible, i.e. for which there is zero probability that they are reached by diffusion from within $(0, 1)$.)

The functions $p(x)$ and $m(x)$ are also central to the calculation of fixation probabilities and stationary distributions respectively, when these are appropriate. We defer consideration of these until we take up specific cases later.

We conclude this section by emphasizing that our main interest is in Markov chain models such as (3.18), and we view diffusion processes mainly as approximations to these. Usually the approximations are excellent, but in some instances (particularly near boundaries) they are less so and some care is needed in proceeding. We take up this matter in more detail in Sect. 5.7.

5.2 No Selection or Mutation

When there is no selection or mutation the model defined by (3.18) and (3.27) reduces to (1.43). Rather complete knowledge of the diffusion approximation to this model is available, and in this section we explore this in some detail. Clearly we have

$$a(x) = 0, \quad b(x) = x(1 - x) \tag{5.9}$$

and the forward equation becomes

$$\frac{\partial f(x; t)}{\partial t} = \frac{1}{2} \frac{\partial^2}{\partial x^2} \{x(1 - x)f(x; t)\}. \tag{5.10}$$

The solution of this equation, and others more complex, was achieved in a series of papers by Kimura (1955a, b, c, 1956, 1957), which heralded a rebirth of the mathematical theory of population genetics. Most of the results in this section were given in these papers. The explicit solution of (5.10), subject to the requirement $x = p$ when $t = 0$, is

$$f(x; t) = \sum_{i=1}^{\infty} \frac{4(2i + 1)p(1 - p)}{i(i + 1)} T_{i-1}^1 (1 - 2p) T_{i-1}^1 (1 - 2x) \exp \{-\tfrac{1}{2} i(i + 1)t\}. \tag{5.11}$$

Here $T_{i-1}^1(x)$ is a Gegenbauer polynomial defined in terms of the hypergeometric function by

$$T_{i-1}^1(x) = \tfrac{1}{2} i (i + 1)F(i + 2, 1 - i, 2, \tfrac{1}{2} (1 - x)),$$

so that in particular

$$T_0^1(x) = 1, \quad T_1^1(x) = 3x. \tag{5.12}$$

The speed measure for the coefficients (5.9) is

$$dm(x) = 2x^{-1}(1-x)^{-1}dx \tag{5.13}$$

and we confirm that (5.11) is of the form defined by (4.86) and (4.91) with

$$\phi_i(y) = 2(2i+1)^{1/2} \{i(i+1)\}^{-1/2} y(1-y)T_{i-1}^1(1-2y). \tag{5.14}$$

The probability $P_0(t)$ and $P_1(t)$ that the diffusion has reached 0 or 1 respectively by time t are

$$P_0(t) = 1 - p + \sum_{i=1}^{\infty} (2i+1)p(1-p)(-1)^i F(1-i, i+2, 2, 1-p) \exp\left(-\tfrac{1}{2}i(i+1)t\right), \tag{5.15}$$

$$P_1(t) = p + \sum_{i=1}^{\infty} (2i+1)p(1-p)(-1)^i F(1-i, i+2, 2, p) \exp\left(-\tfrac{1}{2}i(i+1)t\right). \tag{5.16}$$

The probability of ultimate fixation at $x = 1$ can be found by letting $t \to \infty$ in (5.16) or else by computing (4.18), with $\psi(x)$ defined by (4.17) and (5.9). Evidently $\psi(x) = 1$ and hence

Prob (ultimate fixation at $x = 1$) = p. \tag{5.17}

The mean fixation time can be found by noting from (4.23) and (4.24) that

$$\bar{t}(x;p) = 2(1-p)/(1-x), \quad 0 \leqslant x \leqslant p,$$

$$\bar{t}(x;p) = 2p/x, \qquad\qquad p \leqslant x \leqslant 1. \tag{5.18}$$

Thus the mean absorption time is

$$\bar{t}(p) = -2\{p \log p + (1-p) \log (1-p)\} \tag{5.19}$$

time units or $-4N \{p \log p + (1-p) \log (1-p)\}$ generations. Note that this agrees with the value (3.5) found without recourse to diffusion processes and yields (3.6) and (3.7) as cases of particular interest.

The variance of the absorption time can be found from (4.32) and is

$$4[p \int_p^1 \xi(x)dx - (1-p) \int_0^p \xi(x)dx], \tag{5.20}$$

where

$$\xi(x) = \int^x [(1-y)^{-1} \log y + y^{-1} \log (1-y)] dy. \tag{5.21}$$

The value (5.20) is in terms of (squared) time units and must be multiplied by $4N^2$ to be brought to a (squared) generation basis.

The complete distribution of the absorption time is implicit in (5.15) and (5.16), since

$$\text{Prob \{absorption time} \leqslant t\} = P_0(t) + P_1(t). \tag{5.22}$$

Because of the form of the solutions (5.15) and (5.16), this expression is perhaps of most use when t is large. We shall show later how this solution may be supplemented by an asymptotic expansion the accuracy of which is best for small values of t so as to yield a rather complete picture of the distribution of the absorption time.

What do these diffusion results mean for the Markov chain model (1.43)? The fixation probability (5.17) is exactly correct for this model since we have seen that this value can be reached directly. The mean absorption time approximation has been confirmed. We have, however, arrived at the more detailed information from (5.18) and (4.27), that if the initial number of A_1 genes in the Markov chain model is k, the mean number of generations for which this number assumes the value j, before reaching 0 or 2N, is approximately

$$\overline{t}_{k,j} = 2(2N - k)/(2N - j), \quad j \leqslant k,$$

$$\overline{t}_{k,j} = 2k/j, \quad\quad\quad\quad j \geqslant k. \tag{5.23}$$

The particular case $k = 1$, of particular interest to Fisher and Wright, gives $\overline{t}_{1,j} = 2j^{-1}$, in agreement with (1.51).

We turn now to the spectral expansion (5.11). Recalling the difference in time scale between the Markov chain (1.43) and the diffusion process (5.10), it is clear that the eigenvalue exp $\{-\frac{1}{2} i(i+1)\}$ is the analogue of the Markov chain value

$$\left[\left(1 - \frac{1}{2N}\right)\left(1 - \frac{2}{2N}\right) \ldots \left(1 - \frac{i}{2N}\right)\right]^{2N}$$

$$\approx \exp - \{1 + 2 + \ldots + i\}$$

$$= \exp - \tfrac{1}{2} \{i(i+1)\}.$$

There is also a parallel between the eigenfunctions in (5.11) and the eigenvectors of (1.43). For large t we may write

$$f(x;p, t) = 6p(1-p) \exp(-t) + 30p(1-p)(1-2p)(1-2x) \exp(-3t) + \ldots . \tag{5.24}$$

Thus the function $p(1 - p)$ in the leading term is clearly the analogue of the right eigenvector (1.46). Since this leading term is independent of x, the analogue of the corresponding left eigenvector is l_k = constant. This shows that the asymptotic $(t \to \infty)$ conditional $(x \neq 0, 1)$ distribution of x is uniform over $(0, 1)$, in agreement with the approximation for the model (1.43) noted after equation (1.49). The complete expansion (5.24) shows, as was noted by Kimura, that the extensive attention paid to this distribution was possibly misplaced. The leading term in (5.24) does not dominate the second term until $t \approx 2$ (i.e. $4N$ generations), and the distribution of the fixation time, given by (5.22), shows that fixation of one or other allele is likely to have occurred by this time, especially in the interesting case $p = (2N)^{-1}$.

We note from (4.86) and (4.91) that the eigenfunction solution to equations such as (5.11) appear in the form

$$f(x;p, t) = \sum_n \{\phi_n(p)\} \{\phi_n(x) \; \frac{dm(x)}{dx} \} \exp(\lambda_n t). \tag{5.25}$$

Thus the eigenfunctions corresponding to initial and current point bear a simple relationship to each other. Now in the model (1.43) it is quite easy to find the right eigenvectors exactly but very difficult to find the left eigenvectors, and Eq. (5.25) suggests an approximate relation between the two. Since for the process (5.10)

$$dm(x) = 2x^{-1}(1 - x)^{-1} dx, \tag{5.26}$$

we may make the approximation for the model (1.43)

$$l_{ij} = j^{-1}(2N - j)^{-1} r_{ij}, \tag{5.27}$$

where $l_{ij}(r_{ij})$ is the jth element in the ith left (right) eigenvector of the transition matrix. Since we know $r_{2j} = j(2N - j)$, this rederives the uniform approximation to the leading left eigenvector.

The solution (5.11) is exact for all t but is most useful for large values of t (say $t > 1$). For small values of t (e.g. $t < 0.1$) the infinite series converges slowly, and many terms must be calculated to obtain satisfactory approximations. Fortunately an asymptotic expansion solution of (5.10) (and of more complex equations involving selection and mutation) is available which is most accurate for small t. This and the infinite series (5.11) dovetail nicely near $t = 1$, and this then allows rather complete knowledge of the solution of (5.10). We do not give the derivation of this asymptotic expansion (which is due to Voronka and Keller, 1975; for a wider range of applications see Tier and Keller, 1978) and note only that for $t < 1$,

$$P_1(p;t) \sim \{p(1 - p)/C\}^{1/4} \exp(-2Ct^{-1}), \tag{5.28}$$

where

$$C = \tfrac{1}{4} \{arcos(2p - 1)\}^2.$$

Clearly, by symmetry,

$$P_0(p;t) \sim \{p(1-p)/C'\}^{1/4} \exp\left(-2C't^{-1}\right) \tag{5.29}$$

where

$$C' = \tfrac{1}{4} \{\text{arcos}\,(1-2p)\}^2.$$

These two values can be combined to give an asymptotic expression for the probability of fixation by time t. When $p = \tfrac{1}{2}, t = 0.65$, Eq. (5.28) gives $P_1\left(\tfrac{1}{2}, t\right) \approx 0.119$, whereas the correct value, found after much computation from (5.16), is 0.117. For $t = 1$, (5.28) gives $P_1\left(\tfrac{1}{2}, t\right) \approx 0.232$ whereas the approximation

$$P_1(\tfrac{1}{2};t) \approx \tfrac{1}{2} - \tfrac{3}{4}\,e^{-1},$$

found by taking the two leading terms only in (5.16), gives the value 0.224. (The correct value is 0.223.) Remarks parallel to these apply for the density function $f(x;p,t)$: the asymptotic expansion is very accurate up to $t = 1$, where it agrees with the expression found by taking the three leading terms in (5.11). The latter then provide excellent approximations for larger t values.

We consider now conditional process and suppose for definiteness that $x = 1$ is the absorbing state ultimately reached. Equations (4.54) and (5.11) show that the density function of x at time t is

$$f^*(x;p,t) = \sum_{i=1}^{\infty} \frac{4(2i+1)x(1-p)}{i(i+1)}\, T_{i-1}^1(1-2p)T_{i-1}^1(1-2x)\exp\{-\tfrac{1}{2}\,i(i+1)t\}. \tag{5.30}$$

For large t and small p this gives

$$f^*(x;p,t) \sim 6x \exp\left(-t\right), \tag{5.31}$$

so that

$$\lim_{t\to\infty} f(x|x \neq 0, 1, \text{ eventual fixation at } x = 1) = 2x. \tag{5.32}$$

It is interesting to compare this with (1.81). There is no similarity between the two expressions, and we conclude that this is a case where the natures of the branching process, on which (1.81) is based, and the model (1.43) are sufficiently different so that one gives little information about the other for large t values.

The functions $t^*(x)$ defined in (4.50) and (4.51) now become

$$t^*(x) = 2(1-p)x/\{p(1-x)\}, \quad 0 \leqslant x \leqslant p,$$
$$t^*(x) = 2, \qquad\qquad\qquad\quad p \leqslant x \leqslant 1. \tag{5.33}$$

The conditional mean absorption time, found by integration, becomes

$$t^*(p) = -2p^{-1}(1-p) \log(1-p).$$ (5.34)

In the Markov chain (1.43) this suggests the approximation that, if k is the initial value of the Markov variable,

$$t^*_{k,j} \approx 2(2N-k)j/k(2N-j), \quad j \leqslant k,$$

$$t^*_{k,j} \approx 2, \qquad\qquad\qquad j \geqslant k,$$ (5.35)

and a conditional mean fixation time of $-4Np^{-1}(1-p) \log(1-p)$ generations. Perhaps the most interesting case concerns a unique selectively neutral new mutant destined for fixation. Equation (5.35) suggests that on average it spends two generations at each possible frequency value, leading to a total mean fixation time of $4N-2$ generations. It is instructive to note how easily information about the conditional process can be found from information concerning the unconditional process.

The conditional variance of the absorption time can be found by solving (4.31), subject to appropriate boundary conditions. Here we must use the conditional process drift coefficient

$$a^*(x) = 1 - x$$ (5.35)

rather than the unconditional value. It is found that

$$(\sigma^*)^2(p) = 8[\pi^2/6 + p^{-1}(1-p) \log(1-p)\{1 - (2p)^{-1}(1-p) \log(1-p)\}$$
$$- \sum_{i=1}^{\infty} p^i/i^2].$$ (5.36)

In the limiting case $p \to 0$ this gives

$$(\sigma^*)^2 \approx 8[\pi^2/6 - 1.5] \approx 1.16,$$ (5.37)

or, for the process (1.43), $4.64N^2$ (squared generations). The complete distribution of the conditional absorption time can be found immediately from (5.16). We have

Prob {absorption at $x = 1$ before time t|eventual absorption at $x = 1$}

$$= \frac{\text{Prob \{absorption at } x = 1 \text{ before time } t\}}{\text{Prob \{eventual absorption at } x = 1\}}$$

$$= 1 + \sum_{i=1}^{\infty} (2i+1)(1-p)F(i+2, 1-i, 2, p)(-1)^i \exp\{-i(i+1)t\}.$$ (5.38)

The expressions (5.34) and (5.37) can in principle be found from this distribution, but it is far simpler to arrive at them in the manner we have shown. Kimura (1970) established (5.38) and discussed the nature of the corresponding density function for various p values.

The asymptotic expansion of Voronka and Keller may be used immediately for conditional processes. Thus, for example, (5.28) gives

$$P^*(p; t) = p^{-1} \{p(1 - p)C\}^{1/4} \exp(-2Ct^{-1}),$$
(5.39)

where $P^*(p; t)$ is the conditional probability of fixation at $x = 1$ by time t, given fixation eventually occurs. Similarly the conditional density function $f^*(x; p, t)$ can be accurately approximated for small t, and this leads, as with $f(x; p, t)$, to rather complete knowledge of its nature.

We conclude this section by discussing a model generalizing (1.43). Consider for example the two-sex model introduced in Sect. 3.9. Using the notation of that section, the quantities $k(t)$ (number of A_1 genes among the males in generation t) and $l(t)$ (number of A_1 genes among the females in generation t), are jointly Markovian. We make progress by using the Norman theory of quasi-Markovian variables introduced in Sect. 4.8. The weighted average gene frequency $x(t)$ is

$$x(t) = k(t)(4N_1)^{-1} + l(t)(4N_2)^{-1}$$
(5.40)

and it is easy to check, for the model considered, that

$$E\{x(t + 1) - x(t)|x(t)\} = 0$$

$$\text{Var}\{x(t + 1) - x(t)|x(t) = x\} = x(1 - x)(2N_e)^{-1} + e_{2,t},$$
(5.41)

$$E\{|x(t + 1) - x(t)|^3\} = e_{3,t},$$

where $N_e = 4N_1N_2/(N_1 + N_2)$ and, for large N_1 and N_2, the error terms $e_{i,t}$ can be shown to satisfy (4.99) with $\tau_N = (2N_e)^{-1}$. Thus $x(t)$ is a quasi-Markovian variable, and the conclusions given in Sect. 4.8 for such variables can be applied. In particular the probability of fixation of A_1 is

$$k(0)(4N_1)^{-1} + l(0)(4N_2)^{-1}.$$

Note also that the variance formula confirms the value (3.112) for the variance effective population size.

5.3 Selection

Suppose now that the three genotypes have fitnesses given by (5.1). Assuming no mutation, the drift coefficient (5.6) becomes

$$a(x) = \alpha x(1 - x) \{x + h(1 - 2x)\},$$
(5.42)

and hence the scale function and speed measure are

$$p(x) = \int_{c_1}^{x} \psi(y)dy,$$ (5.43)

$$m(x) = 2 \int_{c_2}^{x} y^{-1}(1-y)^{-1} \{\psi(y)\}^{-1}dy,$$ (5.44)

where

$$\psi(y) = \exp \alpha \{(2h-1)y^2 - 2hy\}.$$ (5.45)

Both boundaries $x = 0$, $x = 1$ are exit, and the probability is unity that one or other boundary is eventually reached. The respective probabilities are given by (4.16) and (4.18) with $\psi(y)$ defined by (5.45).

These expressions simplify significantly only in the case of no dominance ($h = \frac{1}{2}$), for which

$$P_1(p) = \{1 - \exp(-\alpha p)\}/\{1 - \exp(-\alpha)\}.$$ (5.46)

Note that this agrees with the approximation (3.29) found without using diffusion methods. Some numerical values calculated from (5.46) are given in Table 5.1.

The conclusions to be drawn from this table are obvious enough. When N, s and p are jointly sufficiently large, fixation of the favored allele is essentially certain: this occurs roughly when $Nsp > 5$. As N, s or p decreases, the fixation probability decreases, and if $Ns < 0.1$ it does not differ (relatively) by more than 10% from the neutral value p. Perhaps the most striking conclusion is the very strong effect of selection in influencing fixation probabilities: as noted below Eq. (3.29), selective differences far too small to be found in the laboratory can nevertheless have a decisive effect on evolutionary behavior, at least in populations that are not too small. The same conclusion holds, at least qualitatively, when there is dominance (i.e. $h \neq \frac{1}{2}$), although some minor modifications to the numerical values are necessary, especially when dominance is complete ($h = 0$ or $h = 1$). Note that even in the overdominant case ($sh > s > 0$) fixation of one

Table 5.1. Values of $P_1(p)$, for various values of N, s, and p, calculated from (5.32)

		$p = 0.001$			$p = 0.5$		
		$N = 10^4$	$N = 10^5$	$N = 10^6$	$N = 10^4$	$N = 10^5$	$N = 10^6$
	0.01	0.181	0.865	1.000	1.000	1.000	1.000
	0.001	0.020	0.181	0.865	1.000	1.000	1.000
s	0.0001	0.002	0.020	0.181	0.731	1.000	1.000
	0.00001	0.001	0.002	0.020	0.525	0.731	1.000

or other allele is certain although, as we see later, this will normally take an extremely long time, and in practical terms one must then question the appropriateness of the assumptions made, in particular that there is no mutation and that the population size, selective differences, and dominance relationship remain unchanged. So far as fluctuations in population size are concerned it seems likely, for any fixed selection scheme, that (5.46) and (4.18) still apply if N is replaced by the variance effective population size, although the theory for this has not been verified.

The complete solution of the forward equation (4.7), with $a(x)$ and $b(x)$ defined by (5.42) and (5.5), is very complex. Nevertheless solutions were found by Kimura (1955a, b, c; 1957) initially for the no dominance case and subsequently for the general case. Unfortunately the very complexity of the solutions makes examination of their implications difficult, although this does not detract from the influence that the derivation of these solutions has had on population genetics theory. For more details concerning these solutions, see Crow and Kimura, 1970, pp. 396–414.

5.4 Selection: Absorption Time Properties

Despite the very complex form of $f(x;t)$ referred to at the end of the previous section, a rather simple expression exists for the function $\bar{t}(x;p)$, defined in (4.26), and since this function summarizes perhaps the most important features of the transient behavior of the process, we now compute it for the selective models we are considering. All that is required to do this is to substitute (5.42) and (5.5) into the general formulae (4.23) and (4.24). For $h = \frac{1}{2}$

$$\bar{t}(x;p) = 2P_0(p)\{\alpha x(1-x)\}^{-1}\{\exp(\alpha x) - 1\}, \quad 0 \leqslant x \leqslant p$$

$$\bar{t}(x;p) = 2P_1(p)\{\alpha x(1-x)\}^{-1}\{1 - \exp(-\alpha(1-x))\}, \quad p \leqslant x \leqslant 1$$

(5.47)

where $P_1(p)$ is found from (5.46) and $P_0(p) = 1 - P_1(p)$. For the Markov chain defined by (3.18) and (3.27), this implies that the mean number of generations for which there are $j = 2Nx\,A_1$ alleles, given an initial number $k = 2Np$, is approximately

$$\bar{t}_{k,j} = 2\{\exp(-2p) - \exp(-\alpha)\}\{\exp(\alpha x - 1)\}[\alpha x(1-x)\{1 - \exp(-\alpha)\}]^{-1} \quad (j \leqslant k),$$

(5.48)

$$\bar{t}_{k,j} = 2\{1 - \exp(-\alpha p)\}\{1 - \exp-\alpha(1-x)\}[\alpha x(1-x)\{1 - \exp(-\alpha)\}]^{-1} \quad (j \geqslant k).$$

Note, for $k = 1$, that this gives Eq. (1.57) if we make the approximation $1 - \exp(-\alpha/2N) = \alpha/2N$.

The mean time for fixation is given by (4.22) and (5.47), but unfortunately no explicit evaluation of the integral is possible, and numerical computation is necessary. There is, however, one case where useful progress can be made. If α and p are jointly sufficiently large so that fixation of the favored allele can be taken as being certain, we get

$$\bar{t}(x;p) \approx 0, \qquad\qquad\qquad\qquad\qquad 0 \leqslant x \leqslant p$$

$$\bar{t}(x;p) \approx 2\{\alpha x(1-x)\}^{-1}, \qquad\qquad\qquad p \leqslant x \leqslant 1 - 4\alpha^{-1}, \qquad (5.49)$$

$$\bar{t}(x;p) \approx 2\{\alpha x(1-x)\}^{-1}\{1 - \exp - \alpha(1-x)\}, \quad 1 - 4\alpha^{-1} \leqslant p \leqslant 1.$$

The first equation shows that the favored allele's frequency spends negligible time less than its initial value. The second equation is perhaps the most interesting. Converting to generations, it implies that in the Markov chain the mean time spent in the frequency range (x_1, x_2) where $p \leqslant x_1 < x_2 \leqslant 1 - 4\alpha^{-1}$, is approximately

$$\int_{x_1}^{x_2} \{\tfrac{1}{2} sx(1-x)\}^{-1} dx$$

generations. This is identical to the value (1.28) found for the corresponding deterministic process, and we can conclude that the behavior of the process in $(p, 1 - 4\alpha^{-1})$ is "quasi-deterministic". When the frequency exceeds $1 - 4\alpha^{-1}$ the deterministic value no longer gives an adequate guide to the stochastic behavior. In particular, the mean number of generations (in the Markov chain) for which $x = 1 - i(2N)^{-1}$, for small integers i, is equal to the "neutral" value 2. This is severely overestimated by the deterministic formula, and clearly at this stage of the process selective forces have become of secondary importance, and random sampling almost wholly determines the gene frequency behavior.

For general values of h in $(0, 1)$ the expressions (4.23) and (4.24) do not simplify readily. However, the general behavior just noted for the no dominance case continues to apply: quasi-deterministic behavior obtains for sufficiently large p and α, at least until the frequency x of A_1 approaches unity, when selective forces once more can be ignored. The value of x where this occurs will depend to some extent on the level of dominance but will not differ materially from the value $1 - 4\alpha^{-1}$ found in the no dominance case.

The value $k = 1$ is of particular interest. Here we may approximate $P_1\{(2N)^{-1}\}$ by $(2N)^{-1} \left[\int_0^1 \psi(y)dy\right]^{-1}$ and this leads, in the Markov chain, to the approximation (1.55).

The variance of the absorption time is given in principle by (4.32), but evaluation of this will certainly require numerical methods.

The overdominance case presents features of special interest. Here the deterministic theóry gives a stable polymorphism while the stochastic theory predicts eventual loss of one or other allele. On the other hand, it is plausible, in the stochastic case, that long periods of time are spent near the quasi-equilibrium point, at least when selection is strong, so that in some sense the deterministic theory provides a useful guide to the stochastic behavior. (Of course, if mutation is allowed, an entirely different stochastic behavior arises and one which should be well described by the deterministic theory.) We now discuss the stochastic process in more detail and show in particular that the plausibility argument given above does not necessarily apply.

We start from the observation just made, that in the stochastic process fixation of one or other allele is eventually certain if there is no mutation, and ask how much time fixation takes to occur. This is best answered by considering the mean fixation time or, more crudely, the leading eigenvalue in the spectral expansion of the density function $f(x; t)$. The latter approach was taken by Miller (1962) and Robertson (1962) and produced a surprising answer. Define the quasi-equilibrium point by x^* (see Eq. (1.30): we now assume the notation (1.24c)). If x^* is close to 0 or 1 it is possible that this leading eigenvalue λ_1, in an expansion of the form (4.91), is larger in absolute value than for the selectively neutral case: this suggests a more rapid fixation process under overdominance than under neutrality. This perhaps surprising conclusion is explained by noting that if x^* is close to 0 (or 1), selection tends to drive the frequency of A_1 close to 0 (or 1) comparatively rapidly, and then random sampling effects (which, we have noted, play a predominant role near the boundaries) lead to loss or fixation of A_1. The magnitude of this effect can be measured by taking the ratio of the absolute values of the leading eigenvalue and its neutral counterpart. When x^* is close to 0 or 1, this ratio increases with s for values of s of order a few percent (although, for very large s values, the ratio ultimately decreases as s increases). The discussion just given shows why this behavior should occur. For intermediate values of x^* (approximately $0.2 < m < 0.8$) the ratio always decreases as s increases, so that here heterosis always slows down the fixation process. It is clear that, for these values of x^*, selection does not provide a thrust towards the boundaries sufficient to speed up fixation. Tables and graphs illustrating this behavior are given by Robertson (1962); note that Robertson centres attention on retardation behavior and thus considers the reciprocal of the ratio defined above.

A perhaps more complete analysis is provided by considering the mean fixation time although a further degree of complexity arises here since this depends, unlike the eigenvalues, on the initial frequency p of A_1. It is perhaps natural to pay special attention to the case $p = x^*$ although a general analysis is quite straightforward. The mean fixation time is given by (4.23), (4.24), (4.16) and (4.18), where $b(x) = x(1-x)$ and

$$\psi(x) = \exp \{\alpha(2h - 1)(x - x^*)^2\}. \tag{5.50}$$

The retardation factor (corresponding to Robertson's eigenvalue ratio) is the ratio of the mean fixation time so calculated to its neutral theory counterpart (5.19). We do not give here details of the numerical values found (for these see Ewens and Thomson, 1970). The conclusions are in general agreement with those of Robertson, at least for $p = x^*$. For general values of p the behavior can be quite complex, the retardation factor sometimes increasing, then decreasing, then increasing again as s increases.

We turn now to mean absorption times conditional on eventual fixation (or loss) of a specified allele. The formulae appropriate to calculate this are (4.50) and (4.51) or (4.52) and (4.53). Perhaps the case of greatest interest is when $0 < h < 1$ and the condition is made that the favoured allele fixes. When $h = \frac{1}{2}$, (4.50) and (4.51) give

$$t^*(x;p) = 2e^{-\alpha x}\{1 - e^{-\alpha(1-p)}\}\{e^{\alpha x} - 1\}^2 [\alpha x(1 - x)\{1 - e^{-\alpha}\}\{e^{-\alpha p} - 1\}]^{-1}$$

$$0 \leqslant x \leqslant p, \tag{5.51}$$

$$t^*(x;p) = 2\{e^{\alpha x} - 1\}\{e^{\alpha(1-x)} - 1\}[\alpha x(1-x)\{e^\alpha - 1\}]^{-1},$$

$$p \leqslant x \leqslant 1. \tag{5.52}$$

Similarly, if the condition is made that eventually A_1 is lost,

$$t^{**}(x;p) = 2\{e^{\alpha x} - 1\}\{e^{\alpha(1-x)} - 1\}[\alpha x(1-x)\{e^\alpha - 1\}]^{-1}, \quad 0 \leqslant x \leqslant p, \tag{5.53}$$

$$t^{**}(x;p) = 2e^{-\alpha(1-x)}\{1 - e^{-\alpha p}\}\{e^{\alpha(1-x)} - 1\}^2[\alpha x(1-x)\{1 - e^{-\alpha}\}\{e^{\alpha(1-p)} - 1)\}]^{-1},$$

$$p \leqslant x \leqslant 1. \tag{5.54}$$

There are several interesting points about these equations. First, note that $t^*(x;p)$, for $x \geqslant p$, is identical to $t^{**}(x;p)$ for $x \leqslant p$. We will explain why this should be so when considering time-reversal properties (Sect. 5.9). The second point concerns the nature of the formula for $t^*(x;p)$ for very small p (or correspondingly $t^{**}(x;p)$ for very large p), and is relevant when considering a selectively favored new mutant destined for fixation. We observe that $t^*(x;p)$ is symmetric about $x = 0.5$; the mean time spent in any interval $(x, x + \delta x)$ is the same as the mean time spent in $(1 - x - \delta x, 1 - x)$. Even more surprisingly, $t^*(x;p)$ remains unchanged if α is replaced by $-\alpha$, so that a selectively disadvantageous mutant, if destined for fixation, spends as much time, on the average, in any frequency range as a corresponding selectively advantageous mutant destined for fixation. This remarkable fact, noted first in effect by Maruyama (1974), will again be reconsidered later in the light of time-reversal properties. It is indeed easy to see that the entire behavior of the conditional process is independent of the sign of s, since the diffusion coefficient $b^*(x)$, calculated from (4.57) and (5.5), is independent of s while the drift coefficient $a^*(x)$, calculated from (4.56), (5.42), and (5.46) is

$$a^*(x) = \tfrac{1}{2} \alpha x(1-x)/\tanh \left(\tfrac{1}{2}\alpha x\right).$$

Clearly $a^*(x)$ is independent of the sign of α: this more detailed conclusion was first noted by Watterson (1977b). Note, however, that despite the symmetry of $t^*(x)$ around $x = \tfrac{1}{2}$ it is not true that $a^*(x) = a^*(1-x)$.

For arbitrary levels of dominance, (4.51) shows that with $p = (2N)^{-1}$,

$$t^*(x; (2N)^{-1}) = 2 [b(x)\psi(x) \int_0^1 \psi(y)dy]^{-1} \int_0^x \psi(y)dy \int_x^1 \psi(y)dy, \tag{5.55}$$

where $\psi(y)$ is defined by (5.45). If this expression is written more fully as $t^*(x; \alpha, h, (2N)^{-1})$, it may be noted that

$$t^*(x; \alpha, h, (2N)^{-1}) = t^*(1-x; -\alpha, 1-h, (2N)^{-1}). \tag{5.56}$$

This implies that conditional mean fixation time properties for a favored allele are the same as those for the corresponding disadvantageous allele, provided the dominance relation is reversed. This generalizes the conclusion just reached for the case of no dominance.

Using the notation (5.1), the fitnesses will display overdominance if $sh > s > 0$ or underdominance if $sh < s < 0$. In either of these cases the quasi-equilibrium point m, defined above, may be written more fully as $m = m(h) = h/(2h - 1)$. Then Eq. (5.56) may be written as

$$t^*(x, m, (2N)^{-1}) = t^*(1 - x, 1 - m, (2N)^{-1}),$$ (5.57)

an equation first noted in this form by Nei and Roychoudhury (1973).

For general levels of dominance it is no longer true (as it was with no dominance) that $t^*(x, \alpha, h, (2N)^{-1}) = t^*(x, -\alpha, 1 - h, (2N)^{-1})$, nor is it true that $a^*(x, \alpha, h) = a^*(x, -\alpha, 1 - h)$. There is, however, one relation, first noted by Maruyama and Kimura (1974), that does remain true. Keep the fitness scheme (5.1) fixed and consider two cases, one where the initial frequency of A_1 is $(2N)^{-1}$ and the condition is made that A_1 eventually fixes, and the other where the initial frequency of A_1 is $1 - (2N)^{-1}$ and the condition is made that A_1 is eventually lost. By considering A_2 rather than A_1 it is clear that the equation

$$t^*(1 - x, -\alpha, 1 - h, (2N)^{-1}) = t^{**}(x, \alpha, h, 1 - (2N)^{-1})$$ (5.58)

must be true, and this may be used with (5.57) to show that

$$t^*(x, \alpha, h, (2N)^{-1}) = t^{**}(x, \alpha, h, 1 - (2N)^{-1}).$$ (5.59)

(We noted above the special case of this equation when $h = \frac{1}{2}$.) Thus the mean time spent in any frequency range is the same for both processes. On the other hand, it is not true that $a^*(x) = a^{**}(x)$ so that, despite (5.59), the two processes have quite different properties. Again, these perhaps paradoxical conclusions will be reconsidered, and to a large extend resolved, when we consider time-reversal properties of diffusion processes.

5.5 One-Way Mutation

Until now in this chapter we have ignored the possibility of recurrent mutation from one allelic type to another when considering allele-frequency behavior. In some circumstances this might cause little inaccuracy but in general, especially from a macro-evolutionary rather than a micro-evolutionary point of view, it is essential that such mutation be taken into account. In this section we make a start on this by supposing a model such as (3.18) where A_1 mutates to A_2 (at rate u), with no reverse mutation. The drift and diffusion coefficients for the diffusion process approximating this Markov chain are, when there is no selection,

$$a(x) = -\tfrac{1}{2}\theta x, \quad b(x) = x(1 - x),$$ (5.60)

where $\theta = 4Nu$. Clearly A_1 eventually becomes lost from the population and interest centers entirely on properties of the time until such loss occurs. These properties are defined in large measure by the function $t(x; p)$, and insertion of the coefficients (5.60) into (4.39) and (4.40) gives this function immediately. The values so calculated are given in (3.20), where allowance must be made for the fact that the time-scale assumed there assumes unit time for one generation. Perhaps the case of most interest is when $p = (2N)^{-1}$, so that to a close approximation the mean time that A_1 exists in the population is

$$\bar{t} \{(2N)^{-1}\} \approx \int_{(2N)^{-1}}^{1} 2y^{-1}(1-y)^{\theta-1} dy \tag{5.61}$$

generations. Note that this is of order $2 \ln (2N)$ generations for moderate values of θ: a new mutant A_1 cannot on average be expected to remain in the population for very long, or to attain a high frequency, if there is no recurrent mutation $A_2 \rightarrow A_1$.

The process we are considering, since it admits the possibility of only two alleles, is perhaps of limited interest. However, several of its properties throw considerable light on important features of the infinite allele model (3.65). Some of these were already given in (3.82) and (3.83). It is clear that in the infinite allele model we may normally expect several low-frequency alleles in the population (for example if $\theta = 1$, $2N = 10^6$, there will be ten alleles, on average, with frequency less than 0.01). If θ is small enough the most likely situation is where there is one remaining allele at high frequency: thus from (3.82),

Pr (there exists an allele with frequency greater than 0.9)

= mean number of alleles with frequency greater than 0.9

$$= \theta \int_{0.9}^{1} x^{-1}(1-x)^{\theta-1} dx$$

$$> \theta \int_{0.9}^{1} (1-x)^{\theta-1} dx \tag{5.62}$$

$$= (0.1)^{\theta}.$$

For $\theta = 0.1$ this probability is about 0.8. For larger values of θ ($\theta > 1$ approximately) it becomes rather unlikely that such a high-frequency allele will exist, and the most likely configuration is one where a number of alleles exist at low but unequal frequencies. Note that in all cases the least likely situation is one where two, three or four alleles exist with approximately equal frequencies. These arguments suggest an approach to testing whether a model such as (3.65) (and not, for example, one involving selection), is adequate to explain observed allelic frequencies, an observation which we take up at greater length in Chap. 9 when considering tests for neutrality using allele frequency data.

There are two further points that are of interest in considering the model (3.65) and its evolutionary behavior. The first concerns the nature of the boundary $x = 1$ for the two allele model originally considered. Use of (5.60) in (4.73) shows that this boundary is entrance if $\theta \geqslant 1$: this implies that it is impossible to reach this boundary by diffusion from the interior of $(0, 1)$ in this case. It is therefore impossible to consider behavior conditional on the requirement that this boundary is reached, and further it is unnecessary to impose the condition that the boundary is not reached and then consider conditional behavior: this latter condition is already implicit and formulae such as (5.61) apply immediately. When $\theta < 1$ the boundary $x = 1$ is regular and hence attainable and now new behavior arises under the condition that this boundary is not reached. Again assuming $p = (2N)^{-1}$ we find that (5.61) must be replaced, conditional on $x = 1$ not being reached, by

$$\bar{t} \{(2N)^{-1}\} = \int_{(2N)^{-1}}^{1} 2y^{-1}(1-y)^{1-\theta} dy \qquad (5.63)$$

generations. The integrand in (5.63) has the usual interpretation of providing the mean time spent in various frequency ranges.

The second point concerns the frequency of the most frequent allele. The argument that led to (5.62) shows that for $0.5 \leqslant x \leqslant 1$ the probability density function of the frequency of the most frequent allele in the infinite alleles model is, at equilibrium,

$$f(x) = \theta x^{-1}(1-x)^{\theta-1}. \qquad (5.64)$$

For values of x less than 0.5 a deeper argument is clearly required: nevertheless the probability density function of the most frequent allele can be found for these values also (see Watterson and Guess, 1977; Watterson, 1976b). Further details are given in Sect. 5.10. Of particular interest to us now is the use of (5.64) to approximate the value of P_{mono}, defined by

P_{mono} = Prob (only one allele present in the population in any generation
at equilibrium). $\qquad (5.65)$

If we make the approximation

$$P_{\text{mono}} \approx \int_{1-(2N)^{-1}}^{1} f(x)dx, \qquad (5.66)$$

then immediately from (5.64)

$$P_{\text{mono}} \approx (2N)^{-\theta}. \qquad (5.67)$$

This approximation has been made on several occasions in the literature, and in Sect. 5.7 we examine its accuracy. (Note that in the Moran infinite alleles model an exact expression can be given for P_{mono}.) It is of some interest to compare (5.67) to the value found from (3.71) by formally putting $i = 2N$, namely

$$(2N-1)!/[(1+\theta)(2+\theta) \dots (2N-1+\theta)]. \qquad (5.68)$$

The two expressions (5.67) and (5.68) agree at $\theta = 0$ and $\theta = 1$ and are very close for all values of θ less than unity. For $\theta > 1$ the expression (5.67) becomes increasingly smaller than (5.68), and we shall see later that, surprisingly, (5.68) gives a better approximation to P_{mono} than does the more frequently used (5.67).

5.6. Two-Way Mutation

Suppose now in the model (3.18) that mutation both from A_1 to A_2 (at rate u) and from A_2 to A_1 (at rate v) is allowed with no selection, so that π_i is given by (3.21). As we have already noted there will now exist a stationary distribution for the frequency x of A_1, for which we already have an exact expression (3.22) for the mean and an approximate expression (3.23) for the variance. Our aim now is to approximate the entire distribution by diffusion methods. The drift and diffusion coefficients are found from (5.6) (putting $\alpha = 0$) and (5.5) and then (4.46) leads to the stationary distribution

$$f(x) = \frac{\Gamma\{2\beta_1 + 2\beta_2\}}{\Gamma\{2\beta_1\}\Gamma\{2\beta_2\}} x^{2\beta_2-1}(1-x)^{2\beta_1-1}. \tag{5.69}$$

The mean and variance of this distribution are $\beta_1/(\beta_1 + \beta_2)$ and $\beta_1\beta_2/\{(\beta_1 + \beta_2)^2(2\beta_1 + 2\beta_2 + 1)\}$ respectively, and these agree with the exact and approximate values (3.22) and (3.23) once allowance is made for a change of scale. The general shape of the curve (5.69) is clear enough. For small β_i (i.e. small mutation rates and/or population sizes) most of the probability mass is in the extremes of the distribution, so that the most likely situation is one where one or other allele is at a low frequency (or even temporarily absent from the population). For β_i large the variance becomes small and the behavior is "quasi-deterministic": only small deviations are likely from the deterministic theory equilibrium point. This may be illustrated by supposing $u = v$ so that the mean of the stationary distribution is 0.5 irrespective of the population size. Thus supposing $u = v = 2.5 \times 10^{-6}$, the stationary probability that the frequency of A_1 is between 0.4 and 0.6 rises from 0.2 for $N = 10^5$ to 0.8 for $N = 10^6$ and is essentially unity for $N = 10^7$ or more.

   The probability distribution (5.69) allows no atoms of probability at the boundaries $x = 0$, $x = 1$: this is in accordance with the coefficients (5.5) and (5.6), which suggest instantaneous reflection from these boundaries. Nevertheless for the discrete process (3.18) there must exist non-zero stationary probabilities for the states $\{0\}$ and $\{2N\}$, and if u and v are small these probabilities will be quite large. It therefore becomes a matter of some interest to find how these boundary probabilities can be approximated from (5.69). This matter is taken up in the next section.

   When selection is allowed together with two-way mutation there will still exist a stationary distribution although its form is naturally more complicated. Use of the complete expressions (5.5) and (5.6) gives, from (4.46), the formula

$$f(x) = \text{const } x^{2\beta_2-1}(1-x)^{2\beta_1-1} \exp\{2\alpha hx - \alpha(2h-1)x^2\} \tag{5.70}$$

for this distribution, where the constant is a function of β_1, β_2, α, and h and may be found in principle (although rarely in practice) by normalization. Recalling that $\beta_1 = 2Nu$, $\beta_2 = 2Nv$, (5.70) is identical to (1.63). The form of this distribution is of most interest with overdominance or when one allele is at a selective advantage to the other. The former case was discussed in some detail below Eq. (1.63) so we consider here only the latter. Assume for definiteness that $s < sh < 0$ so that A_1 is at a selective disadvantage and consequently usually at a low frequency. Assuming then that x is small we may ignore the term $(1 - x)^{2\beta_1 - 1}$ as well as the term in x^2 in the exponent to get

$$f(x) \approx \text{const } x^{2\beta_2 - 1} \exp (2\alpha h x).$$

Clearly $0 \leqslant x \leqslant 1$ but for values of $|\alpha h|$ sufficiently large ($|\alpha h| > 3$ should normally suffice), $f(x)$ is negligibly small for $x > 1$, and the normalizing constant may be evaluated to a sufficient approximation by supposing $0 \leqslant x < \infty$. This leads to

$$f(x) \approx |2\alpha h|^{2\beta_2} \{\Gamma(2\beta_2)\}^{-1} x^{2\beta_2 - 1} \exp \{-|2\alpha h|x\}. \tag{5.71}$$

From this the mean and variance of the stationary distribution of the frequency of A_1 are found to be

$$\beta_2/|\alpha h|, \quad \tfrac{1}{2} \beta_2/(\alpha h)^2 \tag{5.72}$$

respectively. Allowing for changes in notation, the mean value agrees with the deterministic equilibrium point (1.35), while the variance provides new information and gives some idea of the extend of stochastic variation that can be expected around the mean. Parallel values may be calculated when $s > sh > 0$.

5.7 Diffusion Approximations and Boundary Conditions

The aim of this section is to consider the extent to which various formulae derived or suggested by diffusion methods provide accurate approximations to the true (but unknown) values for the Markov chain specified by (3.18) and (3.27). It must of course be kept in mind that this brings us no closer to an evaluation of how close our results are to "reality": the diffusion process may well provide a better reflection of real processes than does the Markov chain model.

If one adopts the view that the primary process of interest is a Markov chain such as (3.18), approximating diffusion formulae can usually be obtained in two different ways. The first is to approximate the Markov chain process by a diffusion process by finding the appropriate drift and diffusion coefficients (using the theory of Sect. 5.1) to calculate the required quantity for this diffusion process and to use the value so found as an approximation for the Markov chain. The second way is more direct, and was used several times in Chapt. 3: there is no concept of an approximating diffusion process, and the quantity of interest is approximated by considering only the leading terms in a Taylor series expansion. The two approaches give the same formulae (compare, for example, Eqs. (3.5) and (5.19)), and an extension of the second approach, using higher-order Taylor approximations, thus leads to an assessment of the accuracy of diffusion formulae, using standard techniques.

Consider for example the diffusion approximation (5.46) for the probability of fixation of a favored allele A_1 in the absence of dominance and mutation. This equation was also found, as (3.29), without using diffusion theory. Now (3.29) was reached by ignoring certain small order terms in the derivation of (3.28), and a formula somewhat more accurate than (3.28) is found from

$$0 = \sum_{i=1}^{4} E(\delta x)^i \, \pi^{(i)}(x)/i!, \qquad (5.73)$$

as this equation incorporates terms of order N^{-2} as well as terms of order N^{-1}. (Note that an even better approximation would arise by replacing the derivatives in (5.73) by finite differences.) If we now put

$$\pi(x) = \{1 - \exp(-\alpha x)\}\{1 - \exp(-\alpha)\}^{-1} + N^{-1}g(x) \qquad (5.74)$$

in (5.73), a second order differential equation for $g(x)$ will be obtained. Since $g(0) = g(1) = 0$, this equation can be solved for $g(x)$ and thus a correction term to $\pi(x)$ of order N^{-1} obtained. More complete details are given in Ewens (1964).

Similar corrections may be made to the mean absorption times although here difficulties arise for very large or very small values of p, since the higher derivatives of $\bar{t}(p)$, defined by (3.5) or (5.19), become increasingly large for such p values. Nevertheless even for $p = (2N)^{-1}$ the diffusion approximation (3.6) is remarkably accurate: a more precise value, found in Fisher (1958, p. 98) and confirmed by Watterson (1975), is

$$\bar{t}\{(2N)^{-1}\} = 1.355076 + 2 \log 2N \qquad (5.75)$$

generations. An almost identical correction occurs when there is one-way mutation (cf. Eq. (5.61)).

It is also possible to consider corrections to complete distributions arrived at by diffusion theory, in particular to the stationary distribution (4.46). Here it must first be decided in what way the diffusion formula is used as an approximation to a discrete distribution: thus if the stationary distribution $\{\alpha_i\}$ defined below (3.21) is approximated by a continuous distribution $f(x)$, both approximations

$$\alpha_i \approx \text{const} \int_{(i-1/2)/2N}^{(i+1/2)/2N} f(x)dx, \quad \alpha_i \approx \text{const } f(i/2N) \qquad (5.76)$$

could be used. Although in general the two approximations will give similar values for moderate values of i, problems can arise with both definitions at $i = 0$ and $i = 2N$. Thus for the stationary distribution (5.70), the latter definition leads to a zero or infinite value for α_i, both clearly unsatisfactory. However, the first approximation in (5.76) requires adjustments to the terminals of integration for $i = 0$ and $i = 2N$. Further, the integration involved often cannot be completed exactly and numerical methods, which reduce to evaluation of $f(x)$ at discrete point values, are then required. Altogether the best way to view the diffusion approximation is probably to use the second approximation in (5.76) for all values of i other than 0 and $2N$ and to estimate α_0 and α_{2N} through the stationarity equations

$$\alpha_0 = \Sigma \alpha_j p_{j,0}, \quad \alpha_{2N} = \Sigma \alpha_j p_{j,2N}. \tag{5.77}$$

The constant is now chosen so that $\Sigma \alpha_i = 1$. This approach normally leads to quite accurate diffusion approximations to the stationary distribution of Wright-Fisher models.

In considering approximations to α_0 and α_{2N}, Wright (1931, p. 123 and many subsequent papers; see in particular 1969, pp. 356–357) replaced (5.77) by the approximation

$$2Nv\alpha_0 \approx \tfrac{1}{2}\alpha_1, \quad 2Nu\alpha_{2N} \approx \tfrac{1}{2}\alpha_{2N-1}. \tag{5.78}$$

These approximations were suggested by parallel approximations for the asymtotic conditional distribution l_j (see (1.49)), considered in great detail by Wright and Fisher. In subsequent work Wright regarded (5.78) as rough approximations only and paid careful attention to their domains of validity. Unfortunately the approximations (5.78) have often been used uncritically by other authors, and this has led to rather inaccurate expressions for α_0 and α_{2N}. Similar uncritical use of the approximations

$$\alpha_0 \approx \int_0^{(2N)^{-1}} f(x)dx, \quad \alpha_{2N} \approx \int_{1-(2N)^{-1}}^1 f(x)dx \tag{5.79}$$

has also led to estimates of large (relative) error.

This latter point may be illustrated by discussing approximations to the quantity P_{mono}, defined in Sect. 5.5, for the infinite alleles model (3.65). The approximation (5.67) for this quantity can be reached by using (5.79) as a starting point: this was essentially the approach of Kimura (1971) who computed the corresponding value in the K-allele model (3.61) and then let $K \to \infty$. This approach, however, uses diffusion approximations for precisely those values when they are most suspect and a more detailed computation of P_{mono} is possibly needed. This was provided by Watterson (1975) who arrived at the approximation

$$P_{mono} \approx \exp(-0.10030)\Gamma(1 + \theta)(2N)^{-\theta}. \tag{5.80}$$

We have already noted the approximation (5.68) derived by formally putting $i = 2N$ in (3.71). Table 5.2 displays for $2N = 1000$ exact values of P_{mono} found numerically together with the approximations arrived at from (5.80), (5.68), and (5.67).

Table 5.2. Values of P_{mono} for $2N = 1000$ and various θ

θ	0.1	0.5	1	5	10
exact	0.472	0.267×10^{-1}	0.902×10^{-3}	0.669×10^{-13}	0.979×10^{-24}
approximation (5.80)	0.472	0.267×10^{-1}	0.905×10^{-3}	0.727×10^{-13}	1.331×10^{-24}
approximation (5.68)	0.477	0.280×10^{-1}	1.000×10^{-3}	1.188×10^{-13}	3.470×10^{-24}
approximation (5.67)	0.501	0.316×10^{-1}	1.000×10^{-3}	0.010×10^{-13}	1.000×10^{-30}

Clearly (5.80) gives an excellent approximation for all values of θ considered while (5.67) gives a good approximation only when $\theta < 1$.

In considering the above diffusion approximations it has been assumed that the population size, mutation rate, and selective differences are all such that the parameters α, β_1, and β_2 in (5.2) are $O(1)$. Not only is this condition rather imprecise: it may also not apply in several cases of interest. An attempt to overcome this problem has been made by Ethier and Norman (1977) who provide bounds on diffusion approximations irrespective of the order of magnitude assumptions discussed above. More specifically, for the model defined by (3.18) and (3.21), Ethier and Norman provide an explicit upper bound for the difference between the expectation of any infinitely differentiable function of the gene frequency x and the value as calculated from the approximating diffusion process, for any values of N, u and v. This bound is uniform over time and thus applies also to the stationary distribution. For further details see Ethier and Norman (1977), in particular Eq. (7).

An interesting case arises for the heterozygosity measure $2x(1 - x)$. The stationary expectation of this quantity for the diffusion process may be found immediately from (5.69). If we assume for convenience that $2\beta_1 = 2\beta_2 = \theta$, the value found for this expectation is $\theta/(2\theta + 1)$. However, the stationary expectation for the Markov chain defined by (3.18) and (3.21) can be found exactly and is, explicitly,

$$\frac{\theta(1 - u)\{1 - (2N)^{-1}\}}{2\theta - 2u\theta + (1 - 2u)^2},$$

and the difference between this and the diffusion approximation is

$$\frac{u\{2 - 2u + \theta\}}{(2\theta + 1)\{2\theta - 2u\theta + (1 - 2u)^2\}}.$$

For the Ethier and Norman theory the upper bound provided for the error in the diffusion approximation is found by applying their Eq. (7) for the function $2x(1 - x)$. The bound is

$$\max(u, v) + (4N)^{-1} + 27 \max(u^2, v^2) + (7/16)N^{-2}.$$

In our case this may be written

$$u\{1 + \theta^{-1} + 27u + 7u\theta^{-2}\},$$

and it is not hard to verify that this function does indeed bound the exact error in the diffusion approximation given above.

A further remark should be made about order of magnitude assumptions. The diffusions we have considered assume that $E(\delta x)$ for the Markov chain (3.18) is of the same order of magnitude as $Var(\delta x)$, namely N^{-1}. This assumption is not always justified, for example if in the model defined by (3.18) and (3.21) N and u are jointly large enough so that Nu is not small and cannot be taken as of order unity. Suppose, to be specific, that $E(\delta x)$ is of order ϵ and $N\epsilon$ is large. Such processes have been discussed by Karlin and McGregor (1974) and Norman (1974, 1975a). For those processes the gene frequency clusters around its deterministic value given by the "infinite population" theory outlined in Chap. 2. Deviations of gene frequency from this value are, asymptotically, normally distributed with standard deviation of order $(N\epsilon)^{-1/2}$. For certain parameter values this diffusion approximation and the one we have discussed earlier overlap. This situation is analogous to the overlapping domains of applicability of the Poisson and normal approximations to the binomial distribution in statistical theory, and in such cases the two approximations approximate not only the discrete process but also each other.

We note in conclusion that diffusion theory can give a quite false impression not only quantitatively but also qualitatively about the boundary behavior in some Markov chains and illustrate this by considering the stationary distribution (5.70). The criteria (4.73) show that the boundary $x = 0$ is unattainable if $\beta_2 \geqslant \frac{1}{2}$, i.e. if the mutation rate $A_2 \to A_1$ is sufficiently large. Further, this conclusion remains unchanged whatever the selective parameters. Suppose now $\alpha << \alpha h << 0$: the discussion centred around (5.71) shows that the frequency of A_1 will usually be very small, and in a Markov chain model such as that defined by (3.18) and (3.27) there will be a substantial probability that at any time A_1 is absent from the population, in contrast to the diffusion theory prediction. Thus suppose $2N = 10^6, v = \frac{1}{2} \times 10^{-6}, s = -0.2, sh = -0.1$ so that $\beta_2 = 0.5, \alpha = -2 \times 10^5$. Equation (5.72) shows that the mean and standard deviation of the number of A_1 genes in the population at any time are 5 and 5, respectively, and this certainly suggests a non-negligible probability that there are no A_1 genes present. The distribution (5.71) suggests that for small positive integers i, the stationary probability of i A_1 genes is approximately $0.2 \exp(-0.2i)$, and the approximation (5.77) suggests that the stationary probability of the value 0 is about 0.13. This is not negligibly small, and we conclude that the diffusion theory boundary behavior gives a rather misleading picture of the behavior of the Markov chain for very small numbers (including zero) of A_1 genes.

5.8 Random Environments

So far in this book it has been supposed that stochastic changes in gene frequencies have been brought about solely by random sampling effects in finite populations. There are however further sources of stochastic variation, and perhaps the most important of these is that brought about by random temporal changes in the selection parameters, due perhaps to fluctuations in the environment. We will examine this at greater length in Chap. 10 and in particular will then consider the joint stochastic effects of fluctuating environments and finite population sizes. For the moment we limit consideration to the first source of variation only and discuss some simple models which give some idea of the flavor of the conclusions reached.

Suppose the selection parameters w_{ij} in the infinite population model (1.23) are of the form

$$w_{11} = 1, \quad w_{12} = 1 - s, \quad w_{22} = (1 - s)^2 \tag{5.81}$$

where s is a random variable with mean $\mu\delta$, variance $\sigma^2\delta$, and higher moments $O(\delta^2)$ or less. We assume δ is a small parameter, and we subsequently let $\delta \to 0$. Let s_t be the value assumed by the random variable s in generation t. Then with a slight change of notation, (1.23) can be written

$$x_{t+1} = x_t/\{1 - s_t(1 - x_t)\}. \tag{5.82}$$

If $y_t = x_t/(1 - x_t)$, this becomes

$$y_{t+1} = y_t(1 - s_t)^{-1} \tag{5.83}$$

and putting $z_t = \log y_t$, (5.83) leads to

$$z_t = z_0 - \sum_{i=0}^{t-1} \log(1 - s_i). \tag{5.84}$$

Thus apart from the constant z_0, z_t is the sum of independently and identically distributed random variables and thus the central limit theorem may be applied. Since

$$E\{\ln(1 - s)\} \sim E\{-(s + \tfrac{1}{2}s^2)\} = -\delta(\mu + \tfrac{1}{2}\sigma^2),$$

$$\text{Var}\{\ln(1 - s)\} \sim \text{Var}(-s) = \delta\sigma^2,$$

z_t will have an approximate normal distribution with mean

$$z_0 + t\delta(\mu + \tfrac{1}{2}\sigma^2) = \mu_z \tag{5.85}$$

and variance

$$t\delta\sigma^2 = \sigma_z^2. \tag{5.86}$$

A standard statistical transformation now gives the corresponding density function for x_t as

$$f(x;t) = \frac{1}{\sqrt{2\pi\sigma_z x(1-x)}} \exp\left[-\tfrac{1}{2}\{\log\tfrac{x}{1-x} - \mu_z\}^2/\sigma_z^2\right]. \tag{5.87}$$

This conclusion is reached by more or less exact methods, the only approximation involved being the normal distribution assumption as justified by the central limit theorem. We now show how it can be obtained by diffusion methods. From (5.82),

$$E(\delta x) = E\{sx(1-x) + s^2 x(1-x)^2 + 0(s^3)\}$$
$$= \delta x(1-x)\{\mu + \sigma^2(1-x)\} + 0(\delta^2), \tag{5.88}$$
$$E(\delta x)^2 = \delta\sigma^2 x^2(1-x)^2 + 0(\delta^2), \tag{5.89}$$

with higher moments $0(\delta^2)$ or less. Equations (5.88) and (5.89) provide the drift and diffusion coefficients for an approximating diffusion process: note that several authors have incorrectly omitted the term in σ^2 in $E(\delta x)$ and thus have obtained incorrect solutions to the diffusion equation. The equation becomes

$$\frac{\partial}{\delta t} f(x;t) = -\frac{\partial}{\partial x}\{\delta x(1-x)\{\mu + \sigma^2(1-x)\}f(x;t)\}$$
$$+ \tfrac{1}{2}\frac{\partial^2}{\partial x^2}\{\delta\sigma^2 x^2(1-x)^2 f(x;t)\}. \tag{5.90}$$

It may be checked by substitution that the solution of (5.90) is (5.87), and thus the diffusion approximation leads to exactly the same solution as the central limit theorem approximation.

We make several remarks on the solution (5.87). First, if $\mu + \tfrac{1}{2}\sigma^2 > 0$, the density function (5.87) increasingly concentrates near $x = 1$ as $t \to \infty$, while for $\mu + \tfrac{1}{2}\sigma^2 < 0$ it concentrates increasingly near $x = 0$. This behavior has been termed "quasi-fixation" by Kimura (1954) and has been defined more rigorously by Karlin and Liberman (1974) through the concept of "stochastic local stability". We discuss this latter concept at greater length in Chap. 10.

Secondly, note that even when $\mu = 0$, so that "on the average" A_1 and A_2 have equal fitnesses, the density function of the frequency x of A_1 still concentrates increasingly near $x = 1$. This reveals a most important new observation: the variance in fitness is just as important as the (arithmetic) mean fitness in determining evolutionary behavior and, with equal mean arithmetic fitnesses, the allele with the smaller variance in fitness is in effect selectively favored. The true fitness is in fact measured best by the geometric mean fitness which we consider in more detail later and which combines both mean and variance components. In the above case, A_1 and A_2 are selectively equivalent only if $\mu + \tfrac{1}{2}\sigma^2 = 0$, where in this case the left-hand side is the geometric mean selective advantage of A_2.

Finally, note that the solution (5.87), in contrast to other solutions (such as (5.11)) of diffusion equations, does not have the form of an eigenfunction expansion.

This is confirmed by the theory of Sect. 4.7: in the above process application of (4.73) shows that the boundaries $x = 0, x = 1$ are natural, and for such boundaries it is not necessary that the solution be in eigenfunction form.

A second selection scheme perhaps reveals more of the flavor of random environment models. Suppose the fitnesses in each generation are of the form

$$w_{11} = 1 + s, \quad w_{12} = 1, \quad w_{22} = 1 + \alpha s \tag{5.91}$$

where $0 < \alpha \leqslant 1$. This model has been studied in detail by Karlin and Liberman (1974) in discrete time and Levikson and Karlin (1975) in the diffusion case. The conclusions are analogous in the two situations, and we present only several discrete time results of some interest. We again assume s is a random variable having mean $\delta\mu$, variance $\delta\sigma^2$, and higher moments $0(\delta^2)$ or less.

Suppose first that $\alpha = 1$. If $\mu = 0$ the homozygotes have the same arithmetic mean fitness as the heterozygote but a lower geometric mean fitness and thus, from the previous discussion, can be regarded as being at a selective disadvantage. This is confirmed by Karlin and Liberman who show that, with probability one, each trajectory of gene frequency converges to 0.5. This behavior occurs even for positive μ so long as $\mu - \frac{1}{2}\sigma^2 < 0$. When $\mu - \frac{3}{4}\sigma^2 < 0 < \mu - \frac{1}{2}\sigma^2$ convergence of the frequency of A_1 to $0, 0.5$, or 1 can occur, depending on the initial frequency, while if $\mu - \frac{3}{4}\sigma^2 < 0$ the only two limiting behaviors are quasi-fixation of A_1 or A_2. In all cases the actual convergence behavior is not deterministic in the sense that it will depend on the values taken by s in the early generations.

When $\alpha < 1$ the picture is far more complex, and it is then possible in suitable circumstances that a stationary distribution for the frequency of A_1 can arise. The condition for this is that

$$3\alpha/(1 + \alpha) < 2\mu/\sigma^2 < 1 \tag{5.92}$$

and that the initial frequency x_0 of A_1 be between $\alpha/(1 + \alpha)$ and 1. (Note that the condition (5.92) requires $\alpha < \frac{1}{2}$.) Condition (5.92) follows from Eq. (4.8) in Karlin and Liberman (1974) but is rather less general than their condition, applying only when the moments of s have the properties we have assumed. Now the drift and diffusion coefficients for the model (5.91) are

$$a(x) = x(1 - x)\{(1 + \alpha)x - \alpha\}\{\mu - \sigma^2(x^2 + \alpha(1 - x)^2)\},$$
$$b(x) = \sigma^2 x^2(1 - x)^2 \{1 + \alpha)x - \alpha\}^2. \tag{5.93}$$

If these values are formally inserted into (4.46), we get

$$f(x) = \text{const } x^{-2\mu/\alpha\sigma^2}(1 - x)^{-2\mu/\sigma^2} \{x - \alpha/(1 + \alpha)\}^{2(1+\alpha)\mu/\alpha\sigma^2 - 4} \tag{5.94}$$

which, for the parameters specified by (5.92), is integrable over $(\alpha/(1 + \alpha), 1)$. Is it however the required stationary distribution? Tanaka (1957) shows that for diffusions with inaccessible boundaries formal calculation of $f(x)$ in this way does indeed provide the correct stationary distribution, provided that the resulting function is integrable and that $\psi(x)$, defined by (4.17), is non-integrable at both boundaries. We have already checked the former condition and since in this case

$$\psi(x) = \text{const } x^{-2+2\mu/\alpha\sigma^2}(1-x)^{-2+2\mu/\sigma^2}\{x-\alpha/(1+\alpha)\}^{2-2(1+\alpha)\mu/\alpha\sigma^2}$$

we see that the second requirement is also satisfied. Further, (4.73) shows that both $x = 1$ and $x = \alpha/(1+\alpha)$ are inaccessible and thus (5.94) does provide the required stationary distribution.

The distribution (5.94) also applies for the diffusion case (Levikson and Karlin, 1975). We will generalize it in Chap. 10 and conclude now by noting that the diffusion processes of this section possess natural boundaries so that, since regular, exit and entrance boundaries have already been encountered in previous sections, rather simple genetic models can lead to all four of the boundary classifications given in (4.73).

5.9 Time-Reversal and Age Properties

It has been remarked on several occasions earlier that for reversible processes, information about the past behavior can be obtained by determining properties of the future behavior. Since diffusion processes with stationary distributions are reversible we should be able to use some of the conclusions reached above to discuss past behavior and in particular the "age" of an allele.

The time-reversal property states that for any diffusion on [0, 1] admitting a stationary distribution, the probability of any sample path leading from x (at time 0) to y (at time t) is equal to that of the "mirror-image" path leading from y (at time $-t$) to x (at time 0). Unfortunately this observation is not immediately useful for several questions of interest in population genetics since these questions refer to processes for which 0 and/or 1 are accessible absorbing states of the diffusion process, and thus for which no stationary distribution exists. This problem can be overcome in the following way.

Suppose 0 is an absorbing barrier but 1 is not: this will occur in practice, for example, if there is mutation from A_1 to A_2 but no reverse mutation. Now introduce mutation from A_2 to A_1 at rate ϵ: a stationary distribution now exists and reversibility arguments apply. In particular, given a current value x for the frequency of A_1, the distribution of the time (in the future) until 0 is next reached is identical to that of the time (in the past) that it was last left. Now let $\epsilon \to 0$ and note that the distribution of the time (in the future) until the frequency reaches 0 converges to that applying when $\epsilon = 0$. The limiting distribution is then identical to the age distribution of an allele which arose as a unique new mutation and whose current frequency is x. This argument can be made more precise (Watterson, 1977b; Levikson, 1977a, b) by introducing a "return" process whereby the frequency of A_1 is returned from 0 to $\delta(\delta > 0)$ whenever 0 is reached: in practice we put $\delta = (2N)^{-1}$ to correspond to the frequency of a new mutant. We now give some examples of the conclusions reached by this argument.

Consider first the case of no selection or mutation. Assume the allele A_1 arose by a unique mutation in an otherwise pure A_2A_2 population and is now observed with frequency x. The distribution of its age is thus the distribution of its time until loss, conditional that eventual loss does occur. This distribution can be found by centering attention on A_2 (with current frequency $1 - x$) rather than A_1 and is then given by (5.38) with $p = 1 - x$. The mean age can be found either through this distribution or alternatively by replacing p by $1 - x$ in (5.34). This leads to a mean age of

$$-4Nx(1-x)^{-1} \log x \qquad (5.95)$$

generations. The variance of the age is similarly found by putting $p = 1 - x$ in (5.36).

A parallel formula can be found when we assume fitness values $1 + s$ (for A_1A_1), $1 + \frac{1}{2} s$ (for A_1A_2), and 1 (for A_2A_2). Use of (5.53) (with $p = x$) shows that the mean age of A_1, given that it is currently observed with frequency x, is

$$\int_0^x 4N[\alpha\{e^\alpha - 1\}]^{-1} \{e^{\alpha y} - 1\}\{e^{\alpha(1-y)} - 1\}\{y(1-y)\}^{-1} dy$$

$$+ \int_x^1 4N\{1 - e^{-\alpha x}\}[\alpha\{1 - e^{-\alpha}\}\{e^{\alpha(1-x)} - 1\}]^{-1} e^{-\alpha(1-y)} \{e^{\alpha(1-y)} - 1\}^2$$

$$\times \{y(1-y)\}^{-1} dx \qquad (5.96)$$

generations. This converges to (5.95) as $\alpha \to 0$, as we expect, and the form of the integrand indicates the mean time, in the past, that the frequency of A_1 has assumed any arbitrary value y.

Suppose now A_1 mutates to A_2 at rate u with no reverse mutation. If one initial A_1 gene occurred by a unique mutation and the frequency of A_1 is currently observed at x, the mean age of A_1 is, from (3.20),

$$4N(1-\theta)^{-1} \int_0^x y^{-1}\{(1-y)^{\theta-1} - 1\}dx$$

$$+ 4N(1-\theta)^{-1}\{1 - \{(1-x)^{1-\theta}\} \int_x^1 (1-y)^{\theta-1} dy \qquad (5.97)$$

generations. A case of particular interest is when $x = 1$ corresponding to temporary fixation of A_1: this evaluation is allowed only when $\theta < 1$.

It is also possible to consider the mean age of A_1 conditional on the requirement that the frequency of A_1 was never unity in the past. This is identical to the mean time for loss of A_1 given that its future frequency never achieves the value unity and will be given by (5.97) for $\theta > 1$ (since then the condition that the frequency of A_1 never reaches unity is automatically satisfied). For $\theta < 1$ the probability that the frequency of A_1 never reaches unity given a current value x is found from (4.16) to be $(1-x)^{1-\theta}$. Use of (4.52) and (4.53) then shows that the conditional mean age of A_1 is

$$\int\limits_{0}^{x} 4Ny^{-1}(1-\theta)^{-1}\{1-(1-y)^{1-\theta}\}dy$$

$$+ \int\limits_{x}^{1} 4Ny^{-1}(1-\theta)^{-1}(1-y)^{1-\theta}\{(1-x)^{\theta-1}-1\}dy. \tag{5.98}$$

This reduces to (5.63) for $x = (2N)^{-1}$. Equation (5.98) was first given by Kimura and Ohta (1973), using a quite different approach.

The conclusions reached above can be reached in a different way by the method due to Sawyer (1977) referred to in Sect. 3.10. If the Markovian random variable is approximated by a diffusion variable the appropriate drift and diffusion coefficients are

$$a(x) = 1 - x - \tfrac{1}{2}\theta x, \quad b(x) = x(1-x). \tag{5.99}$$

This diffusion process has the further property that a random times the diffusion variable is moved immediately to the boundary $x = 0$. These shifts occur at times independent of the current value of the variable and the times between consecutive shift and independent random variables with density function

$$f(t) = \tfrac{1}{2}\theta \exp\left(-\tfrac{1}{2}\theta t\right), \quad 0 \leqslant t < \infty. \tag{5.100}$$

This form of behavior arises because if a given gene in the ancestral line considered is a new mutant, the frequency of its allelic type in the population is $(2N)^{-1}$ when it first occurs, irrespective of the frequency x of the allelic type of its parent gene in the previous generation. Here the boundary $x = 1$ is not absorbing while the boundary $x = 0$ is not accessible, i.e. cannot be reached by drift from the interior of $(0,1)$, (although we have seen this boundary can be reached because of the discrete shifts in the diffusion). The frequency x of the allelic type in the ancestral line thus has a stationary distribution which is found by Sawyer using renewal theory arguments. It turns out that

$$\lim_{t \to \infty} \text{Prob } \{x_t \leqslant p\} = 1 - (1-p)^{\theta/2}, \tag{5.101}$$

and the reversibility argument shows immediately that the distribution of frequencies seen in the past is identical to that to be seen in the future, given a current frequency x. From this observation we can rederive (5.97) as well as find further age properties of the process.

We conclude this section by discussing the various symmetry properties in Sect. 5.4 in the light of time-reversal arguments. We take as our starting point the observation that any two of Eqs. (5.56), (5.58) and (5.59) imply the remaining equation. Now Eq. (5.58) is true by direct argument so the observation of prime interest is that (5.56) implies (5.59) and vice versa. Our starting point in Sect. 5.4 was that (5.56) was true by computation of both sides in the formula, and this then implies (5.59). Our starting point here is that time-reversal arguments imply the truth of (5.59). This is true because the reflection of a path leading from ϵ (ϵ small) to $1 - \epsilon$ is a path leading from

$1 - \epsilon$ to ϵ. The identity of the probability of the two paths leads directly to (5.59), and we conclude that time-reversibility implies (5.59) and hence (5.56) and (5.57). The further facts that $t^*(x; p)$ defined in (5.52) is symmetric about 0.5 and independent of the sign of α do not appear to follow directly from time-reversal arguments although use of (5.56) (which we have seen can be arrived at through time-reversal arguments) with $h = \frac{1}{2}$ shows that either property implies the other.

5.10 Multi-allele Processes

In this section we consider diffusion approximations to K-allele models of the form (3.61).

The simplest version of (3.61) arises when $\psi_i = X_i/2N$. It is clear immediately that in this Markov chain model the probability of fixation of any allele is initial frequency, and we also have the eigenvalue formula (3.62) concerning the rate of decrease of the probability that j or more alleles exist at time t. To obtain further results we turn to the diffusion approximation to (3.61).

Put $x_i = X_i/2N$ $(i = 1, 2, ..., K - 1)$: then elementary theory shows that, given x_1, ..., x_{K-1},

$$E(\delta x_i) = 0, \quad \mathrm{Var}(\delta x_i) = (2N)^{-1} x_i(1 - x_i), \quad \mathrm{Covar}(\delta x_i, \delta x_j) = (2N)^{-1} x_i x_j.$$

These parameters, in conjunction with (4.95), lead to the following partial differential equation for the joint density function of $x_1, ..., x_{K-1}$ at time t (unit time corresponds to $2N$ generations):

$$\frac{\partial}{\partial t} \{f(x_1, ..., x_{K-1}; t)\} = \frac{1}{2} \sum_{i=1}^{K-1} \frac{\partial^2}{\partial x_i^2} \{x_i(1 - x_i) f(x_1, ..., x_{K-1}; t)\}$$

$$- \sum_{i<j} \sum \frac{\partial^2}{\partial x_i \partial x_j} \{x_i x_j f(x_1, ..., x_{K-1}; t)\}.$$

This is a generalization of (5.10) and admits an eigenfunction solution generalizing (5.11) (see Littler and Fackerell, 1975; Griffiths, 1978). Writing $f = f(x_1, ..., x_{K-1}; t)$, the corresponding backward equation is

$$\frac{\partial f}{\partial t} = \frac{1}{2} \sum_i p_i(1 - p_i) \frac{\partial^2 f}{\partial p_i^2} - \sum_{i<j} \sum p_i p_j \frac{\partial^2 f}{\partial p_i \partial p_j},$$

where p_i is the initial value of x_i. This equation may be used to find various fixation probabilities. Thus the probability π $(= \pi(p_1, p_2, ..., p_{K-1}))$ of any fixation event satisfies

$$\tfrac{1}{2} \Sigma\, p_i(1 - p_i) \frac{d^2\pi}{dp_i^2} - \Sigma\Sigma_{i<j} p_i p_j \frac{d^2\pi}{dp_i dp_j} = 0, \tag{5.102}$$

subject to the appropriate boundary conditions. Thus the probability that A_i eventually fixes satisfies (5.102) together with the boundary conditions

$$\pi(p_1, \ldots, p_{K-1}) = 1 \quad \text{if} \quad p_i = 1,$$

$$\pi(p_1, \ldots, p_{K-1}) = 0 \quad \text{if} \quad p_j + p_m + \ldots + p_u = 1 \; (j, m, \ldots, u \neq i).$$

The solution of these equations is $\pi = p_i$ which we know also to be exactly correct for the model (3.61) with $\psi_i = X_i/2N$. Suppose now we wish to find the probability π that ultimately A_i and A_j are the last two alleles to exist. Here the boundary conditions are

$$\pi(p_1, \ldots, p_{K-1}) = 1 \quad \text{if} \quad p_i + p_j = 1,$$

$$\pi(p_1, \ldots, p_{K-1}) = 0 \quad \text{if} \quad p_m + p_s + \ldots + p_u = 1 (m, s, \ldots, u \neq i, j),$$

and the solution of (5.102) satisfying these is

$$\pi = p_i p_j \{(1 - p_i)^{-1} + (1 - p_j)^{-1}\}.$$

Note that in the case $K = 3$ this shows, for example, that the probability that A_1 is the first allele lost is

$$p_2 p_3 \{(1 - p_2)^{-1} + (1 - p_3)^{-1}\}.$$

Similar probabilities may be found for other fixation events.

We turn now to questions concerning the time until various fixation events occur. The development is easiest when $K = 3$, so we follow the development in this case and merely quote results for larger values of K. Complete details are available in Littler (1975).

Define T_i as the time required until exactly $i(i = 1, 2)$ alleles exist in the population. We first find an expression for $E(T_1)$. Conditional on the event that A_1 is the last remaining allele, the mean of T_1 is

$$E(T_1 | A_1) = 2p_1^{-1} (1 - p_1) \log (1 - p_1),$$

from (5.34). Since the probability is p_1 that indeed A_1 is the last remaining allele we have

$$E(T_1) = -2[(1 - p_1) \log (1 - p_1) + (1 - p_2) \log (1 - p_2) + (1 - p_3) \log (1 - p_3)].$$

Clearly this value can be extended immediately to the case of K alleles to get

$$E(T_1) = -2 \, \Sigma \, (1 - p_i) \log (1 - p_i). \tag{5.103}$$

It is equally straightforward to use the analysis leading to (5.36) to find an expression for the variance of T_1.

In the three-allele case we find $E(T_2)$ as follows. The event $T_2 \leqslant t$ implies that, at time t, at least one p_i value is zero. Standard probabilistic formulae for unions of events give

$$E(T_2) = -2[\Sigma\, p_i \log p_i + \Sigma\, (1 - p_i) \log (1 - p_i)]. \tag{5.104}$$

In the particular case $p_i = 1/3$ these formulae give

$$E(T_1) \sim 3.2N \text{ generations}, \quad E(T_2) \sim 1.6N \text{ generations}.$$

Littler (1975) shows that in the K-allele case,

$$E(T_i) = -2[\sum_{s=1}^{i} (-1)^{i-s}\binom{K-1-s}{i-s}\{\Sigma\,(1 - p_{i_1} - \dots - p_{i_s}) \log (1 - p_{i_1} - \dots - p_{i_s})\}] \tag{5.105}$$

where the inner sum is taken over all possible values $1 \leqslant i_1 < i_2 < \dots < i_s \leqslant K$. This reduces to (5.103) when $i = 1$ and generalizes (5.104) to arbitrary K when $i = 2$. It is of some interest to note that if $p_i = K^{-1}$,

$$\lim_{K \to \infty} E(T_j) = 2/j, \quad j = 1, 2, \dots . \tag{5.106}$$

This conclusion may be compared to that deriving from (3.62), viz.

Prob (at least j allelic types remain present after $2Nt$ generations)
$\sim$ const exp $\{-\frac{1}{2} j(j - 1)t\}$.

Evidently the eigenvalues give a very poor indication of the way in which $E(T_j)$ changes as a function of j.

A further asymptotic result of some interest is also given by Littler (1975). If $M^{(t)}$ is the number of alleles present at time t in a K-allele model with $p_i = K^{-1}$,

$$E(M^{(t)}) \sim 1 + 3e^{-1} + 5e^{-3t} + 7e^{-6t} + \dots + (2j + 1e)^{-\frac{1}{2}j(j+1)t} + \dots$$

as $K \to \infty$.

Suppose now A_i mutates to A_j at rate u_{ij} $(i \neq j)$. It follows that

$$E(\delta x_i) = -x_i \sum_j u_{ij} + \sum_j x_j u_{ji}$$

$$= (2N)^{-1} m_i(x_1, \dots, x_{K-1})$$

say, where

$$m_i(x_1, \ldots, x_{K-1}) = -x_i \Sigma \beta_{ij} + \Sigma x_j \beta_{ji}$$

and $\beta_{ij} = 2Nu_{ij}$. If each $u_{ij} > 0, (i \neq j)$, there will exist a stationary distribution for the joint frequency of $A_1, \ldots, A_{K-1}$. This distribution has not been found in general although it must clearly satisfy the stationarity equation

$$0 = - \Sigma \frac{\partial}{\partial x_i} \{m_i(x_1, \ldots, x_{K-1}) f(x_1, \ldots, x_{K-1})\} + \frac{1}{2} \Sigma \frac{\partial^2}{\partial x_i^2} \{x_i(1 - x_i) f(x_1, \ldots, x_{K-1})\}$$

$$- \Sigma \Sigma_{i<j} \frac{\partial^2}{\partial x_i \partial x_j} \{x_i x_j f(x_1, \ldots, x_{K-1})\}. \tag{5.107}$$

Fortunately, in one case of special interest, (5.107) can be solved explicitly, namely the equi-mutation model for which u_{ij} is independent of i and j. Suppose

$$u_{ij} = u/(K - 1)$$

and put $\theta = 4Nu$ (note that u is the total mutation rate for any gene). Then the appropriate solution of (5.105) is

$$f(x_1, \ldots, x_{K-1}) = \frac{\Gamma\{K\epsilon\}}{\{\Gamma(\epsilon)\}^K} \{x_1 x_2 \ldots x_K\}^{\epsilon-1}, \tag{5.108}$$

where we write for convenience $x_K = 1 - x_1 - x_2 \ldots - x_{K-1}$ and $\epsilon = \theta/(K - 1)$. This is the so-called Dirichlet distribution, which arises in several areas of statistics and whose properties are well known. It may be used to rederive various formulae already found by other methods and, as we soon note, to find new ones. Thus the probability that two genes drawn at random are of the same allelic type is

$$\sum_{i=1}^{K} \int \ldots \int x_i^2 f(x_1, \ldots, x_{K-1}) dx_{K-1}$$

and this leads, after some calculation, to (3.63). Suppose now the gene frequencies are arranged in decreasing order

$$x_{(1)} \geqslant x_{(2)} \geqslant x_{(3)} \ldots \geqslant x_{(K)} \geqslant 0. \tag{5.109}$$

These frequencies are called the order statistics. Their joint distribution is, directly from (5.108),

$$f(x_{(1)}, \ldots, x_{(K-1)}) = \frac{K! \Gamma\{K\epsilon\}}{\{\Gamma(\epsilon)\}^K} \{x_{(1)} x_{(2)} \ldots x_{(K)}\}^{\epsilon-1}. \tag{5.110}$$

From this distribution the joint distribution of the first j order statistics $x_{(1)}, x_{(2)}, \ldots,$ $x_{(j)}$ may be found although we do not give the formula here.

The limiting case $K \to \infty$ is of special interest. While the joint distribution (5.108) has no non-trivial limit, Kingman (1975, 1977b) has shown that the distribution of the first j order statistics *does*, for any j, converge to a non-trival limit, namely the Poisson-Dirichlet distribution and that this concides with the joint distribution of the first j order statistics in the infinite allele model. This remarkable result is most important as it allows us, so long as we concentrate on order statistics and functions derived from them, to move freely between the K-allele model of Sect. 3.5 and the infinite allele model of Sect. 3.6, so that we may approach certain problems in the infinite allele model either directly or through the K-allele model, whichever we prefer. To illustrate this, the probability that two genes at random in the infinite allele model are of the same allelic type has been computed directly in (3.67) or may be found by letting $K \to \infty$ in (3.63). The reason why the two approaches yield the same value is that the probability in question can be expressed in terms of the order statistics as

$$E(F) = E\,(x_{(1)}^2 + x_{(2)}^2 + x_{(3)}^2 + ...),\qquad(5.111)$$

where F is the (random) population homozygosity. The eigenvalue set (3.81) for the infinite allele model can also be found through a limiting process from the K-allele case.

The expression for the Poisson-Dirichlet limiting order statistics distribution is rather complex. The distribution of $x_{(1)}$, at least over the range $(0.5, 1)$, has already been noted in (5.64). For $(2N)^{-1} \leqslant x_{(1)} \leqslant 0.5$, Watterson and Guess (1977) show that the density function of $x_{(1)}$ is of the form

$$f(x_{(1)}) = \Gamma(\theta + 1)e^{\gamma\theta}x_{(1)}^{\theta - 2}g((1 - x_{(1)})/x_{(1)}),\qquad(5.112)$$

where $g(\cdot)$ is a complicated function which is best defined through the Laplace transform equation

$$\int_0^\infty e^{-tz}g(z)dz = \exp\,\{\theta \int_0^1 y^{-1}(e^{-tz} - 1)dy\}.\qquad(5.113)$$

More generally the joint density function of $x_{(1)}, x_{(2)}, ..., x_{(r)}$ is

$$f(x_{(1)}, ..., x_{(r)}) = \theta^r\Gamma(\theta)e^{\gamma\theta}g(y)\,\{x_{(1)}x_{(2)} ... x_{(r)}\}^{-1}x_{(r)}^{\theta - 1}\qquad(5.114)$$

where $y = (1 - x_{(1)} - x_{(2)} - ... -x_{(r)})/x_{(r)}$. These conclusions are noted by Watterson (1976). The expression (5.114) simplifies, when $x_{(1)} + ... + x_{(r-1)} + 2x_{(r)} \geqslant 1$, to

$$f(x_{(1)}, ..., x_{(r)}) = \theta^r\,\{x_{(1)}x_{(2)} ... x_{(r)}\}^{-1}(1 - x_{(1)} - ... - x_{(r)})^{\theta - 1}.\qquad(5.115)$$

One interesting conclusion to be reached from (5.112) concerns the age of the most frequent allele currently observed in the population. In particular, the following question can be asked: "What is the probability that the most frequent allele in the population is also the oldest?" Time-reversal arguments show that this is identical to the probability that the most frequent allele will last longest into the future. Given the current allele frequency configuration this is, by symmetry, the current frequency $x_{(1)}$

of the most frequent allele, so the probability in question is just the mean value of $x_{(1)}$. This is

$$(2N)^{-1} \int_{0}^{1} x_{(1)} f(x_{(1)}) dx_{(1)} \tag{5.116}$$

where $f(x_{(1)})$ is given by (5.112). The simple form for this density in (0.5, 1) shows that a lower bound to this probability is

$$\theta \int_{0.5}^{1} x_{(1)} \{x_{(1)}^{-1}(1 - x_{(1)})^{\theta - 1}\} dx_{(1)} = (\tfrac{1}{2})^{\theta}, \tag{5.117}$$

and Watterson and Guess (1977) compute a more accurate value using (5.112) and (5.116). In a similar way the probability that the ith most frequent allele is the oldest can be computed from the expected value of the ith order statistic.

A further use for (5.114) is in deriving the distribution of the population hetero-zygosity $F = x_1^2 + x_2^2 + \dots$. As we have seen, F can equally well be expressed as the sum of squares of ordered frequencies, so its distribution for the K-allele model converges to that for the infinite allele model as $K \to \infty$. The complete distribution of F in the K-allele model is not simple. Nevertheless it is comparatively easy to find the mean and variance of F from (5.107): we obtain

$$E(F) = (1 + \epsilon)/(1 + K\epsilon) \tag{5.118}$$

$$\mathrm{Var}(F) = 2\theta(1 + \epsilon)/[(1 + K\epsilon)^2(2 + K\epsilon)(3 + K\epsilon)], \tag{5.119}$$

where $\epsilon = \theta/(K - 1)$. The value for the mean is in effect computed by (5.109) and is identical to that reached in (3.63) by different methods. By letting $K \to \infty$ and using the convergence theory, the infinite allele model mean and variance become

$$E(F) = (1 + \theta)^{-1}, \quad \mathrm{Var}(F) = 2\theta/[(1 + \theta)^2(2 + \theta)(3 + \theta)]. \tag{5.120}$$

Again the value for the mean is a standard result, cf. (3.67).

Suppose now in the K-allele model that selective differentials exist. Let the fit-ness of $A_i A_j$ be $1 + s_{ij}$ and suppose s_{ij} is of order N^{-1}. Writing $\alpha_{ij} = 2Ns_{ij}$ and assuming mutation as above, the joint density function of $x_1, \dots, x_{K-1}$ can be found by solving (5.107) where now $m_i(x_1, \dots, x_{K-1})$ contains a further turn due to selective differences. The explicit solution (Wright, 1949, p. 383) is

$$f(x_1, \dots, x_{K-1}) = \mathrm{const}\ \exp\ \{\sum_{i=1}^{K} \sum_{j=1}^{K} x_i x_j \alpha_{ij}\} \{x_1 \dots x_K\}^{\epsilon - 1}. \tag{5.121}$$

In particular, if all heterozygotes have equal fitness $1 + s$ ($s > 0$) and all homozygotes have fitness 1, and if $\alpha = 2Ns$, (5.121) becomes

$$f(x_1, \dots, x_{K-1}) = \mathrm{const}\ \exp\ \{-\alpha F\} \{x_1 \dots x_K\}^{\epsilon - 1}. \tag{5.122}$$

We have noted that the limiting ($K \to \infty$) behavior can be analysed only by transferring attention to the order statistics $x_{(1)}, x_{(2)}, \ldots, x_{(j)}$. Unfortunately the joint distribution of the order statistics is, in general, too complicated to yield useful information. Considerable progress can be made, however, when the α_{ij} are small, and thus for the rest of this chapter we assume this is the case. (Formulae applying for larger values of α_{ij} will be given in Chap. 9.) We thus put in (5.121)

$$\exp \{\textstyle\sum \sum x_i x_j \alpha_{ij}\} = 1 + \sum \sum x_i x_j \alpha_{ij} + 0(\alpha_{ij}^2).$$

The joint distribution of the order statistics may be found by summing in (5.121) over all possible permutations of $1, 2, \ldots, K$. This yields

$$f(x_{(1)}, x_{(2)}, \ldots, x_{(K)}) = \frac{K! \Gamma \{K\epsilon\}}{\{\Gamma(\epsilon)\}^K} [1 + A \{F - \frac{\epsilon + 1}{\theta + 1}\} + 0(\alpha_{ij}^2)]$$

$$\times \{x_{(1)} x_{(2)} \cdots x_{(K)}\}^{\epsilon - 1}, \tag{5.123}$$

where

$$A = K^{-1} \sum \alpha_{ii} - \{K(K-1)\}^{-1} \sum_{i \neq j} \sum \alpha_{ij}, \quad F = \sum x_{(i)}^2.$$

We are interested in two particular fitness schemes. The first (or "heterotic") scheme has just been noted: all heterozygotes have fitness $1 + s$ and all homozygotes, fitness 1. Here $\alpha_{ii} = 0$, $\alpha_{ij} = \alpha = 2Ns$ so that

$$f(x_{(1)}, \ldots x_{(K)}) = \frac{K! \Gamma \{K\epsilon\}}{\{\Gamma(\epsilon)\}^K} [1 - \alpha \{F - \frac{\epsilon + 1}{\theta + 1}\} + 0(\alpha_{ij}^2)] \{x_{(1)} \cdots x_{(K)}\}^{\epsilon - 1}. \tag{5.124}$$

In the second selective model (the "deleterious alleles" model) a fraction γ of the K alleles are deleterious: individuals carrying i deleterious genes ($i = 0, 1, 2$) have fitness $1 - is$. Here $A = 0$ and the terms in α_{ij}^2 in (5.123) must be computed. There results

$$f(x_{(1)}, \ldots, x_{(K-1)}) = \frac{K! \Gamma \{K\epsilon\}}{\{\Gamma(\epsilon)\}^K} [1 + 2\alpha^2 \gamma (1 - \gamma) \{F - \frac{\epsilon + 1}{\theta + 1}\} + 0(\alpha_{ij}^2)]$$

$$\{x_{(1)} \cdots x_{(K)}\}^{\epsilon - 1}. \tag{5.125}$$

From (5.124) and (5.125) the joint distribution of $x_{(1)}, \ldots, x_{(j)}$ may, in principle, be found. From Kingman's theory this joint distribution will converge, as $K \to \infty$, to a non-trivial limit, namely that of the first j order statistics in the infinite allele heterosis and deleterious alleles models. (In the latter model γ is defined as the probability that any new mutant allele is deleterious.) We do not pursue these distributions here other than to note that, since to the order of approximation considered (5.125) can be obtained from (5.124) by replacing α by $-2\alpha^2 \gamma (1 - \gamma)$, the same must be true in the limiting order statistics distribution and any function derived from it. The information of most use to us may be found from one such function, namely the frequency spectrum of the two models considered. The frequency spectrum $\phi(x)$ was introduced in Sect. 3.6: it has the interpretation that $\phi(x)\delta x$ is the mean number of alleles in the population with frequency in $(x, x + \delta x)$. From this frequency spectrum may be deduced three quantities of some interest in theoretical population genetics:

$$\int_0^1 \phi(x)dx = \text{mean number of alleles in the population,} \qquad (5.126)$$

$$\int_0^1 x^2\phi(x)dx = \text{mean population homozygosity,} \qquad (5.127)$$

$$\int_0^1 \{1 - (1 - x)^r\}\phi(x)dx = \text{mean number of alleles in a sample} \qquad (5.128)$$
$$\text{of size } r \text{ from the population.}$$

The mean population homozygosity, $E(F)$, is the probability that two genes taken at random from the population are of the same allelic type. For the neutral infinite alleles model,

$$\phi(x) \approx 2N\theta, \quad 0 \leqslant x \leqslant (2N)^{-1},$$
$$\phi(x) \approx \theta x^{-1}(1 - x)^{\theta - 1}, \quad (2N)^{-1} \leqslant x \leqslant 1. \qquad (5.129)$$

Use of (5.129) in (5.126), (5.127) and (5.128) rederives the quantities (3.82), (3.67) and (3.79) found in Chap. 3 by other methods. Our aim is to consider the corresponding three values in the "heterotic" and "deleterious alleles" selection models.

Watterson (1977a) has shown that in the heterotic model the frequency spectrum $\phi(x)$ becomes, for $(2N)^{-1} \leqslant x \leqslant 1$,

$$\phi(x) = \theta x^{-1}(1 - x)^{\theta - 1}[1 + \alpha x \{2 - (2 + \theta)x\} (1 + \theta)^{-1} + 0(\alpha^2)]. \qquad (5.130)$$

It follows immediately from (5.126)–(5.130) that the mean number of alleles in the population exceeds its neutral theory value by an amount

$$\alpha\theta(1 + \theta)^{-2} + 0(\alpha^2), \qquad (5.131)$$

that

$$E(F) = (1 + \theta)^{-1} - 2\alpha\theta \{(1 + \theta)^2(2 + \theta)(3 + \theta)\}^{-1}, \qquad (5.132)$$

and that the mean number of alleles in a sample of r genes exceeds its neutral theory value by

$$\alpha\theta(1 + \theta)^{-2} - 2\theta(\theta + 2r)\{(1 + \theta)(r + \theta)(r + 1 + \theta)\}^{-1}. \qquad (5.133)$$

Clearly, for large r, this approximates (5.131) as would be expected.

We have noted that for the deleterious alleles model the parallel quantities may be found by replacing α by $-2\alpha^2\gamma(1 - \gamma)$. Thus for this model the mean number of alleles in the population falls short of the neutral value by

$$2\theta\alpha^2\gamma(1 - \gamma)(1 + \theta)^{-2}, \tag{5.134}$$

the mean homozygosity is given by

$$\mathrm{E}(F) = (1 + \theta)^{-1} + 4\alpha^2\gamma(1 - \gamma)\theta\,\{(1 + \theta)^2(2 + \theta)(3 + \theta)\}^{-1} \tag{5.135}$$

while the mean number of alleles in a sample of r falls short of its neutral theory value by

$$2\alpha^2\gamma(1 - \gamma)\theta(1 + \theta)^{-2} - 2\alpha^2\gamma(1 - \gamma)\theta(\theta + 2r)\,\{(1 + \theta)(r + \theta)(r + 1 + \theta)\}^{-1}. \tag{5.136}$$

All of these conclusions will be of use to us in Chap. 9 when we consider tests of neutrality based on gene frequency data. They all assume α_{ij} small and generalizations of them for large α_{ij} will also be noted in Chap. 9.

6. Two Loci

6.1 Introduction

Most of the theory in this book so far has assumed that the fitness of any individual depends on his genetic make-up at a single locus. Although for certain specific purposes this assumption may give reasonable approximations, it is in general a gross simplification, in particular when epistatic (i.e. interactive) effects arise between loci. In this chapter we suppose that the fitness of any individual depends on his genetic constitution at two (or sometimes three) loci. Although this assumption is hardly less realistic than the previous one, it does allow substantial advance to be made, as has been noted in Sect. 2.9, on assessing the evolutionary effect of recombination between loci and also on judging the extent to which two-locus behavior is predictable from combining single-locus analyses. We will note later in this chapter that it also allows an investigation of the effects of modifier genes.

The model we use is that described in Sect. 2.9. We assume viability selection only (with fitness scheme given by (2.69) or (2.70)), random mating, no fitness differentials between sexes and a discrete-time parameter. Thus the recurrence relations (2.73) describe the evolution of the frequencies of the gametes A_1B_1, A_1B_2, A_2B_1 and A_2B_2 (i.e. gametes of types 1, 2, 3, 4 respectively).

We have already noted (in Sect. 2.9) two major consequences of these recurrence relations. The first is that the Fundamental Theorem of Natural Selection (that mean fitness increases, or at worst remains stable, from one generation to the next) is no longer true as a mathematical theorem, while the second is that the equilibrium points of the recurrence system can depend on the recombination fraction between the loci. We start by examining these conclusions in greater detail.

6.2 Evolutionary Properties of Mean Fitness

Our aim in this section is to discuss the implications of the recurrence relations (2.73) for the evolution of mean fitness, defined by (2.72). Fisher (1958, see in particular pp. 30–37) attempted to prove that the Fundamental Theorem holds when fitness depends on an arbitrary number of loci, but his analysis is superficial and incorrect. The first substantial analysis of multi-locus mean fitness behavior was given by Kimura (1958) although here we adopt a rather different approach than Kimura's. A second important early discussion of the question, not perhaps sufficiently highly appreciated, is that of Kojima and Kelleher (1961): in particular, the fact that mean fitness can decrease in two-locus systems was first explicitly mentioned in this paper.

If there is linkage disequilibrium at any stable equilibrium point (as is true, for example, at the equilibria (2.76) or (2.77)), it is always possible to find a neighborhood of the equilibrium point such that, starting from any point in this neighborhood, th mean fitness decreases. Thus the Fundamental Theorem cannot be true as a mathematical theorem. Karlin (1975) indeed asserts that it "usually fails" in the sense that for almost all fitness configurations it is possible to find a set of gamete frequencies for which the mean fitness decreases, at least for a few generations.

These considerations, however, are probably of lesser importance from a practical standpoint. We are mainly interested in the behavior of mean fitness during those generations when substantial changes in gene frequency occur, and here it is possible to rescue in large part the spirit of the Fundamental Theorem, at least in a wide variety of cases of practical interest. That such an attempt is not over-optimistic is supported by the observations of Karlin and Carmelli (1975) who noted that when the entries in the fitness matrix (2.69) are chosen randomly, the mean fitness increases for most generations for most fitness configurations. It is important, therefore, to emphasize that our aim is not to prove a mathematical theorem but to determine circumstances in which increases in mean fitness tend to occur.

The first attempt along these lines was through the introduction of the principle of quasi-linkage equilibrium (QLE) (Kimura, 1965). The essence of this principle is as follows. If we define (in the notation of Sect. 2.9) a quantity Z by

$$Z = c_1 c_4 / c_2 c_3, \tag{6.1}$$

then to a first order of approximation the change in the value of log Z between consecutive generations is

$$\Delta \log Z \approx c_1^{-1}\Delta c_1 + c_4^{-1} \Delta c_4 - c_2^{-1}\Delta c_2 - c_3^{-1}\Delta c_3. \tag{6.2}$$

The values of Δc_i can be found from (2.73) as

$$\Delta c_i = \bar{w}^{-1} \{c_i(w_i - \bar{w}) + \eta_i R w_{14}(c_2 c_3 - c_1 c_4)\} \tag{6.3}$$

and substitution into (6.2) eventually leads to the approximation

$$\bar{w}\Delta \log Z \approx \bar{\epsilon} - Rw_{14}(Z - 1)\{c_2 + c_3 + Z^{-1}(c_1 + c_4)\}, \tag{6.4}$$

where

$$\bar{\epsilon} = w_1 - w_2 - w_3 + w_4.$$

Suppose now $Z > 1$. If $\bar{\epsilon}$ can be treated roughly as a constant, there will be a tendency for Z to decrease (at least for values of Z sufficiently large compared to $\bar{\epsilon}$). Similarly when $Z < 1$ there will be a tendency for Z to increase, at least for very small Z. We may thus hope that Z rapidly reaches a constant value at which

$$\Delta \log Z = 0. \tag{6.5}$$

The change in mean fitness may be approximated, ignoring small order terms, by

$$\Delta \bar{w} \approx 2 \Sigma w_i \Delta c_i$$

and substitution from (6.3) then gives

$$\Delta \bar{w} \approx 2 \bar{w}^{-1} \{ \Sigma \, c_i w_i (w_i - \bar{w}) + R w_{12} (c_2 c_3 - c_1 c_4) \bar{\epsilon} \}. \tag{6.6}$$

If now Eq. (6.5) holds we must have, from (6.4),

$$R w_{14} (c_2 c_3 - c_1 c_4) = - \bar{\epsilon} \, (\Sigma \, c_i^{-1})^{-1}$$

and substituting this into (6.6) we find

$$\Delta \bar{w} \approx \bar{w}^{-1} \{ 2 \, \Sigma \, c_i (w_i - \bar{w})^2 - 2 \bar{\epsilon}^2 (\Sigma \, c_i^{-1})^{-1} \}. \tag{6.7}$$

The two terms in the brackets on the right-hand side have arisen in our previous discussion. The first is the total gametic variance defined in (2.88), and the second is the epistatic gametic variance defined in (2.89). The difference between the two must be the additive gametic variance (or equivalently, in view of the discussion in Sect. 2.9, the additive genetic variance), and we conclude that whenever (6.5) holds

$$\Delta \bar{w} \approx \bar{w}^{-1} \sigma_A^2. \tag{6.8}$$

Provided the above reasoning is satisfactory, we may therefore expect the system to evolve rapidly to a state where (6.8) is true, and if this is so we have succeeded in our aim of rescuing the Fundamental Theorem as a reasonable general principle.

The above reasoning, however, requires much closer examination. Clearly any well-behaved function Y of gamete frequencies will converge to its equilibrium as the system itself approaches its equilibrium state, and at the moment we have no reason to prefer Z to any other such function. But the value of $\Delta \bar{w}$ assuming $\Delta Y = 0$ will be different from that assuming $\Delta Z = 0$. Indeed numerical simulations (Kimura, 1965; Ewens, 1976) show that mean fitness usually changes at a much slower rate than Z, and it is therefore unreasonable to consider changes in mean fitness assuming $\Delta Z = 0$: it would be more reasonable to consider changes in Z assuming $\Delta \bar{w} = 0$.

Despite these comments it is possible to arrive at (6.8) by a deeper argument, at least in cases of biological interest. This has been done by Nagylaki (1976: see also Hoppensteadt, 1976; Conley, 1972), and we here outline the main points of Nagylaki's argument.

Suppose that fitness differences in the system (2.69) are small so that $w_{ij} = 1 + sa_{ij}$, where s is a small parameter and the a_{ij} are moderate or small. We consider the linkage disequilibrium measure D $(= c_1 c_4 - c_2 c_3)$. From the recurrence relations (2.73)

$$\Delta D = - RD + sf(c_i, a_{ij}) \tag{6.9}$$

where the exact form of the function f is not important. This leads to

$$D(t) = (1 - R)^t D(0) + s(1 - R)^t \sum_{u=1}^{t} (1 - R)^{-u} f(c_i(u - 1), a_{ij}) \tag{6.10}$$

where $D(t)$ is the value of D in generation t. Clearly, for R moderate, there exists a time t_1 for which $D(t_1)$ is of order s so we can write, for $t \geqslant t_1$, $D(t) = s D^*(t)$ and hence, from (6.9),

$$\Delta D^*(t) = RD^*(t) + f(c_i(t), a_{ij}). \tag{6.11}$$

We also know, from (2.73), that Δc_i is of order s at most and hence

$$f(c_i(t), a_{ij}) - f(c_i(t - 1), a_{ij}) \tag{6.12}$$

is of order s at most. Hence, from (6.11) and (6.12),

$$\Delta D^*(t) - R^{-1} \{f(c_i(t), a_{ij}) - f(c_i(t - 1), a_{ij})\}$$
$$= R \{D^*(t) - R^{-1} f(c_i(t), a_{ij})\} + 0(s),$$

or

$$\Delta G(t) = - RG(t) + 0(s), \tag{6.13}$$

where

$$G(t) = D^*(t) - R^{-1} f(c_i(t), a_{ij}). \tag{6.14}$$

Now there will exist a time t_2 such that

$$G(t_1)(1 - R)^{t_2 - t_1} < s$$

and from (6.14) and (6.11) it then follows that for $t \geqslant t_2$,

$$\Delta Z(t) = 0(s^2). \tag{6.15}$$

Since $D = c_2 c_3 (Z - 1)$ it follows that

$$\Delta Z = 0(s^2) \quad \text{for } t \geqslant t_2. \tag{6.16}$$

We turn now to changes in mean fitness. A more exact computation than that leading to (6.8) yields

$$\Delta \bar{w} = [\sigma_A^2 - 2RD\bar{\epsilon} + 2\,\bar{\epsilon}^2(\Sigma c_i^{-1})^{-1}] + 0(s^3) \tag{6.17}$$

for $t \geqslant t_1$. Further, a more exact computation than that leading to (6.4) gives

$$Z^{-1}\Delta Z = \bar{\epsilon} - RD\Sigma c_i^{-1} + 0(s^2). \tag{6.18}$$

Using this equation, (6.17) becomes

$$\Delta \bar{w} = [\sigma_A^2 + 2\bar{\epsilon}(\Sigma c_i^{-1})^{-1}Z^{-1}\Delta Z] + 0(s^3). \tag{6.19}$$

Now $\bar{\epsilon}$ is of order s and (6.16) shows that for $t \geqslant t_2$, ΔZ is of order s^2. Thus the second term in the brackets on the right-hand side of (6.19) is $0(s^3)$ for $t \geqslant t_2$. Since in general σ_A^2 is $0(s^2)$, it follows that for $t \geqslant t_2$,

$$\Delta \bar{w} \approx \sigma_A^2 \tag{6.20}$$

during those epochs of the process for which the various order of magnitude arguments hold. Note that near an equilibrium point σ_A^2 is very close to zero so that it is then possible that the second term in brackets in (6.19) dominates the first term, leading to possible decreases in mean fitness. These are probably small and of little evolutionary consequence.

The correct statement of the QLE principle is thus this: for small selective differences (of order s) and loose linkage, a state soon arises where the change in mean fitness is given by (6.19), where ΔZ is of order s^2 and $\bar{\epsilon}$ is of order s. Since σ_A^2 is usually of order s^2 to a leading order of approximation, Eq. (6.20), embodying the main concepts of the Fundamental Theorem, is usually true. It is not correct to say, as some versions of the QLE principle claim, that (6.20) is true because changes in Z are smaller than changes in $\bar{w}$ and can be ignored: it is possible (see Kimura, 1965) that $\Delta Z \to \infty$ and yet (6.20) still holds.

It is of some interest to illustrate this conclusion by some numerical examples. Table 6.1 shows the values of $\Delta \bar{w}$, ΔZ, σ_A^2 and $\bar{\epsilon}$ for the evolution of a system with fitness matrix (2.75) with $R = 0.5$ and various initial gamete frequencies. Also tabulated is the value of $P = \sigma_A^2 + 2\bar{\epsilon}(\Sigma c_i^{-1})^{-1}Z^{-1} \Delta Z$ (see Eq. (6.19)) which we expect to provide a close approximation to $\Delta \bar{w}$. Note that for these fitness values we may take $s \sim 0.02$.

The values in the table illustrate the various points made above. First the mean fitness may decrease in the early generations, but eventually it begins to increase. Second, the value ΔZ and σ_A^2 are, after the early generations, both of order s^2 while $\bar{\epsilon}$ is of order s. Finally, and most important, values of $\Delta \bar{w}$ are closely approximated by the quantity P throughout the process and, after the initial generations, by σ_A^2. The only exception to this rule arises in case 3 where the population starts in extreme linkage disequilibrium. Here the early generations of the process show large values for ΔZ and $\Delta \bar{w}$ and exceptionally large values of P. The additive genetic variance, however, is

Table 6.1. Parameters associated with the evolution of the two-locus system with fitness matrix (2.75); $R = 0.5$

Generation	$\Delta \bar{w}(\times 10^4)$	$\Delta Z(\times 10^4)$	$\sigma_A^2(\times 10^4)$	$P(\times 10^4)$	$\bar{\epsilon}(\times 10^2)$
Case 1: Initial gamete frequencies 0.11, 0.16, 0.39, 0.34					
1	−1.67	1,750.0	0.532	−2.12	0.793
2	−0.462	1,020.0	0.496	−0.541	0.728
5	0.376	140.0	0.450	0.350	0.676
10	0.386	3.39	0.402	0.400	0.690
20	0.310	−0.526	0.322	0.322	0.719
30	0.246	−0.401	0.254	0.255	0.744
40	0.193	−0.301	0.200	0.201	0.758
50	0.152	−0.224	0.157	0.157	0.775
100	0.045	−0.046	0.046	0.046	0.805
200	0.005	0.002	0.005	0.005	0.807
Case 2: Initial gamete frequencies 0.42, 0.09, 0.11, 0.38					
1	−8.26	128,000.0	0.096	−6.27	0.073
2	1.23	15,600.0	0.054	−0.334	−0.384
5	0.693	777.0	0.034	0.648	−0.773
10	0.047	19.9	0.026	0.047	−0.821
20	0.019	0.032	0.020	0.020	−0.822
30	0.014	0.043	0.014	0.015	−0.818
40	0.010	0.038	0.011	0.011	−0.821
50	0.008	0.034	0.008	0.008	−0.818
100	0.002	0.015	0.002	0.002	−0.810
Case 3: Initial gamete frequencies 0.00001, 0.48386, 0.51612, 0.00001					
1	−78.3	1,180.0	0.000	1,130,000.0	−2.288
2	−23.6	2,560.0	0.002	−47.2	−1.533
5	− 1.36	970.0	0.004	− 1.51	−0.895
10	− 0.030	32.4	0.004	− 0.030	−0.812
20	0.003	0.042	0.003	0.003	−0.808
30	0.002	0.013	0.002	0.002	−0.808
40	0.002	0.010	0.002	0.002	−0.807
50	0.001	0.009	0.002	0.002	−0.897

quite small. This is clearly an extreme case where the behavior of the system before time t_1 cannot be predicted from our analysis.

In his original discussion of the QLE principle, Kimura (1965) provided an example in which (6.20) holds to a good approximation after a number of generations has passed, even though $\Delta Z \to \infty$ as $t \to \infty$. The state of QLE does not arise for some time (approximately 100 generations) due to the very low value of R assumed (0.001). In this example the fitness matrix (2.70) is

$$
\begin{array}{ccc}
1.00 & 1.00 & 0.95 \\
1.00 & 1.00 & 0.95 \\
0.95 & 0.95 & 1.10
\end{array}
$$

and the initial gamete frequencies are $c_i = 0.25$. In generation 100, $\Delta Z \approx 400$ and yet $\Delta \bar{w} = 11.59 \times 10^{-5}$, $\sigma_A^2 = 11.01 \times 10^{-5}$. Clearly we cannot say $\Delta \bar{w} \approx \sigma_A^2$ because $\Delta Z \approx 0$. Despite the large value of ΔZ, the quantity $P \,(= \sigma_A^2 + 2\bar{\epsilon}(\Sigma c_i^{-1})^{-1} Z^{-1} \Delta Z)$ takes the value 12.6×10^{-5} (because of the extremely small values of $\bar{\epsilon}$ and Z^{-1}) and clearly Eq. (6.19) holds and the QLE principle applies.

There are several points one can make in conclusion. First we have considered only two alleles at the loci in question: the extension of the above arguments to many alleles has been made by Nagylaki (1977). It is also of some interest to calculate the total change in mean fitness from the initial point to the equilibrium point: this may sometimes be negative because large decreases in mean fitness during the early generations outweigh the consistent but small increases during later generations. Nagylaki (1977) examines this question and suggests that this happens comparatively seldom: in most cases the total course of the evolutionary process increases the mean fitness. Finally one may ask whether there are any special classes of fitness matrices for which the mean fitness always increases. One class of fitness matrices where this is the case has been provided by Ewens (1969a, b). Suppose that the fitness matrix (2.70) is in the "additive" form

	$B_1 B_1$	$B_1 B_2$	$B_2 B_2$	
$A_1 A_1$	$\alpha_1 + \beta_1$	$\alpha_1 + \beta_2$	$\alpha_1 + \beta_3$	(6.21)
$A_1 A_2$	$\alpha_2 + \beta_1$	$\alpha_2 + \beta_2$	$\alpha_2 + \beta_3$	
$A_2 A_2$	$\alpha_3 + \beta_1$	$\alpha_3 + \beta_2$	$\alpha_3 + \beta_3$	

The fitness of any individual is here the sum of two components, one characterizing the genotype at the A locus and the other, the genotype at the B locus. For this fitness matrix the mean fitness $\bar{w}$ becomes

$$\bar{w} = \alpha_1(c_1 + c_2)^2 + 2\alpha(c_1 + c_2)(c_3 + c_4) + \alpha_3(c_3 + c_4)^2$$
$$+ \beta_1(c_1 + c_3)^2 + 2\beta_2(c_1 + c_3)(c_2 + c_4) + \beta_3(c_2 + c_4)^2. \qquad (6.22)$$

Suppose the gamete frequencies c_1, c_2, c_3, c_4 take any arbitrary values in generation t. The mean fitness in generation $t + 1$ is found by replacing c_i by c_i' in (6.22). Now (6.22) depends on the gamete frequencies only through the gene frequencies $c_1 + c_2$, $c_3 + c_4$, $c_1 + c_3$, and $c_2 + c_4$: however, from (2.73), the gene frequencies $c_1' + c_2'$, $c_3' + c_4'$, $c_1' + c_3'$, and $c_2' + c_4'$ are independent of R once the c_i are given and thus, in particular, are the same as for the special case $R = 0$. But when $R = 0$ the system (2.73) is identical to a four-allele single-locus system, and then Kingman's theorem (see Sect. 2.4) shows that mean fitness is non-decreasing. It follows that in the two-locus system (6.21) the mean fitness is non-decreasing. Note that while we have used single-locus theory to obtain this result it is not true that the complete evolution of the system is identical to that of any four-allele single-locus system: we have used the parallel with the latter merely to assert that mean fitness is non-decreasing. Note that this argument can be extended immediately to cover an arbitrary number of alleles at each of the two loci and indeed an arbitrary number of loci with an arbitrary recombination pattern. Later we will use this conclusion to derive properties of additive fitness models and will later introduce a second class of fitness matrices where the mean fitness is non-decreasing.

6.3 Equilibrium Points

In the previous section we have been concerned with a specific property of the dynamics of the recurrence system (2.73). In the present (brief) section we turn to static properties and introduce the machinery whereby we will examine the equilibrium properties of this system.

If we write Eqs. (2.73) in the form

$$c_i' = \phi_i(c_1, c_2, c_3, c_4, w_{ij}, R) \tag{6.23}$$

the point $c_i = c_i^*$ is an equilibrium point if the c_i^* satisfy the equations

$$c_i^* = \phi_i(c_1^*, c_2^*, c_3^*, c_4^*, R). \tag{6.24}$$

It is clear that the system (2.73) may possess several equilibrium points and further that these points may depend on the value of R. We discuss these observations in more detail later. When the fitness matrix (2.69) possesses special properties the equilibrium equation (6.24) can often be solved explicitly, but in general this cannot be done and numerical methods are required. In the next section we present examples of equilibria found by both methods and discuss their properties in some detail.

We have already noted several connections between mean fitness and the equilibrium points. Thus any unique stable equilibrium point when $R = 0$ corresponds to a maximum of mean fitness while if the coefficient of linkage disequilibrium is non-zero at any stable equilibrium for $R > 0$, the equilibrium mean fitness must be less than this maximum. Further there are gamete frequency trajectories near such equilibria along which mean fitness is decreasing, at least over some generations.

Suppose an equilibrium point of the system (2.73) has been found. It is then necessary to examine its stability behavior since unstable equilibrium points are of little interest. The local linear stability of the system is tested by standard methods which we here outline. Suppose in any generation $c_i = c_i^* + \delta_i$, $(\Sigma \delta_i = 0)$, where the δ_i are small deviations from the equilibrium value. If the corresponding deviations in the following generation are δ_i', then from (2.73)

$$c_i^* + \delta_i' = \phi_i(c_1^* + \delta_1, c_2^* + \delta_2, c_3^* + \delta_3, c_4^* + \delta_4)$$

$$= \phi_i(c_1^*, c_2^*, c_3^*, c_4^*) + \sum_i \delta_j \left[\frac{\partial \phi_i}{\partial c_j} \right]^* + 0(\delta^2). \tag{6.25}$$

Here $[f]^*$ means the function f evaluated at the equilibrium point. Ignoring small-order terms, this gives

$$\delta' = A\delta \tag{6.26}$$

where A is the 4×4 matrix with the $i - jth$ term $[\partial\phi_i/\partial c_j]^*$. Since

$$\delta^{(n)} = A^n\delta^{(0)} \tag{6.27}$$

it is clear that the equilibrium point is locally linearly stable if and only if all eigenvalues of A are less than unity in absolute value. These eigenvalues can be evaluated numerically by standard computer subroutines or, in special cases, found algebraically. The condition $\Sigma \delta_i = 0$ can be used to simplify the calculations under both approaches.

Our definition of stability is a local one only. Questions concerning global stability and domains of attraction are far harder to answer and, especially concerning the latter, little is known about them.

6.4 Special Models

Historically the first analyses, both static and dynamic, of the recurrence system (2.73) assumed special forms for the fitness matrix (2.69). While this often allowed explicit expressions for the equilibria and explicit criteria for their stability which suggested rather general conclusions for other fitness matrices, there was no certainty that the conclusions reached were not an artifact of the simple forms assumed and thus no certainty of the generality of the conclusions reached. In this section we attempt to overcome these difficulties by presenting explicit conclusions for certain special matrices and numerical conclusions found from a number of more or less arbitrary fitness matrices.

We have already noted the additive fitness model (6.21). For the multiplicative fitness model the matrix (2.70) appears in the form

$$
\begin{matrix}
\alpha_1\beta_1 & \alpha_1\beta_2 & \alpha_1\beta_3 \\
\alpha_2\beta_1 & \alpha_2\beta_2 & \alpha_2\beta_3 \\
\alpha_3\beta_1 & \alpha_3\beta_2 & \alpha_3\beta_3
\end{matrix}
\tag{6.28}
$$

while a third special class of fitness matrices is given by the "symmetric viability" model for which the fitnesses are

$$
\begin{matrix}
1 - \delta & 1 - \beta & 1 - \alpha \\
1 - \gamma & 1 & 1 - \gamma \\
1 - \alpha & 1 - \beta & 1 - \delta
\end{matrix}
\tag{6.29}
$$

In this model we usually assume $\alpha, \beta, \gamma, \delta > 0$ and do so here throughout. These models are not mutually exclusive: thus if $\beta + \gamma = \alpha = \delta$ the symmetric viability model is also an additive model. Several models which do not initially appear in one of these forms can in fact be so written by suitable reparameterization. Thus the model

$1 + s$	$1 + t$	$1 - s$
1	$1 + t$	1
$1 - s$	$1 + t$	$1 + s$

(6.30)

of Kimura (1956b) can be cast in the form (6.29) by putting $\alpha = (s + t)/(1 + t)$, $\beta = 0$, $\gamma = t/(1 + t)$, $\delta = (t - s)/(1 + t)$. (Note that Kimura's analysis of the model (6.30) marked the beginning of the mathematics of two-locus models as discussed in this book.) However, a second model studied by Kimura (1956b), for which the fitness matrix is

$1 + s$	$1 + s + t$	$1 - s$
$1 + s$	$1 + s + t$	$1 - s$
$1 - s$	$1 - s + t$	$1 + s$

(6.31)

cannot in general be cast in any of the three forms above. Similarly the model (1.87) does not fall into any of these forms, whereas the models (1.88) and (1.89) do. In examining in this section both static and dynamic properties of the additive, multiplicative and symmetric viability models and of various numerical models we assume throughout that $0 < R \leqslant 0.5$: the case $R = 0$ reduces to a one-locus four-allele model to which the theory of Sect. 2.4 can in large part be applied, while cases with $R > 0.5$ are not of biological interest.

We take first the additive fitness scheme (6.21). Using the fact that mean fitness is non-decreasing in this model, Karlin and Feldman (1970b) have shown in the case $\alpha_2 > \alpha_1, \alpha_3$ and $\beta_2 > \beta_1, \beta_3$ that there exists a unique internal equilibrium point at

$$
\begin{aligned}
c_1 &= (\alpha_2 - \alpha_3)(\beta_2 - \beta_3)/[(2\alpha_2 - \alpha_1 - \alpha_3)(2\beta_2 - \beta_1 - \beta_3)], \\
c_2 &= (\alpha_2 - \alpha_3)(\beta_2 - \beta_1)/[(2\alpha_2 - \alpha_1 - \alpha_3)(2\beta_2 - \beta_1 - \beta_3)], \\
c_3 &= (\alpha_2 - \alpha_1)(\beta_2 - \beta_3)/[(2\alpha_2 - \alpha_1 - \alpha_3)(2\beta_2 - \beta_1 - \beta_3)], \\
c_4 &= (\alpha_2 - \alpha_1)(\beta_2 - \beta_1)/[(2\alpha_2 - \alpha_1 - \alpha_3)(2\beta_2 - \beta_1 - \beta_3)],
\end{aligned}
$$

(6.32)

which is globally stable. Note that this equilibrium is what would be expected by a composition of single-locus analyses, using Eq. (1.30). In particular the equilibrium (6.32) is independent of R and further the coefficient of linkage disequilibrium is zero at the point (6.32). Since also mean fitness is non-decreasing one might be tempted to conclude that for this model the loci are practically "independent" and for practical purposes can be studied separately. This conclusion, however, is not quite true: thus if the coefficient of linkage disequilibrium is zero in any generation it does not necessarily remain zero in future generations, and this can prevent gene frequencies converging monotonically to their equilibrium values implied by (6.32).

If $\alpha_1 > \alpha_2 > \alpha_3$ the frequency of A_1 converges to unity. If at the same time $\beta_2 > \beta_1, \beta_3$ a polymorphism will be maintained at the B locus in accordance with (1.30). This case and the corresponding cases where fixation occurs at the B locus or at both loci essentially reduce, so far as equilibrium properties are concerned, to single-locus systems; we consider them no further here. Note that the criteria for such fixation depend only on the selective parameters α_i, β_i and are independent of the recombination fraction R.

An important question concerning the concept of marginal fitnesses arises with respect to the equilibrium (6.32). Suppose we observe the A locus only and compute the marginal fitnesses defined by (2.78). These reduce to $\alpha_1 + \beta$, $\alpha_2 + \beta$, $\alpha_3 + \beta$ (for some β), as we might expect. A parallel observation applies for the B locus. The important observation for us is that, since in computing (6.32) it has been assumed that $\alpha_2 > \alpha_1, \alpha_3$ and $\beta_2 > \beta_1, \beta_3$, the marginal fitnesses will also exhibit overdominance. The question of whether marginal overdominance always applies for stable internal two-locus equilibria is an interesting one which we will return to on several occasions.

We turn now to the multiplicative system (6.28). The properties of this model are more complex than those for the additive system. We again suppose that $\alpha_2 > \alpha_1, \alpha_3$ and $\beta_2 > \beta_1, \beta_3$. Then there exists an equilibrium point at (6.32) for all values of R. However, this equilibrium is stable only if R is large enough: Bodmer and Felsenstein (1967) show that stability applies if and only if

$$R > \frac{(\alpha_2 - \alpha_1)(\beta_2 - \beta_1)(\alpha_2 - \alpha_3)(\beta_2 - \beta_3)}{(2\alpha_2 - \alpha_1 - \alpha_3)(2\beta_2 - \beta_1 - \beta_3)\alpha_2\beta_2} . \tag{6.33}$$

This is a condition for "local" stability. Moran (1968) has shown that a sufficient condition for global stability of (6.32) is

$$R > \min \left| \frac{(\alpha_2 - \alpha_3)(\alpha_2 - \alpha_1)}{\alpha_2(2\alpha_2 - \alpha_1 - \alpha_3)}, \frac{(\beta_2 - \beta_3)(\beta_2 - \beta_1)}{\beta_2(2\beta_2 - \beta_1 - \beta_3)} \right| . \tag{6.34}$$

It is not known how close this is to being a necessary condition.

The requirement (6.33) has been generalized in an elegant way to an arbitrary number of alleles at the two loci by Roux (1974). Suppose we associate a multiplicative fitness component a_{ij} to the genotype A_iA_j at locus A and a multiplicative component b_{ij} to B_iB_j at locus B. Assuming that treated as single-locus systems each of these has an internal equilibrium where

$$\text{freq } (A_i) = p_i, \quad \text{freq } (B_j) = q_j, \tag{6.35}$$

the two-locus system will have an equilibrium point at which freq $(A_iB_j) = p_iq_j$. Now define the A and B locus equilibrium mean fitnesses as

$$\bar{w}_A = \Sigma \Sigma p_ip_ja_{ij}, \quad \bar{w}_B = \Sigma \Sigma q_iq_jb_{ij}, \tag{6.36}$$

put

$$c_{ij} = a_{ij}\sqrt{p_ip_j}/\bar{w}_A, \quad d_{ij} = b_{ij}\sqrt{q_iq_j}/\bar{w}_B, \tag{6.37}$$

and let λ, ψ be the largest non-unit eigenvalues of the matrices $\{c_{ij}\}$, $\{d_{ij}\}$. Then Roux shows that the condition for the equilibrium where freq $(A_iB_j) = p_iq_j$ to be stable is

$$R > \lambda\psi/\{(1 - \lambda)(1 - \psi)\}. \tag{6.38}$$

In the case of two alleles at each locus, we may adopt our previous notation and put, fo the A locus,

$$a_{11} = \alpha_1, \quad a_{12} = a_{21} = \alpha_2, \quad a_{22} = \alpha_3.$$

Further,

$$p_1 = (a_2 - a_3)/\{2\alpha_2 - \alpha_1 - \alpha_3\} = 1 - p_2 \tag{6.39}$$

and hence

$$\overline{w}_A = (\alpha_2^2 - \alpha_1\alpha_3)/(2\alpha_2 - \alpha_1 - \alpha_3). \tag{6.40}$$

It follows from (6.37) – (6.40) that

$$c_{11} = \frac{\alpha_1(\alpha_2 - \alpha_3)}{\alpha_2^2 - \alpha_1\alpha_3}, \quad c_{12} = c_{21} = \frac{\alpha_2 \{(\alpha_2 - \alpha_3)(\alpha_2 - \alpha_1)\}^{1/2}}{\alpha_2^2 - \alpha_1\alpha_3},$$

$$c_{22} = \frac{\alpha_3(\alpha_2 - \alpha_1)}{\alpha_2^2 - \alpha_1\alpha_3}.$$

The eigenvalues of the matrix $\{c_{ij}\}$ are easily found to be 1 and

$$\lambda = (\alpha_2 - \alpha_3)(\alpha_2 - \alpha_1)/(\alpha_2^2 - \alpha_1\alpha_3).$$

In a similar way we find for the B locus

$$\psi = (\beta_2 - \beta_3)(\beta_2 - \beta_1)/(\beta_2^2 - \beta_1\beta_3).$$

Insertion of these values into (6.38) gives precisely the condition (6.33).

Suppose now the condition (6.33) is not met. In the symmetric case $\alpha_1 = \alpha_3$, $\beta_1 = \beta_3$ there will now exist two stable equilibria, both possibly exhibiting large numerical values of linkage disequilibrium. Thus, for example, for the fitness parameters (6.34) and $R = 0.000001$ the two stable equilibria are

c_1	c_2	c_3	c_4	
0.45	0.05	0.05	0.45	(6.41)
0.05	0.45	0.45	0.05	

Since such equilibria arise in multiplicative models only for very tightly linked loci, they may be regarded as being of limited interest. However, we shall note in the next chapter that for interactive systems involving many loci, rather less stringent conditions on the value of the recombination fraction still lead to equilibria of the form (6.41) (i.e. with $D \neq 0$).

As with the additive case, whenever the conditions $\alpha_2 > \alpha_1$, α_3 and $\beta_2 > \beta_1$, β_3 do not both hold, "edge" or "corner" equilibria may arise. These are of little interest to us now and we consider them no further.

We turn now to the mean fitness. First, since equilibria of the form (6.41) have non-zero linkage disequilibrium values, mean fitness cannot be maximized at them. The equilibrium (6.32) is perhaps of greater interest. Surprisingly, the mean fitness is not maximized at this print: rather, the fitness surface has a saddle point at (6.32). These conclusions imply that mean fitness can be decreasing in the neighbourhood of equilibrium points.

One special property of multiplicative fitness schemes concerns the linkage disequilibrium function D. If in any generation we have $D = 0$, then $D = 0$ for all future generations. (This may be checked directly from the recurrence relations.) In such a case the mean fitness must be non-decreasing since the recurrence relations then reduce to single-locus equations. More generally, for $R \leqslant 0.5$, the sign of D will not change throughout the entire evolution of the recurrence process (Karlin, 1975).

We consider next the value of the equilibrium mean fitness as a function of R. When the inequality (6.33) obtains the equilibrium point (6.32) is independent of R and thus so is the equilibrium mean fitness. When (6.33) does not hold the equilibrium point depends on R. However, since the equilibrium mean fitness decreases with R for $R \approx 0$, we may suspect that it is non-increasing with R for all R, and this has been confirmed for this model by Karlin (1975).

We note finally that if $\alpha_2 > \alpha_1$, α_3 and $\beta_2 > \beta_1$, β_3 and an internal equilibrium point, either (6.32) or of the form typified by (6.41), exists, the marginal fitness values as computed from (2.79) exhibit overdominance. This is easily checked for (6.32) and may also be verified for the linkage-disequilibrium equilibrium points.

The last special case we consider in some detail is the symmetric viability model (6.29). This model possesses several unusual features not shared by additive or multiplicative schemes. Perhaps the most important is the unexpected existence of so-called asymmetric equilibria. The fitness matrix (6.29) possesses certain symmetry properties and because of this one might expect that at any equilibrium point,

$$c_1^* = c_4^*, \quad c_2^* = c_3^*. \tag{6.42}$$

Strangely, while equilibria satisfying (6.42) usually do exist, there is a further class of equilibria for which (6.42) does not hold. This fact was discovered by Karlin and Feldman (1970a) and revealed hitherto unthought of complexities in the model (6.29). Suppose for example that (6.29) takes the form

$$
\begin{matrix}
0.97 & 0.96 & 0.98 \\
0.96 & 1.00 & 0.96 \\
0.98 & 0.96 & 0.97
\end{matrix}
\tag{6.43}
$$

and that $R = 0.04$. Then the recurrence system (2.73) admits five equilibrium points and at only one of these (the first noted below) does (6.42) hold. The five equilibria are

c_1	c_2	c_3	c_4	
0.238	0.262	0.262	0.238	
0.154	0.656	0.036	0.154	
0.154	0.036	0.656	0.154	(6.44)
0.889	0.050	0.050	0.011	
0.011	0.050	0.050	0.889	

For smaller values of R the number of symmetric equilibria for which (6.42) holds increases to three. Thus for $R = 0.004$, apart from four asymmetric equilibria, there are equilibria at

c_1	c_2	c_3	c_4	
0.0034	0.4966	0.4966	0.0034	
0.2734	0.2266	0.2266	0.2734	(6.45)
0.4950	0.0050	0.0050	0.4950	

This does not automatically occur for every choice of α, β, γ, δ. Thus if $\beta + \gamma > \alpha$, δ do not both hold, there will be only one equilibrium of the form (6.42).

It is of particular importance for this model to ask two questions concerning stable equilibrium points. As noted above, up to seven internal equilibria can exist for certain parameter choices in the model (6.29). First, how many of these can be stable and second, will there always be at least one stable internal equilibrium, at least for certain values of R?

Karlin (1975) has stated that irrespective of R there can never be more than two stable internal equilibria for the system (6.29). Thus if seven such equilibria exist, at least five must be unstable. Can we be guaranteed that at least one internal stable equilibrium exists? Karlin and Feldman (1970a, b) show that this is so, at least provided R is sufficiently small (and, as we are assuming, that α, β, γ, $\delta > 0$). For larger values of R quite complex behavior is possible, and this behavior can be described explicitly in the special case $\delta = \alpha$. We restrict our attention to symmetric equilibria of the form (6.42), since Karlin and Feldman (1970a) show that the asymmetric equilibria are unstable in this case. Here the equilibrium point $c_i^* = 0.25$ exists for all R but is stable only if

$$R > \tfrac{1}{4}(\beta + \gamma - \alpha) \qquad (6.46)$$

and

$$\alpha > |\beta - \gamma|. \qquad (6.47)$$

If the right-hand side in (6.46) is negative and (6.47) holds, this of course implies stability for all R. When the inequality (6.46) does not hold there will exist two equilibria of the form

$$c_1^* = c_4^* = 0.25 \pm 0.25 \{1 - 4R/(\beta + \gamma - \alpha)\}^{1/2}, \tag{6.48}$$

$$c_2^* = c_3^* = 0.25 \pm 0.25 \{1 - 4R/(\beta + \gamma - \alpha)\}^{1/2}.$$

The condition (on the recombination fraction R) that these be stable is that

$$\phi(R) = 4R^2(\beta + \gamma - \alpha) + 2R\{2\alpha^2 - \beta^2 - \gamma^2 - \alpha(\beta + \gamma)\}$$
$$+ \alpha(\beta + \gamma - \alpha)^2 > 0. \tag{6.49}$$

This equation is always satisfied for R sufficiently close to zero: thus the equilibra (6.48) are always stable if linkage is sufficiently tight. When $R = \frac{1}{4}(\beta + \gamma - \alpha)$, the upper limit for R for which equilibria of the form (6.48) exist, condition (6.49) reduces to (4.37). If we denote the solutions of the equation

$$\Phi(R) = 0 \tag{6.50}$$

by R_1 and R_2 ($R_1 < R_2$) it follows that (6.48) will be stable in the following cases:

1) if $R_1 > 0, |\beta - \gamma|\alpha$, then (6.48) is stable for $0 < R < R_1$;
2) if $R_1 > \frac{1}{4}(\beta + \gamma - \alpha)$, then (6.48) is stable for $0 < R < \frac{1}{4}(\beta + \gamma - \alpha)$;
3) if $0 < R_1 < R_2 < \frac{1}{4}(\beta + \gamma - \alpha)$, then (6.48) is stable for $0 < R < R_1$ and for $R_2 < R < \frac{1}{4}(\beta + \gamma - \alpha)$.

Note that in the latter case there exists a gap of instability (when $R_1 < R < R_2$) for which there are no internal stable equilibria. Clearly the stability behavior, even of the symmetric equilibria, is not straightforward.

Before leaving symmetric viability models we note that the model (1.89) can be cast in the form (6.29) by suitable renormalization and parameterization. If there exists an equilibrium of the form (6.42) it is easy to see that $c_1^* = c$ satisfies (1.91). Further properties of this model follow from the general properties we give for symmetric viability models.

Given a point of stable equilibrium, what can be said about the behaviour of the equilibrium mean fitness as a function of R? Karlin (1975) asserts that this mean fitness is non-increasing in R so that the behavior here agrees with that for additive and multiplicative models. Further, it can be shown (Karlin, 1975) that marginal over-dominance holds: if the marginal fitnesses are calculated as in (2.78) for any stable equilibrium point of a symmetric viability system, the marginal heterozygote fitness, at both loci, will exceed that of the corresponding homozygotes.

If is convenient now to list the general conclusions and impressions drawn from our analyses of additive, multiplicative, and symmetric models.

(i) Only in one class of models (the additive class) does the Fundamental Theorem of Natural Selection hold. In other models violations of this theorem are possible.

(ii) When the double heterozygote is the most fit individual ($\alpha_2 > \alpha_1, \alpha_3$ and $\beta_2 > \beta_1$, β_3 in additive and multiplicative models, $\alpha, \beta, \gamma, \delta > 0$ in the symmetric model) there always exists a stable internal equilibrium point for sufficiently small values of R. At least for the symmetric viability model, the existence of stable equilibria for larger values of R depends on quite complicated criteria involving R and the selective parameters.

(iii) Stable equilibria with small R often exhibit linkage disequilibrium whereas stable equilibria with large R do not.

(iv) For any given value of R there can be at most two stable equilibria.

(v) The value of the equilibrium mean fitness decreases (or at worst remains stable) as R increases, in accordance with the argument of Fisher outlined in Sect. 1.5.

(vi) Whenever a stable equilibrium exists, the marginal fitnesses, computed from (2.78), exhibit induced overdominance.

How general are these conclusions? It may be argued, since they have been derived from fitness matrices possessing special properties, that they are artifacts of the special features assumed and do not apply generally to arbitrary fitness matrices. To check this we present conclusions derived algebraically from other fitness matrices and also numerically from arbitrary numerical fitness matrices. The latter have been considered in particular by Karlin (1975) and Karlin and Carmelli (1975), and we draw heavily on their conclusions in our analysis.

We consider first the Fundamental Theorem. We have noted that if the coefficient of linkage disequilibrium D is non-zero at a stable equilibrium point, the Fundamental Theorem cannot be true. Now (2.73) shows that the requirement $D = 0$ at equilibrium implies $w_i = \overline{w}(i = 1, 2, 3, 4)$ at that equilibrium. These equations imply certain constraints on the fitness parameters which will only hold in special cases, and we can conclude that in this sense, as a mathematical statement, the Fundamental Theorem is false. However, it is far more important to note, from our analysis of QLE, that the result implied by the theorem is "usually" true.

Are there any general fitness schemes (other than the additive scheme) for which the Fundamental Theorem is necessarily true? Karlin (1975) has shown that for the symmetric viability model with fitness matrix

$$\begin{matrix} 0 & 1-\beta & 0 \\ 1-\gamma & 1 & 1-\gamma \\ 0 & 1-\beta & 0 \end{matrix} \qquad (6.51)$$

with $\beta + \gamma < 1$, the Fundamental Theorem is necessarily true. This model (implying lethality of all double homozygotes) may be regarded as an unusual one, and since no other general class of models implying the mathematical truth of the Fundamental Theorem has been found, we may conclude that in practice the additive model is the only important example where this theorem holds.

We turn next to the question of the existence of stable equilibria when the double heterozygote is the most fit genotype. In all cases that we considered above a stable equilibrium was guaranteed for small R and we now ask whether this is true for arbitrary fitness matrices. Karlin (1975) shows that this is not so: even if the double heterozygote is the most fit genotype it is possible, for some fitness schemes, that no internal stable equilibrium point exists for any positive recombination rate. One fitness scheme possessing this property is, in the notation of (2.70),

0.98	0.998	0.98
0.976	1	0.965
0.97	0.995	0.96

$$(6.52)$$

Are there special conditions on the fitness values that ensure a stable internal equilibrium, at least for small R? When all four double homozygotes are more fit than any single heterozygote but less fit than the double heterozygote, there are two such equilibria for small R. Perhaps of greater interest is the case where the double homozygotes are the least fit, with single heterozygotes of intermediate fitness (and the double heterozygote most fit). Here there also exists at least one stable internal equilibrium for small values of R. All these results are given by Karlin (1975), from which reference further details on the method of proof of these statements may be obtained.

We turn next to the conclusion that equilibria for small R usually exhibit linkage diequilibrium, whereas for large R the linkage disequilibrium is small or even zero. Of course this conclusion is not uniformly true: for additive models, for example, the (unique) internal equilibrium (6.32) obtains for all R and exhibits zero linkage disequilibrium. Nevertheless, as a broad statement, the above conclusion is largely correct. Thus for $R = 0.5$ the fitness matrix (2.75) possesses a stable equilibrium at the point (2.77), and at this point the linkage disequilibrium is quite small (-0.000935). When $R = 0.001$ there exist two stable equilibria, at the points

$$c_1 = 0.447, \quad c_2 = 0.030, \quad c_3 = 0.022, \quad c_4 = 0.500$$

and

$$c_1 = 0.015, \quad c_2 = 0.503, \quad c_3 = 0.469, \quad c_4 = 0.013.$$

At these two points the linkage disequilibrium D is 0.223 and -0.236 respectively, and clearly this is far larger than at the point (2.77).

Despite these remarks, the relationship between R and the equilibrium value of D is not quite clear-cut. Franklin and Feldman (1977) provide an example, quite unexpected in view of our previous conclusions, of a fitness matrix for which, with certain values of R, there exist two stable equilibria, at one of which $D = 0$ while at the other $D \neq 0$. An example of a fitness matrix, written in the form (2.70), for which this occurs is

0.78	0.82	0.77
0.82	0.79	0.805
0.77	0.805	0.795844 .

$$(6.53)$$

Here the equilibria c_1 = 0.29989, c_2 = 0.143184, c_3 = 0.143184, c_4 = 0.683744 (for which D = 0) are stable for all R greater than 0.05. At the same time here exists, for each R, a second stable equilibrium for which $D \neq 0$. The location of this equilibrium depends on R: thus for R = 0.1 it is at (0.446233, 0.223923, 0.223923, 0.105920) and here D = −0.002877, whereas at R = 0.5 it is at (0.443863, 0.222814, 0.222814, 0.110508) for which D = 0.000596. Even more unexpectedly, similar behavior can arise under multiplicative fitnesses (Karlin and Feldman, to appear).

Next we consider the question of the maximum number of stable internal equilibria. Karlin (1975) proves that for small values of R no more than two such equilibria are possible. For larger R values no mathematical result is available although the many simulations of Karlin and Carmelli (1975) and others, for arbitrary fitness matrices, suggest strongly that no more than two stable equilibria are possible for any value of R.

The next point concerns the behavior of the equilibrium mean fitness $\bar{w}*$ as a function of R. In the special cases examined above, $\bar{w}*$ is non-increasing with R, an this accords with the verbal discussion of Fisher (1958). Unfortunately this behavior does not occur for arbitrary fitness matrices. While $\bar{w}*$ is locally (and indeed globally) maximized at R = 0, it is possible that, for certain ranges of R values, $\bar{w}*$ increases as R increases. An example o a fitness scheme, in the notation (2.70), for which this occurs is provided by Karlin and Carmelli (1975):

$$
\begin{array}{lll}
0.462245 & 0.403142 & 0.188776 \\
0.136754 & 0.481281 & 0.391682 \\
0.182915 & 0.245957 & 0.182463
\end{array}
\qquad (6.54)
$$

The equilibrium mean fitness, as a function of R, is shown in Table 6.2. When $\bar{w}*$ increases over certain R values, the behavior of $\bar{w}*$ as a function of R is often of the form displayed in this table: an initial decrease in $\bar{w}*$ is followed by an increase over a small range of R values with an ultimate flattening out of the values of $\bar{w}*$ as R increases to 0.5. Of course if two stable equilibria exist for certain R values there will exist

Table 6.2. Values of equilibrium mean fitness for the viability matrix (6.54), for various values of R

R	Equilibrium mean fitness
0	0.463385
0.005	0.462887
0.010	0.462480
0.015	0.462168
0.020	0.461958
0.027	0.461845
0.030	0.461866
0.035	0.462003
0.037	0.462095
$\geqslant 0.042$	0.462245

two curves of $\overline{w}^*$ against R, and it is possible that both such curves exhibit the form of behavior just described.

The form of fitness scheme where this behavior tends to arise is typified by (6.54). One double homozygote and the double heterozygote have larger fitnesses than the remaining genotypes, with the double heterozygote having the largest fitness. This ensures a stable equilibrium at $R = 0$ with two gametes only, and by continuity, fo R small, a stable equilibrium exists with these gametes predominating. For large values of R the frequency of the favored double homozygote is unity with a consequent fairly high mean fitness. For intermediate values of R all genotypes exist at positive frequency and because of the lower fitness of most genotypes the equilibrium mean fitness here takes its minimum value. The conditions required for this behavior are perhaps rather special and, at least in the numerical example given, the effect of R on $\overline{w}^*$ is quite small. Nevertheless it is of some interest to note that, in principle at least, this curious behavior can occur.

We turn finally to the question of induced overdominance. In all cases considered above it can be proved that, if a stable internal equilibrium exists, the marginal mean fitness must exhibit overdominance. Although no numerical counter-example for arbitrary fitness matrices has yet been found, no proof exists that this behavior applies generally. Note also that the converse to this proposition is false: there may exist an internal equilibrium exhibiting marginal overdominance which is *unstable*. Thus starting near this equilibrium the gamete frequencies will change and this can produce behavior, if one of the two loci only and its marginal fitnesses are observed, contrary to that of single-locus theory. An example of this is given by Ewens and Thomson (1978). For the symmetric fitness matrix

0.79	0.60	0.79
0.90	1.00	0.90
0.79	0.60	0.79

$$(6.55)$$

written in the form (2.70), the equilibria (6.48) become

$$c_1^* = c_4^* = 0.25 \pm 0.25 \ \{1 - R/0.0725\}^{1/2},$$
$$c_2^* = c_3^* = 0.25 \mp 0.25 \ \{1 - R/0.0725\}^{1/2}.$$

These exist whenever $R < 0.0725$ and are stable, from (6.49), whenever $R < 0.05756$. The marginal fitnesses at the B locus always exhibit overdominance, but those at the A locus do only if $R < 0.05971$. Thus whenever $0.05756 < R < 0.05971$, both marginal fitnesses exhibit overdominance at the equilibrium and yet the equilibrium is not stable.

It will be clear that difficulties arise in enunciating general principles for equilibria of two-locus systems. While several conclusions, as discussed above, are generally true, counter-examples are usually possible. These sometimes refer to cases of unlikely biological interest, and it remains a challenge to discover a principle, like QLE, that indicates normal behavior in biologically realistic circumstances.

6.5 Modifier Theory

One of the most interesting applications of two- (or multi-) locus theory arises when one of the loci considered is a modifier locus, i.e. the genes present at that locus modify in some way the values of various genetic parameters (e.g. mutation rate, linkage) at or between other loci. These other loci are called the primary loci, and our main interest is to consider the way in which evolutionary processes at these loci are affected by the existence of the modifier locus.

Two general classes of modifier locus theory may be defined. In the first class it is supposed that an individual's fitness depends in part on his genetic constitution at the modifier locus. The modification scheme for the evolution of dominance, mentioned already (see (1.87)) and discussed in more detail below, falls in this class. In the second class of modification schemes the fitness of any individual is assumed independent of his genetic constitution at the modifier locus. Any evolution at the modifier locus is then a result of its interaction with the primary loci and may then be described (see, for example, Feldman and Krakauer, 1976, p. 561) as due to secondary selection. This class of modification schemes was introduced into the literature by Nei (1967), who considered a modifier controlling the recombination fraction between two primary loci.

Our first example is of the former class and concerns Fisher's (1928a, b, 1930b, 1931, 1934) theory of the evolution of dominance through the action of a modifier locus. We have already proposed in (1.87) a fitness scheme for this situation. A proper description of the joint evolution at A and M loci requires consideration of the frequencies of the four gametes A_1M_1, A_1M_2, A_2M_1 and A_2M_2, but, if s is small and the recombination fraction between primary and modifier locus is not small, no serious error is made by making the approximation that linkage equilibrium obtains throughout. Thus the frequency of any gamete may be written as the product of the frequencies of the two constituent alleles: (a more refined analysis (Feldman and Karlin, 1971) confirms that the error in this approximation is negligible).

Suppose that the frequency of A_1 is close to unity and that A_1 mutates to A_2 at rae u. Then under our simplifying assumptions, if x is the frequency of A_1 and y that of M_1,

$$\Delta y = sx(1 - x)y(1 - y)\{4ky + 2h - 2hy - 2k\}, \tag{6.56}$$

$$\Delta x = sx(1 - x)\{1 - x + (2x - 1)(1 - y)(2ky + h - hy)\} - ux. \tag{6.57}$$

The final term in (6.57) does not of course come from (2.73) but arises from the recurrent mutation from A_1 to A_2. Clearly both x and y change and our aim is to use (6.56) and (6.57) to find an expression for Δy that is independent of x. Before doing this we note from (1.33) (with a slight change of notation) that the equilibrium value of x when $y = 0$ is $1 - u/sh$ and when $y = 1$ is $1 - (u/s)^{1/2}$. Thus $x(1 - x)$ always lies in the interval

$$(u/sh)(1 - u/sh), \quad (u/s)^{1/2}\{1 - (u/s)^{1/2}\} \tag{6.58}$$

and is consequently always bounded above by $(u/s)^{1/2}$. This implies that Δy is always bounded above by

$$(us)^{1/2} y(1-y)\{4ky + 2h - 2hy - 2k\}. \tag{6.59}$$

For $k = \frac{1}{2} h$ this yields the upper bound

$$\Delta y \leqslant (us)^{1/2} hy(1-y) \tag{6.60}$$

while for $k = 0$,

$$\Delta y \leqslant 2h(us)^{1/2} y(1-y)^2. \tag{6.61}$$

One way of obtaining a more accurate assessment of the value of Δy would be to use (6.56) and (6.57) to form the differential equation

$$\frac{dy}{dx} = \frac{s(1-x)y(1-y)\{4ky + 2h - 2hy - 2k\}}{s(1-x)\{1-x + (2x-1)(1-y)(2ky + h - hy)\} - u}. \tag{6.62}$$

If this equation could be solved for x (as a function of y) the resulting solution could be inserted in (6.56) to obtain a very accurate value for Δy. Unfortunately no solution of (6.62) has yet been found and the best that appears possible in this direction is to solve (6.62) numerically: this would, however, be equivalent to a joint numerical solution of (6.56) and (6.57).

A slightly less accurate method is to argue as follows. For any fixed value of y there will exist an equilibrium value of x, somewhere in the interval (6.58). This value is found by solving from (6.57) the equation $\Delta x = 0$. Although we cannot expect x to have reached this equilibrium value for any current value of y, a reasonable approximation is obtained by assuming that it has. The solution of the equation $\Delta x = 0$, or

$$(1-x)^2 + (1-x)(2x-1)(1-y)(2ky + h - hy) = u/s \tag{6.63}$$

is not as straightforward as might appear. Since $x \approx 1$ one might be tempted to ignore the term in $(1-x)^2$ and, putting $2x - 1 \approx 1$, to write

$$1 - x \approx (u/s)[(1-y)(2ky + h - hy)]^{-1}.$$

Insertion of this in (6.56) gives, for the important particular case $k = h$,

$$\Delta y \approx uy, \tag{6.64}$$

which is the value given by Fisher (1928b). If we write $\Delta y = y(1-y)\phi(y)$, this gives

$$\phi(y) \approx u(1-y)^{-1} \tag{6.65}$$

so that $\phi(y) \to \infty$ as $y \to 1$. This conclusion was stressed by Fisher (1928c) as an essential point of his argument.

The above analysis, however, is incorrect: this can be seen immediately by noting that it leads to a violation of the upper bound (6.60). For $y \approx 1$ the term $(1 - x)^2$ becomes the dominant factor on the left-hand side of (6.63). It is, however, possible to replace the term $2x - 1$ in (6.63) by 1 with little loss of accuracy, since the error involved in doing this is of order $(1 - x)^2(1 - y)$ and is thus always extremely small. Solving the resulting equation gives

$$1 - x = \tfrac{1}{2} [-(1 - y)(2ky + h - hy) + \{(1 - y)^2(2ky + h - hy)^2 + 4u/s\}^{1/2}]$$

and insertion in (6.56) shows that, to a very close approximation,

$$\Delta y \approx \tfrac{1}{2} sy(1 - y)[4ky + 2h - 2hy - 2k][-(1 - y)(2ky + h - hy)$$
$$+ \{(1 - y)^2(2ky + h - hy)^2 + 4u/s\}^{1/2}]. \qquad (6.66)$$

For $k = h$ this becomes

$$\Delta y \approx \tfrac{1}{2} shy(1 - y)[-h(1 - y) + \{h^2(1 - y)^2 + 4u/s\}^{1/2}]. \qquad (6.67)$$

Note that since

$$-h(1 - y) + \{h^2(1 - y)^2 + 4u/s\}^{1/2} \leqslant (4u/s)^{1/2} ,$$

(6.67) gives

$$\Delta y \leqslant (us)^{1/2} hy(1 - y) \qquad (6.68)$$

in agreement with (6.60). Defining $\phi(y)$ as above we get

$$\phi(y) \approx (us)^{1/2} h. \qquad (6.69)$$

Parallel calculations give

$$\Delta y \leqslant y(1 - y)\{1.24 h^{1/2} u^{3/4} s^{1/4}\} \qquad (6.70)$$

for $k = 0$ with similar expressions for other values of k. Numerical calculations, proceeding from the recurrence relations governing gamete frequencies, show (Ewens, 1967) that (6.67)–(6.70) are very accurate and confirm our belief that linkage equilibrium can be assumed to a close approximation throughout this process.

What value do the above calculations have for the evolution of dominance question? It is clear that the "selective intensity in favor of the modifier", $\phi(y)$ in the above notation, is always very small and certainly does not become infinitely large, as Fisher's analysis claims. It is essentially because of this observation that Wright (1929a, b) originally cast doubt on Fisher's theory. We have noted in Chap. 1 Wright's emphasis on the pleiotropic effects of genes. If, in line with his view, the modifier gene is subject to a primary selection pressure quite independent of its modification action, the very small selective pressure due to dominance modification cannot control its evolution and is essentially irrelevant. Fisher (1929) resists this viewpoint, but, quite apart from the bias in his argument induced by the mathematical error noted above, this author finds his position unconvincing.

The above analysis has assumed that from the start the favored primary allele A_1 is always at a high frequency and the small effect of dominance modification is in large part due to the very low frequency of heterozygotes A_1A_2 throughout the process. In some cases, for example with the classic evolution of the melanic form in the moth *Biston betularia* due to industrial pollution, the eventually favoured form starts at a low frequency and thus, during the course of its frequency increase, many heterozygotes will appear upon which the force of dominance modification could act. The extent to which the frequency of the modifier is changed in this way depends on the degree of linkage between primary and modifying loci: if the two loci are closely linked there is some possibility for a substantial increase in the frequency of the modifier and this tendency is magnified the larger the selective differences are at the primary locus. At the same time, once the frequency of A_1 reaches a high value the induced selective force on the modifier becomes very weak and the argument of Wright outlined above will again prevail.

We turn now to other ways in which modifier loci can act, considering in particular modification of linkage and mutation rate. In this way we wish to give an explanation for the evolution of these characteristics which is independent of arguments involving interpopulation competition at least implicit in the early discussions of them. The classic papers introducing modifiers which do not change fitnesses were those of Nei (1967, 1969), who considered modification of linkage. We follow here, however, the discussion of this topic given by Feldman (1972).

Consider two primary loci A and B and a modifier locus M the effect of which is to influence the recombination fraction between A and B loci. Let these loci lie on a chromosome in the order MAB with recombination values R between M and A loci, R_{ij} between A and B loci and $R + R_{ij} - 2RR_{ij}$ between M and B loci. Here R_{ij} depends on the genotype M_iM_j at the modifier locus.

Our mode of approach is the following. Suppose the various genotypes at A and B loci have fitnesses specified by (2.69). If all individuals at the M locus are M_1M_1 the recombination fraction between A and B loci is R_{11}, and we suppose that the population has reached a stable equilibrium point (for this recombination fraction) with gamete frequencies c_1^*, c_2^*, c_3^* and c_4^*. Suppose that the allele M_2 is now introduced at the M locus. The frequencies of the gamete $M_1A_1B_1, M_1A_1B_2, M_1A_2B_1$ and $M_1A_2B_2$ now become $c_1^* + \delta_1, c_2^* + \delta_2, c_3^* + \delta_3, c_4^* + \delta_4$, and we suppose the frequencies of $M_2A_1B_1, M_2A_1B_2, M_2A_2B_1$ and $M_2A_2B_2$ are $\delta_5, \delta_6, \delta_7, \delta_8$, with $\sum_{i=1}^{8} \delta_i = 0$. We may set

up recurrence relations extending (2.73) for the eight gamete frequencies which, if terms in $\delta_i^2, \delta_i\delta_j$ are ignored, become linear recurrence relations in $\delta_5, \delta_6, \delta_7$ and δ_8. If on the one hand the eigenvalues of the matrix governing this recurrence system are less than unity, then $\delta_i \to 0$ and the system returns to its original equilibrium with M_2

absent. If on the other hand at least one eigenvalue is greater than unity in absolute value the frequency of M_2 will increase and the recombination fraction between A and B loci, for those individuals carrying the M_2 gene, will change. Clearly our objective is to find the circumstances (in terms of these eigenvalues) which lead to this increase.

In general this proves to be rather difficult, although we present later a general argument that at least suggests the behavior of these eigenvalues. In the additive, multiplicative and symmetric viability models, Feldman shows that provided $c_1^* c_4^* \neq c_2^* c_3^*$, the frequency of M_2 will increase if and only if $R_{12} < R_{11}$, that is if and only if the modifier heterozygote $M_1 M_2$ leads to tighter linkage between A and B loci than does $M_1 M_1$.

There is no reason to suppose that this conclusion does not apply for general fitness matrices. That this is so is suggested by a conclusion of Feldman and Krakauer (1976). Let the fitness matrix (2.69) at A and B loci be arbitrary and suppose, in the above notation, that $R_{12} < R_{11}, R_{22}$. Now define

$$m_1 = (R_{22} - R_{12})/(R_{11} + R_{22} - 2R_{12}) = 1 - m_2,$$
$$R^* = m_1^2 R_{11} + 2m_1 m_2 R_{12} + m_2^2 R_{22}.$$

Suppose now that $(c_1^*, c_2^*, c_3^*, c_4^*)$ is a solution of the equilibrium equations (2.74) if $R = R^*$. Then there is an equilibrium of the three-locus system where

$$\text{freq} (M_i A_1 B_1) = m_i c_1^*, \quad \text{freq} (M_i A_1 B_2) = m_i c_2^*,$$
$$\text{freq} (M_i A_2 B_1) = m_i c_3^*, \quad \text{freq} (M_i A_2 B_2) = m_i c_4^*,$$

for $i = 1, 2$. The stability properties of this equilibrium have not yet been determined but, if this equilibrium is stable at least for certain recombination values, it strongly suggests the evolution of tighter linkage between A and B loci by secondary selection. Note in passing the curious resemblance between the formula for m_1 and the equilibrium gene frequency (1.30) in a one-locus selective scheme.

We turn now to modification of mutation rate. Suppose the fitnesses at the primary locus A are given by (1.24b) with $s > sh > 0$ (cases with complete dominance in fitness can be considered similarly). Suppose that the mutation rate $A_1 \rightarrow A_2$ is controlled by the genes present at a modifier locus and is u_{ij} for individuals of genotype $M_i M_j$. We may suppose that initially the frequency of M_1 is unity and that the frequency of A_1 is at the mutation-selection equilibrium value $1 - u_{11}/s(1 - h)$. The population mean fitness is now $1 - 2u_{11}$. Suppose now M_2 is introduced at a low frequency. By considering linearized recurrence relations for the four gametic types it is found that the frequency of M_2 increases if and only if $u_{12} < u_{11}$. More generally if $u_{22} < u_{12} < u_{11}$ the frequency of M_2 will steadily increase to unity so that the mutation rate $A_1 \rightarrow A_2$ becomes u_{22} and the population mean fitness, $1 - 2u_{22}$. If $u_{12} < u_{11}, u_{22}$ a polymorphism is established at the M locus. All these conclusions are true irrespective of the linkage arrangement between primary and modifier loci. Again, we attempt to restate these conclusions later as particular applications of a general modifier principle.

Of course the question of the establishment of a optimal mutation rate requires arguments more complex than these and must take into account the need for long-term flexibility perhaps enhanced by a higher mutation rate. These matters were discussed in Chap. 1, and our interest here is simply that interpopulation selection arguments are not required to arrive at an agency reducing the mutation rate.

It is also possible to discuss the dynamics of modifiers of sex-ratio, migration and selfing (Feldman and Krakauer, 1976; Karlin and McGregor, 1974). We turn now rather to the question of whether there exists a general principle for modifier loci embracing the conclusions just reached as particular cases. Such a general principle has been proposed by Karlin and McGregor (1974) which does not have the status of a mathematical theorem but nevertheless applies widely for modifier loci. Consider a primary locus (or loci) with evolution determined at least in part by a parameter θ (e.g. a mutation rate). Suppose the value of θ is determined by a selectively neutral modifier locus and that for individuals of genotype $M_i M_j$ ($i, j = 1, 2$) the parameter takes the value θ_{ij}. Assume that random mating obtains, and let $\bar{w}(\theta_{ij})$ be the mean fitness of the primary system at a stable equilibrium when $\theta = \theta_{ij}$. Then if

$$\bar{w}(\theta_{11}) < \bar{w}(\theta_{12}) < \bar{w}(\theta_{22}) \tag{6.71}$$

the allele M_2 will become fixed at the modifier locus. If the inequalities in (6.71) are reversed, M_1 will become fixed, while if $\bar{w}(\theta_{12}) > \bar{w}(\theta_{11})$, $\bar{w}(\theta_{22})$ a stable polymorphism will arise at the modifier locus. Thus this principle essentially asserts that the evolution at the modifier locus is such as to maximize mean fitness, and this has been observed in the specific examples above. We have proved earlier that in the multiplicative and symmetric viability models the mean fitness is non-increasing in R, and this principle then indicates that modifiers decreasing the recombination value become fixed. This is in agreement with the conclusion reached by Feldman (1972) noted above and generalizes that conclusion by not restricting attention to the frequency of the modifier M_2 when it is small. Note that this principle suggests that there a circumstances where *increased* recombination between loci is favored, as recent (unpublished) work verifies.

6.6 Two-Locus Diffusion Processes

In this section we consider multidimensional diffusion analogs to the two-locus two-allele Markov chain models (3.85) and (3.86), using throughout this section the same notation as for Sect. 3.9. The evolution of the models (3.85 and 3.86) can be described by considering the linearly independent frequencies c_1, c_2 and c_3. However, there is an obvious asymmetry in doing this and in any event we find it more convenient to work with the variables $p = c_1 + c_2$, $q = c_1 + c_3$ and $D = c_1 c_4 - c_2 c_3$. To use the diffusion theory of Sect. 5.1 we must assume that means, variances and covariances of the changes of these quantities between consecutive generations are of order N^{-1}. This will require us to assume in particular that R is $0(N^{-1})$.

We suppose first that there is no mutation and consider initially the RUZ model. Then from (3.93),

$$E\{D(t+1) - D(t)|c_i(t)\} = -(2N)^{-1}\{1 + 2NR\}D(t). \tag{6.72}$$

Assuming R is of order N^{-1} this gives $(1 + 2NR)D$ as the drift coefficient corresponding to D and, if R is still assumed to be $0(N^{-1})$, the same value applies for the RUG model. Indeed it is found that all the coefficients for the diffusion process approximating the RUG model are the same as for the RUZ model so that the diffusion approximations to the two processes are identical.

Using (6.72) and the corresponding values for p and q, as well as the variance and covariance terms for these quantities, it is found for both models that if we scale time so that unit time T corresponds to N generations of the Markov chain, the backward Kolmogorov equation for the joint density of p, q and D, when there is no selection, becomes (Ohta and Kimura, 1969a)

$$\frac{\partial f}{\partial t} = \frac{1}{4} p(1-p) \frac{\partial^2 f}{\partial p^2} + \frac{1}{4} q(1-q) \frac{\partial^2 f}{\partial q^2} + \frac{1}{2} D \frac{\partial^2 f}{\partial p \partial q} + \frac{1}{2} D(1-2p) \frac{\partial^2 f}{\partial p \partial D}$$

$$+ \frac{1}{2} D(1-2q) \frac{\partial^2 f}{\partial q \partial D} + \frac{1}{4} \{pq(1-p)(1-q) + D(1-2p)(1-2q) - D^2\} \frac{\partial^2 f}{\partial D^2}$$

$$- \frac{1}{2} (1 + 2NR) \frac{\partial f}{\partial D}. \tag{6.73}$$

Here f is the joint density function of the frequency of A_1, the frequency of B_1, and the linkage disequilibrium at time t, given initial values p, q and D. Our aim is to use this equation, in conjunction with the theory of Sect. 4.10, to find diffusion analogs for various quantities established for the Markov chain models of Sect. 3.8. We focus here on the diffusion process eigenvalues and the diffusion process expectation of $\{(D(t)\}^2$: these have been found by Ohta and Kimura (1969a), and we follow their analysis closely.

Our point of departure for both problems is to consider the expectations of the three quantities $p(1-p)q(1-q)$, $D(1-2p)(1-2q)$, and D^2 in generation t. Simultaneous equations for these can be found by using (4.106). By inserting trial eigenvalue solutions with undetermined coefficients, Ohta and Kimura show that the expectations of these quantities converge to zero at rate $\exp(\lambda_1 t)$, where λ_1 is the largest root of the equation

$$4\lambda^3 + (20 + 12\gamma)\lambda^2 + (27 + 38\gamma + 8\gamma^2)\lambda + (9 + 26\gamma + 8\gamma^2) = 0. \tag{6.74}$$

In this equation $\gamma = NR$ and from our assumptions is $0(1)$. While an explicit solution is possible, it is perhaps preferable to solve (6.74) numerically for selected values of γ, and some specific solutions are given in Table 6.3.

It is naturally of interest to compare these values with those in Table 3.1. Since unit time in the diffusion corresponds to N generations in the Markov chain, we should compare $\exp \lambda_1 t$ with μ^{Nt} (where μ is the Markov chain eigenvalue), or equivalently μ with $\exp(\lambda_1/N)$. Values of this expression are also given in Table 6.3 for $N = 50$ and can be compared to the final line in Table 3.1. It will be noted that the agreement is excellent, thus showing that the leading eigenvalue for the Markov chain is closely approximated by that for the diffusion process.

Table 6.3. The largest solution (λ_1) of (6.74) for various values of γ

γ	0.5	5	10	25
λ_1	−0.813	−0.9926	−0.9978	−0.9996
$\exp(\lambda_1/50)$	0.984	0.980	0.980	0.980

The value of E $\{D(t)\}^2$ turns out to be a complicated expression involving all three eigenvalue solutions of (6.74). We do not give an explicit expression here: it is sufficient to note that the value obtained is in excellent agreement with the Markov chain solution given in Fig. 3.1. Note in particular with both conclusions that the requirement that R be $0(N^{-1})$ does not appear to be necessary for the agreement between Markov chain and diffusion solutions. This no doubt occurs because in any expression where it occurs R is invariably multiplied by the coefficient of linkage disequilibrium which, when R is not small, is usually small itself.

Suppose now that mutation exists so that there is a stationary distribution for D. Of particular interest to us is the expectation E (D^2) for this stationary distribution. We suppose the mutation rates are u_1 (from A_1 to A_2), v_1 (from A_2 to A_1), u_2 (from B_1 to B_2) and v_2 (from B_2 to B_1), and that all mutation rates are $0(N^{-1})$. The drift and diffusion coefficients for changes in p, q and D (now including terms in the mutation rates) can be inserted in (4.107) to find the stationary expectation of any function $g(p, q, D)$ of these variables. The equation so obtained (Ohta and Kimura, 1969b) is

$$
E\left[p(1-p)\frac{\partial^2 g}{\partial p^2} + q(1-q)\frac{\partial^2 g}{\partial q^2} + 2D\frac{\partial^2 g}{\partial p \partial q} + 2D(1-2p)\frac{\partial^2 g}{\partial p \partial D} \right.
$$

$$
+ 2D(1-2q)\frac{\partial^2 g}{\partial q \partial D} + \{pq(1-p)(1-q) + D(1-2p)(1-2q) - D^2\}\frac{\partial^2 g}{\partial D^2}
$$

$$
+ 4N\{v_1 - (u_1 + v_1)p\}\frac{\partial g}{\partial p} + 4N\{v_2 - (u_2 + v_2)q\}\frac{\partial g}{\partial q}
$$

$$
\left. - D(2 + 4NK)\frac{\partial g}{\partial D} \right] = 0, \tag{6.75}
$$

where $K = R + u_1 + u_2 + v_1 + v_2$ and the drift coefficients implied by the mutation rates are displayed in the equation. The expectation is with respect to the joint stationary distribution of p, q and D.

Our aim is to make suitable choices for g so that a set of three simultaneous equations arises from which E (D^2) can be obtained. Three choices of g which do this are

$$
g = pq(1-p)(1-q) = g_1 \tag{6.76}
$$

$$
g = D(1-2p)(1-2q) = g_2 \tag{6.77}
$$

$$g = D^2 = g_3. \tag{6.78}$$

Thus, for example, inserting $g = g_3$ from (6.78) into (6.75),

$$E[g_1 + g_2 - g_3(3 + 4NK)] = 0. \tag{6.79}$$

The remaining two equations yield expectations for g_1, g_2 and g_3 in terms of expectations for "lower-order" quantities such as $p(1 - p)$, pq, etc. The expectations of the latter may be found from (6.75) by choosing $g = p(1 - p)$, etc. in (6.75). Joint solution of the resulting equations and (6.79) gives

$$E(D^2) = \frac{\{2.5 + N(K + U)\}NA}{(1 + 2NU)(3 + 4NK)(2.5 + NU + NK) - 3 - 4NU} , \tag{6.80}$$

where

$$U = u_1 + u_2 + v_1 + v_2 \quad \text{and}$$

$$A = \frac{8Nu_1 u_2 v_1 v_2}{(u_1 + u_2)(v_1 + v_2)} \left[\frac{1}{4N(u_1 + v_1) + 1} + \frac{1}{4N(u_2 + v_2) + 1} \right] . \tag{6.81}$$

We may usually assume mutation rates are sufficiently small so that NU is moderate. Unless the two loci are very closely linked, NR and NK will both be large and in this case we may write

$$E(D^2) \sim A/\{4R(1 + 2NU)\}. \tag{6.82}$$

This expression is of the same order of magnitude as the mutation rate and will thus usually be very small. This reinforces the conclusion we have reached above that random processes in finite populations is of minor importance in causing non-zero values of linkage disequilibrium.

Some interest also attaches to the standardized linkage disequilibrium σ_D^2, defined by

$$\sigma_D^2 = E(D^2)/\{Epq(1 - p)(1 - q)\}. \tag{6.83}$$

Both expectations can be computed using (6.75) – (6.78) and we find if NR is large that

$$\sigma_D^2 \sim (4NR)^{-1}. \tag{6.84}$$

This is again small. If the two loci are tightly linked so that NR is not large a more accurate expression (Ohta and Kimura, 1969b) is

$$\sigma_D^2 = [3 + 4NK - 2\{2.5 + NK + NU\}^{-1}]^{-1}. \tag{6.85}$$

6.7 Associative Overdominance and Hitch-hiking

We consider in this section two concepts, of potential practical interest, which arise with non-zero values of linkage disequilibrium in finite populations.

The first concept is that of associative overdominance. This was introduced by Frydenberg (1963) to explain secular gene frequency changes in certain experimental *Drosophila* populations. The essence of the notion is that, whereas the genes at the locus of interest may be selectively equivalent, they appear to exhibit overdominance by being linked (with non-zero values of linkage disequilibrium) to a locus where true overdominance does occur. The most detailed theoretical treatment of this concept is by Kimura and Ohta (1971a, pp. 110–116).

Consider two loci A and B for which true overdominance occurs at the A locus while the B locus is selectively neutral, so that the fitnesses of the various genotypes are

$$A_1A_1B - B -   \quad A_1A_2B - B -   \quad A_1A_2B - B -$$
$$1 - s_1 \qquad\qquad 1 \qquad\qquad 1 - s_2.$$

Let the frequencies A_1 and B_1 be x and y respectively. Then the frequencies of the gametes A_1B_1 and A_2B_1 are $xy + D$ and $(1 - x)y - D$, and thus the marginal fitness of B_1B_1 individuals (see (2.78)) is

$$y^{-2}[(xy + D)^2(1 - s_1) + 2(xy + D)\{(1 - x)y - D\} + (1 - s_2)\{(1 - x)y - D\}^2]$$
$$= 1 - s_1(x + y^{-1}D)^2 - s_2(1 - x - y^{-1}D)^2. \tag{6.86}$$

Similarly the marginal fitness of B_1B_2 individuals is

$$1 - s_1(x - y^{-1}D)\{x - (1 - y)^{-1}D\} - s_2\{1 - x + (1 - y)^{-1}D\}(1 - x - y^{-1}D), \tag{6.87}$$

and of B_2B_2 individuals,

$$1 - s_1\{x - (1 - y)^{-1}D\}^2 - s_2\{1 - x + (1 - y)^{-1}D\}^2. \tag{6.88}$$

The apparent selective advantage of B_1B_2 over B_1B_1 is

$$s_1D(x + y^{-1}D)\{y(1 - y)\}^{-1} - s_2D(1 - x - y^{-1}D)\{y(1 - y)\}^{-1} \tag{6.89}$$

and over B_2B_2 is

$$-s_1D\{x - (1 - y)^{-1}D\}\{y(1 - y)\}^{-1} + s_2\{1 - x + (1 - y)^{-1}D\}\{y(1 - y)\}^{-1}. \tag{6.90}$$

There is one case of these formulae of special interest. If the selection at the A locus is so strong that we may assume $x = x^* = s_2/(s_1 + s_2)$, these apparent selective advantages become

$$(s_1 + s_2)D^2/\{y^2(1-y)\} \quad \text{and} \quad (s_1 + s_2)D^2/\{y(1-y)^2\} \qquad (6.91)$$

respectively. These are non-negative so that in this case, if there is non-zero linkage disequilibrium between selected and neutral loci, apparent (or associative) over-dominance will exist at the neutral locus. Clearly the extent of this effect will depend on the value of D^2 (or, in the more general case where we cannot assume $x = s_2/(s_1 + s_2)$, on D and D^2). We now discuss how large this effect might be when linkage disequilibrium is generated from stochastic processes in finite populations.

The formula we have derived for $E(D^2)$ in the previous section assumes no selective effects at either locus and must therefore be generalized to cover the present model. This has been done by Ohta and Kimura (1970) assuming selection is so strong that $x = x^*$. It is found that

$$E(D^2) = \frac{x^*(1-x^*)E\{y(1-y)\}}{1 + 4N(R+u+v) + \dfrac{(1-2x^*)^2}{x^*(1-x^*)} \dfrac{N(R+2u+2v)}{1+N(R+2u+2v)}} . \qquad (6.92)$$

Here u and v are the mutation rates at the B locus. If R is not extremely small this expression is $0(N^{-1})$ and hence very small. We may thus expect little effect of associative overdominance in this case. Similarly, when there is no mutation and fixation at both loci eventually occurs, the effect of linkage disequilibrium, while perhaps initially non-trivial, eventually becomes negligibly small so that associated overdominance is, in this case, a transient phenomenon. Extensions of these conclusions to the case where several overdominant loci are linked to the neutral locus are given by Ohta and Kimura (1970).

We turn now to the concept of hitch-hiking. Hitch-hiking occurs when the gene frequencies at a neutral locus are affected by those at a linked selected locus where a favorable allele is proceeding towards fixation. As the name implies, we are mainly interested in the extent to which the frequency of one allele at the neutral locus increases through linkage to the favored allele. Aspects of this possibility have been examined by Maynard Smith and Haigh (1974), Haigh and Maynard Smith (1976), Ohta and Kimura (1975, 1976) and Thomson (1977).

Haigh and Maynard Smith consider a somewhat different question than do Ohta and Kimura. They assume an initial polymorphism at the neutral locus and, supposing that a substitution then occurs at the selected locus, focus attention on the expected final value of heterozygosity at the neutral locus when the substitution ceases. Ohta and Kimura, however, imagine a new mutation to arise at a neutral locus while a selected locus is substituting and consider the effect on the expected total heterozygosity at the neutral locus of the selected substitution. Which facet is the more relevant biologically is not clear, and both sets of authors argue for their own viewpoint.

The purely mathematical discussion is less controversial, and we consider first the analysis of Ohta and Kimura. In order to have a standard of reference we consider the model (1.43) which concerns a selectively neutral locus without reference to any linked loci. If a single A_1 mutant arises in an otherwise pure A_2A_2 population, the number of A_1 genes will be j on an average for $2j^{-1}$ generations (see Eq. (1.51)). This means, on average, that the total heterozygosity created by this mutation is

$$\sum_{j=1}^{2N-1} 2j(2N - j)(2N)^{-2} 2j^{-1} = 2.$$ (6.93)

We take this value as the standard against which values computed under hitch-hiking may be compared.

Suppose the A locus is selectively neutral while, at the B locus, the favored allele B_1 is steadily replacing B_2. We may assume to a reasonable approximation that this replacement is deterministic so that the frequency y of B_1 satisfies the differential equation

$$\frac{dy}{dt} = sy(1 - y).$$ (6.94)

(Note that, for convenience, this assumes no dominance at the B locus.) We denote by x_1 the frequency of A_1 among B_1 chromosomes and by x_2 the frequency of A_1 among B_2 chromosomes. The total frequency of A_1 is thus $x = yx_1 + (1 - y)x_2$, and our aim is to compute the expectation of

$$H = \int_0^\infty 2x(1 - x)dt.$$ (6.95)

Ohta and Kimura approach this problem by using (4.106). Differential equations for the expected value of x_1^2, $x_1 x_2$, and x_2^2 are found from (4.106) by successively using these functions for $g(\cdot)$. These equations must be solved numerically and the solutions inserted in (6.95) and will be given in terms of the initial values of x_1 and x_2, the selective coefficient s, the recombination fraction R between A and B loci and the value y_0 assumed for y when the initial mutation at the A locus takes place. Ohta and Kimura (1975) give separate values for $E(H)$ depending on whether the initial mutant lies on a B_1 or B_2 chromosome. For our purposes it is probably convenient to consider the weighted average

$$\overline{E}(H) = y_0 E_1(H) + (1 - y_0)E_2(H),$$ (6.96)

where $E_i(H)$ is the expected value of H assuming the initial mutant is on a B_i chromosome. In Table 6.4 we give values (computed from those of Ohta and Kimura, 1975) of $\overline{E}(H)$ for various values of s and y_0 for the values $R = 0.1, N = 100$. It will be seen that $\overline{E}(H)$ does not differ substantially from the value 2 (computed without taking linked loci into account), and from this point of view we may conclude, with Ohta

Table 6.4. Values of $\bar{E}(H)$ for various values of s and y_0 with $R = 0.1, N = 100$[a]

y_0	0.1	0.2	0.5
s 0.05	1.97	1.97	1.97
0.10	1.94	1.94	1.96

[a] See (6.96) and text for definitions.

and Kimura in this case, that hitch-hiking is of comparatively small importance in altering the value of total mean heterozygosity at the neutral locus. Although we have considered only one value for N and one value for R in Table 6.4, the general conclusion reached applies for a very wide range of R and N values and is appreciably in error only when $NR < 5$ and $Ns < 100$. The minimum expected value of H is about 1.2 and occurs when $NR \sim 0, Ns \approx 5$.

Maynard Smith and Haigh consider it biologically more relevant to consider the effect on an existing neutral polymorphism of a selective substitution and do so by comparing the expected final heterozygosity in this case with that existing before the selective substitution starts. They show that if $R \ll s \ll 1$ the ratio of the final heterozygosity H_∞ (when the selective substitution has taken place) to the initial heterozygosity H_0 is of the form

$$\text{const } Rs^{-1}, \tag{6.97}$$

where the constant depends on the initial gametic configuration. Under the assumptions made the quantity (6.97) is quite small. However, Ohta and Kimura (1975) compute H_∞/H_0 for a far wider range of R and s values and conclude that unless Ns is small (< 100, approximately) then $H_\infty/H_0 \sim 1$. In particular this is true if $R > s$, and thus in this case the effect of hitch-hiking is negligible.

6.8 The Evolutionary Advantage of Recombination

We have noted above that the mean fitness of a random-mating population is maximized when the recombination fraction between the loci we consider is zero. (This conclusion may be generalized to cover an arbitrary number of loci with an arbitrary number of alleles at each locus.) Why then have populations not evolved so that recombination does not exist? Even if we were to discount the use of mean fitness as a measure of success in intergroup competition and, further, assert that in any event recombination is determined more by the evolution of modifier genes than by such competition, we must still find an answer to this question since in the analyses we have so far mentioned, such modifiers often act so as to reduce recombination.

Our aim then is to find what advantage the existence of recombination might be for a population. In this context we shall mean sexual recombination: we do not consider the possible advantages of recombination in asexual populations. Thus, in what follows we assume the existence of two sexes with identical fitness patterns: a generalization of the theory of Sect. 2.3 shows that the frequency of any gamete will be the same in males and females so that in the quantitative discussion of recombination no explicit recognition of the existence of the two sexes is necessary.

The classical argument for the existence of sexual recombination is that of Fisher (1930a) and Muller (1932), namely that recombination favors the incorporation into the population of favorable new alleles arising at different loci, since recombination is more efficient in allowing such favored genes to occur in the same individual. A verbal discussion proceeds as follows. Suppose a favorable mutation A_1 arises at a locus A and begins to spread throughout a population. If a favorable mutation B_1 subsequently arises at a locus B, then without recombination A_1 and B_1 cannot both become simultaneously fixed unless the initial B_1 mutation happens to arise on an A_1 chromosome. This is unlikely to occur until the frequency of A_1 is substantial, and thus either the evolution at other loci is slowed down by the evolution of the A locus or the favorable mutation A_1 is lost through the increase in frequency of B_1 (and hence the linked allele A_2) at the B locus. With recombination, both A_1 and B_1 genes can eventually arise on the same chromosome so that evolution, under this argument, proceeds more rapidly than with no recombination.

This argument clearly assumes that ultimately the advantage to the population with recombination will arise through intergroup competition. (Recall that Fisher's original argument for decreased recombination also makes this assumption.) It would be fitting, in line with our discussion of the evolution of modifier genes, to attempt to produce an argument that does not rely ultimately on such an assumption: we mention such arguments later. We emphasize again that our aim is to compare systems with no recombination ($R = 0$) to those with positive recombination ($R > 0$): this is a different question to comparing two populations with positive recombinations R_1, R_2 respectively. It may well be that, since high recombination breaks up favorable gene complexes as well as creating them, the incorporation of favorable new mutants proceeds best, at least in some circumstances, in populations with low but positive recombination. To repeat, this is not the comparison that is being made.

The first attempt to quantify the Fisher-Muller theory was by Crow and Kimura (1965). Crow and Kimura assume a population in which favorable new mutations (with initial frequency N^{-1}) arise in a population of size $\frac{1}{2} N$ at rate NU per generation. The new favorable mutations are all at different loci, and the mutation is nonrecurrent. We may suppose for convenience that each new mutant has selective advantages with no dominance. While (see Sect. 1.4) most favorable new mutations will be lost from the population, we may expect an equal fraction to be lost with and without recombination so that such random loss can be ignored and all processes treated as deterministic. Suppose finally in a population without recombination that on average g generations pass between the occurrence of a favorable new mutation and the occurrence of a second favorable mutation in a descendant of the first. Then in such a population favorable new mutations are incorporated into the population at a rate of one per g generations.

In a population with recombination all favorable new mutations during g generations can be incorporated, and hence since NU favorable mutations arise per generation, NUg arise during g generations. As far as the rate of incorporation of favorable new mutations is concerned, then, the advantage of recombination is $NUg: 1$, and in order to discuss this ratio more usefully, it remains to find a formula for g in terms of N, U and s.

We have assumed no dominance and a selective advantage s to single mutants. The frequency x of individuals carrying a favorable new mutant is then given by (1.26) if we put $h = \frac{1}{2}$ and replace s by $2s$. Under the initial condition $x = N^{-1}$ the solution of (1.26) is clearly

$$x = [1 + (N - 1)e^{-st}]^{-1}. \tag{6.98}$$

Thus, in the first i generations after the occurrence of this mutation the total number of its descendents is

$$N \int_0^i x dt = Ns^{-1} \log \{(N - 1 + e^{si})/N\}. \tag{6.99}$$

The total number of favorable new mutations in these descendents is found by multiplying this quantity by U, and thus g is found as the solution of the equation

$$1 = NUs^{-1} \log \{N - 1 + e^{sg})/N\}.$$

This gives immediately

$$g = s^{-1} \log \{N(e^{s/NU} - 1) + 1\},$$

and thus the rate of incorporation of advantageous mutations in populations with recombination to those without, under this analysis, becomes

$$NUs^{-1} \log \{N(e^{s/NU} - 1) + 1\} : 1. \tag{6.100}$$

Several limiting cases of this formula are of interest. Suppose U is extremely small, so that favorable new mutations arise very rarely. We may then expect that each favorable mutation is incorporated in the population before the next arises, and in this case there is no advantage to recombination. This argument is confirmed by noting that the ratio (6.100) approaches unity as $U \to 0$. Similarly for very large s the incorporation of each new favorable mutant should be very rapid, and we again confirm that the ratio (6.100) approaches unity as $s \to \infty$. Clearly the situation when recombination is most favored is when U/s is large and N is large. Crow and Kimura (1965) give a table of values of (6.100) for various combinations of N, U and s values which document this.

This conclusion was challenged by Maynard Smith (1968) who produced a "counter-example" in which the existence of recombination made no difference to the rate at which two favorable alleles increased in frequency. Maynard Smith considered unfavorable alleles at two loci which are maintained at low frequency in a population through recurrent mutation and showed that the gamete frequencies would then be in linkage equilibrium. Suppose now that the environment alters and that both rare alleles are favored, with a multiplicative fitness scheme of the form (6.28), and steadily increase in frequency. We have shown that with a multiplicative fitness scheme a population having zero linkage disequilibrium initially will persist in a state of zero linkage disequilibrium. In this case the value of the recombination fraction R is irrelevant to the rate of increase of frequency of the two alleles since, as Eqs. (2.73) show, R appears only as a multiplier of the coefficient of linkage disequilibrium. Hence the rate of incorporation of the rare alleles is unaffected by the existence of recombination.

As pointed out by Maynard Smith (1968) and Crow and Kimura (1969), there is an essential difference between the assumptions made in the two analyses. Maynard Smith assumes that gametes carrying both initially unfavored alleles exist at positive frequency. Crow and Kimura's analysis assumes zero initial frequency for such gametes and, as is clear from (2.73) with $R = 0$ or by general reasoning, such gametes can never arise in a population without recombination if their initial frequency is zero. More generally, Maynard Smith asserts that the essential difference between the two analyses arises because in his analysis favorable new alleles arise in a recurrent process whereas in that of Crow and Kimura they arise uniquely. Clearly if favorable new alleles do arise recurrently at a sufficiently high frequency, then even without recombination a favored new mutation at the B locus can arise on a chromosome carrying a favored mutation at the A locus in the course of fixation.

Crow and Kimura (1969) state that their argument does not assume unique favorable mutants, but rather that these occur sufficiently rarely so that double mutants arise very seldom (or, more generally, if many loci are substituting simultaneously, that n-tuple mutants arise with completely negligible frequency). Thus the real essence of their argument remains unchanged. Maynard Smith (1971) carries out an analysis dropping the assumption of zero initial linkage disequilibrium but incorporating a recurrent mutation rate of favorable mutations and concludes that for moderate populations ($N \approx 10^6$) there is little advantage to recombination, while for large populations ($N \approx 10^{10}$) populations with recombination can incorporate favorable new mutations about four or five times faster than populations without recombination.

It is clear that the final answers to these questions rely on biological arguments concerning the most likely circumstances from which a micro-evolutionary process begins and, more generally, on the main nature of evolution. If evolution is mainly of the "shifting balance" type of Wright, outlined in Chap. 1, where gene frequencies are high throughout, the Crow-Kimura argument does not apply. If, however, evolution depends more on the incorporation of rare favorable mutations, their argument is much more important. The extent to which this is so will depend on the rate of occurrence of favorable new mutations, the population size and the selective advantage of the new mutant.

A number of interesting quantitative conclusions have been found when mutation to favorable new alleles is recurrent. Thus when the double mutant is more fit than multiplicative fitnesses would imply, Eshel and Feldman (1970) show that under no recombination the frequency of the double mutant gamete is always larger than it is with positive recombination, provided that the initial disequilibrium $c_1 c_4 - c_2 c_3$ is non-negative (here c_1 is the initial frequency of the double mutant gamete). They further show that when single mutants are deleterious but the double mutant advantageous, for suitable fitness values and sufficiently low mutation rate the two mutants will increase in frequency only if the linkage between the two loci is sufficiently tight. (This conclusion is essentially given also by Crow and Kimura, 1965.) Karlin (1973) considers stochastic versions of the process, paying particular attention to the mean time until the first double mutant gamete is formed and the mean time until fixation of the double mutant gamete.

All the above arguments concern long-term optimization and ultimately rely on intergroup competition for the establishment of the population with the optimal recombination value. Short-term arguments have been offered by Williams (1966, 1975) and Williams and Mitton (1973). These center around the claim that under intense selection, populations having high recombination have an immediate advantage over populations with no recombination because they produce more high-fitness genotypes (which are assumed to be the only genotype to survive). This argument has been contested by Maynard Smith (1971). Felsenstein and Yokoyama (1976) introduce a locus which modifies recombination between primary loci and discuss verbally and by simulation the fate of the allele causing no recombination as favorable mutations arise and become fixed at the primary loci. This argument is in the spirit of Sect. 6.5 and avoids group-competition arguments. Unfortunately the complexities of the argument make a mathematical analysis well-nigh impossible. The argument relies on the existence of randomly generated linkage disequilibrium in finite populations, and thus the analysis of Sect. 6.6, which suggests that unless population sizes are small such linkage disequilibrium is rarely large, becomes relevant.

6.9 Summary

In the preceding sections we have outlined several conclusions concerning the joint evolution at two linked loci. A number of topics have not been discussed. These include the effect of population subdivision (Feldman and Christiansen, 1975; Nei and Li, 1973), the effect of allowing several alleles at one locus (Feldman et al., 1975) and the effect of different recombination fractions in the two sexes (Strobeck, 1974). We cannot hope to cover here all possible extensions and generalizations. Far and away the most important question concerns the degree of linkage disequilibrium in populations. We have noted above (and will note again in the next chapter) that if linkage disequilibrium can normally be taken as negligibly small, a great simplification can be made to the theory. Loci can essentially be examined one by one with interactive effects between loci being of minor importance. Ohta and Kimura (1975), for example, claim that linkage disequilibrium in nature is comparatively rare and that such simplifying assumptions, which allow us to carry the theory a considerable distance, can reasonably be made. Thus they assert that "for large and stable populations the concept of quasi-linkage equilibrium together with the single locus theory is sufficient to treat most problems realistically." Lewontin (1974), on the other hand, emphasizes the role of linkage and linkage disequilibrium in evolution, and Wright, as we have noted, also emphasizes interactive effects of loci. Under the latter view evolution is far more complex and its quantitative assessment extremely difficult. It cannot be claimed that a weight of evidence has yet accumulated on either side. In the next chapter we carry the theory to many loci and note the additional complications (compared to those in a two-locus analysis) that then arise.

7. Many Loci

7.1 Introduction

Our aim in this chapter is to outline certain properties of populations when it is assumed that the various characteristics (and in particular the fitness) of any individual depend on his genetic constitution at an arbitrary (and often large) number of loci. As far as is possible, this will be done without making special assumptions about the numerical values of these characteristics or the recombination pattern between the loci determining them. However, more progress can be made when particular forms are assumed for these values, so we consider here general and specific forms and base our overall conclusions on the results deriving from both.

This chapter has two intertwining themes. The first concerns the relationship between properties of the entire multi-locus system considered and those of the various subsystems (in particular single-locus subsystems) that this defines. The second concerns linkage disequilibrium and its effect on static and dynamic properties of the population. The former theme is of interest because, while many properties of individuals depend on genes at a number of loci, experiments often involve one or a small number of loci, and it is clearly important to assess the extent to which valid inferences can be made from the loci investigated to the entire system involved. We shall find that to a great extent the validity of these inferences depends on the amount of linkage disequilibrium: for highly "architectured" systems possessing high linkage disequilibrium such inferences can be of dubious value, while for "unarchitectured" systems with little or no disequilibrium the inferences are more likely to be valid.

A large literature on multi-locus theory is now becoming available. We do not aim to cover this here: for generalizations and extensions of the theory covered here and for many further multi-locus results see Karlin (1979).

7.2 Notation

Since the notation for multi-locus systems can become remarkably confusing, we collect here most of the notation that will be used in this chapter: note that this notation does not necessarily conform to that used in other chapters.

We suppose the entire genetic system considered to consist of K loci, the generic symbol used for a locus being k (and on occasions where two loci are considered simultaneously, k and l). Thus k takes the values $1, 2, ..., K$: as far as possible we use upper-case symbols for fixed quantities (such as the number of loci in a system or the number of gametes these loci define) and the corresponding lower-case symbol for generic or typical values. We suppose there are I_k alleles possible at locus k, so that there will exist $I = I_1 I_2 \ldots I_K$ different K-locus gametes, which are assumed labelled in some agreed fashion $1, 2, \ldots, i, \ldots, I$.

We are also interested in one-, two-, and in general G-locus subsystems of the entire K-locus system. The alleles at locus k are labelled $A_{k1}, A_{k2}, \ldots, A_{kI_k}$. Two-locus gametes are described by the alleles at the two loci they define, for example (A_{ku}, A_{lv}). For general G-locus systems there will exist $Q = \Pi I_k$ G-locus gametes, where the product is taken over all G loci in the subsystem. These are also assumed to be labelled in some agreed fashion $1, 2, \ldots, q, \ldots, Q$.

We now turn to the frequencies of the various alleles and gametes and here consistently use the notation x for gene frequencies, y for frequencies of two-locus gametes, z for frequencies of G-locus gametes and c for frequencies of K-locus gametes. More explicitly we have

$$
\begin{aligned}
x_{ku} &= \text{frequency of the allele } A_{ku}, \\
y_{ku,lv} &= \text{frequency of the two-locus gamete } (A_{ku}, A_{lv}). \\
z_q &= \text{frequency of the } q\text{th } G\text{-locus gamete}, \\
c_i &= \text{frequency of the } i\text{th } K\text{-locus gamete}.
\end{aligned}
\tag{7.1}
$$

When referring to two or more G-locus gametes we use suffixes p, q and r; when referring to two or more K-locus gametes we use suffixes h, i and j. Naturally, the frequencies x_{ku}, $y_{ku,lv}$ and z_q can be found by appropriate summation of the c_i: thus, for example,

$$
z_q = \sum_{i \in S_q} c_i,
\tag{7.2}
$$

where S_q is the set of all K-locus gametes containing the same alleles at the loci of the G-locus system as the qth G-locus gamete.

The concept of linkage disequilibrium was introduced in Chap. 2 for two-locus, two-allele systems. We use the symbol D for two-locus disequilibria: thus

$$
D_{ku,lv} = y_{ku,lv} - x_{ku} x_{lv}.
\tag{7.3}
$$

Higher-order linkage disequilibria have been defined by Geiringer (1944), Bennett (1954) and Slatkin (1972). Although linkage disequilibrium is a major concern of this chapter, we will not introduce these measures here. We note, however, that if the frequency of every K-locus gamete is the product of the frequencies of its constituent alleles, all these measures are zero and that large linkage disequilibrium values imply that gametic frequencies cannot be found, even approximately, as products of the corresponding allele frequencies.

Each pair of K-locus gametes defines a genotype: that defined by gametes i and j is denoted (i, j). The value of some measured characteristic of an (i, j) individual is written m_{ij}: note that it is assumed that there is no environmentally caused variation in this measurement. A particular case is the fitness of such an individual, given the special notation w_{ij}. The marginal value m_i of gamete i for the character in question is defined by

$$m_i = \sum_j c_j m_{ij}, \qquad (7.4)$$

and the mean value $\bar{m}$ for the entire population is

$$\bar{m} = \sum c_i m_i = \sum \sum c_i c_j m_{ij} \qquad (7.5)$$

assuming (as we do) random mating. The total genetic variance σ_T^2 for the character is

$$\sigma_T^2 = \sum \sum c_i c_j (m_{ij} - \bar{m})^2, \qquad (7.6)$$

and the "additive" component of this, defined in the following section, is denoted σ_A^2.

Similarly, each pair of G-locus gametes defines a G-locus genotype, the (marginal) value m_{pq} for an individual (p, q) being defined, as in (2.78), by averaging over K-locus genotypes. Explicitly,

$$\bar{m}_{pq} = \sum_{i \in S_p} \sum_{j \in S_q} c_i c_j \, m_{ij}/z_p z_q. \qquad (7.7)$$

The marginal value for gamete p is

$$\bar{m}_p = \sum_q \bar{m}_{pq} = \sum_{i \in S_p} \sum_j c_i c_j m_{ij}/z_p \qquad (7.8)$$

and it follows that

$$\sum_p z_p \bar{m}_p = \sum_i \sum_j c_i c_j m_{ij} = \bar{m}. \qquad (7.9)$$

The total variance in the character for the G-locus subsystem is

$$\sum \sum z_p z_q (\bar{m}_{pq} - \bar{m})^2, \qquad (7.10)$$

and this can also be divided into additive and non-additive components. This is true in particular when $G = 1$ so that the z_p are the gene frequencies $x_{k1}, x_{k2}, \ldots$ at a single locus. Since the single-locus case is particularly important, we now exhibit the variance partition for it. The marginal value for the genotype $A_{ku}A_{kv}$ is defined (as in 7.7) by

$$\bar{m}_{ku,kv} = \sum_{i \in S_{ku}} \sum_{j \in S_{kv}} c_i c_j m_{ij}/x_{ku} x_{kv}. \qquad (7.11)$$

From this we may compute the marginal value for A_{ku} as

$$\bar{m}_{ku} = \Sigma \, \bar{m}_{ku,kv} x_{kv} \qquad (7.12)$$

and the additive effect of A_{ku} by

$$a_{ku} = \bar{m}_{ku} - \bar{m}. \qquad (7.13)$$

Then the total k-locus variance for the character, namley

$$\Sigma \Sigma \, x_{ku} x_{kv} (\bar{m}_{ku,kv} - \bar{m})^2, \qquad (7.14)$$
$$u \quad v$$

may be partitioned into an additive component $\sigma_A^2(k)$ and a dominance component $\sigma_D^2(k)$ defined by

$$\sigma_A^2(k) = 2\Sigma \, x_{ku} a_{ku}^2, \qquad (7.15)$$
$$u$$

$$\sigma_D^2(k) = \Sigma \Sigma x_{ku} x_{kv} d_{k,uv}^2 \qquad (7.16)$$
$$u \quad v$$

where

$$d_{k,uv} = \bar{m}_{ku,kv} - \bar{m}_{ku} - \bar{m}_{kv} + \bar{m}. \qquad (7.17)$$

For $G > 1$, additive x additive and other higher-order components of variance arise: these are of interest only in Sect. 7.6, and we defer further discussion of them until that section.

7.3 Recurrence Relations for Gametic Frequencies

The genetic evolution of a random-mating population is described most efficiently by recurrence relations for the various K-locus gamete frequencies. These relations generalize (2.73) and depend on genotypic fitnesses and the recombination pattern between loci. We define the latter through the function $f(i, j \rightarrow h)$, defined as the probability that a randomly chosen one of the two gametes formed by recombination in gametes i and j is gamete h. Then if c_i, c_i' are the frequencies of gamete i in consecutive generations,

$$\bar{w} c_i' = w_i c_i - \Sigma^{(1)} w_{ij} c_i c_j f(i, j \rightarrow h) + \Sigma^{(2)} \, w_{hj} c_h c_j f(h, j \rightarrow i). \qquad (7.18)$$

Here $\Sigma^{(1)}$ denotes summation over all gametes h and j with $i, j \neq h$ and $\Sigma^{(2)}$ denotes summation over all gametes h and j with $h, j \neq i$. Summation of (7.18) over all i in S_p yields

$$\bar{w} z_p' = \bar{w}_p z_p - \Sigma^{(1)} \bar{w}_{pq} z_p z_q f(p, q \rightarrow r) + \Sigma^{(2)} \bar{w}_{rq} z_r z_q f(r, q \rightarrow p) \qquad (7.19)$$

where $\Sigma^{(1)}$ and $\Sigma^{(2)}$ have meanings parallel to those just given. The similarity in form between (7.18) and (7.19) shows that the recurrence relations for G-locus gametes written down by formal analogy with (7.18) do indeed provide the correct recurrence values, but with one restriction: the fitnesses $\bar{w}_{pq}$ (unlike the w_{ij}) are not fixed but normally change from generation to generation, so that (7.19) can be used to predict G-locus gametic frequencies only one generation in advance. For long-term predictions the full system (7.18) must be used.

We have noted in Chap. 2 that for systems where fitness depends on the alleles at two loci, decreases in mean fitness can occur and the Fundamental Theorem is not valid. If, however, the loci involved are in linkage equilibrium, this cannot happen: setting $c_1 c_4 - c_2 c_3 = 0$ in (2.73) yields the "single-locus type" Eq. (2.7), and this implies non-decrease in mean fitness, at least for one generation. The same holds for many loci: if all linkage disequilibria of all orders are zero, (7.18) will also reduce to single-locus type equations and the same conclusion holds. We emphasize that this is a condition on gamete frequencies, not the fitnesses w_{ij}.

We turn now to equilibrium behavior. Note that $c_i' = c_i$ implies $z_q' = z_q$, a conclusion that is put more usefully in contrapositive form: if $z_q' \neq z_q$ then the full K-locus system cannot be in equilibrium. Note that this is a conclusion concerning the full system from a subset of loci. In particular if $G = 1$ and only two alleles A_{k1} and A_{k2} can occur at locus k, the full system cannot be in equilibrium unless the marginal fitness of the heterozygote $A_{k1} A_{k2}$ is outside the range of those of the homozygotes $A_{k1} A_{k1}$ and $A_{k2} A_{k2}$ and unless further the frequency of A_{k1} is at the value predicted by an equation of the form of (1.30).

However, the equations $z_q' = z_q$ $(q = 1, 2, ..., Q)$ do not necessarily imply that the full K-locus system is in equilibrium. Further, if indeed $c_i' = c_i$, the stability of this equilibrium cannot necessarily be gauged by the $\bar{w}_{pq}$: it is possible for the $\bar{w}_{pq}$ to suggest stability of the K-locus equilibrium and yet for that equilibrium to be unstable. For examples see Ewens and Thomson (1977): clearly subsystem behavior does not give necessarily correct information about the full system.

7.4 Components of Variance

In this section we compute the "additive genetic" and the "additive gametic" components of the variance (7.6). We show that the two are identical and then consider their relation to the sum of the single locus additive variances.

The additive genetic variance was defined for one locus (and two alleles) in (1.9) and for two loci in (2.84), while the additive gametic variance was defined after (2.88). We consider first the additive gametic variance. For an arbitrary number of loci, the total gametic variance for any character is defined as

$$2 \Sigma c_i(m_i - \bar{m})^2 \tag{7.20}$$

and is a measure of the differences in marginal gametic values for this measurement. The additive gametic variance measures the extent to which this variance can be accounted for by additive effects of genes. We attach an additive parameter α_{ku} to the allele A_{ku} where for each k the α_{ku} are subject to the constraints

$$\sum_u \alpha_{ku} x_{ku} = 0. \tag{7.21}$$

Subject to (7.21) we now minimize

$$2 \sum_i c_i \{m_i - \bar{m} - \sum \alpha_{ku}\}^2, \tag{7.22}$$

the inner summation being over all alleles in gamete i. The minimizing values $\hat{\alpha}_{ku}$ of the α_{ku} satisfy the equations

$$\Sigma^{(ku)} c_i (m_i - \bar{m}) = x_{ku} \hat{\alpha}_{ku} + \sum_{\substack{l,v \\ l \neq k}} y_{ku,lv} \hat{\alpha}_{lv}. \tag{7.23}$$

where $\Sigma^{(ku)}$ implies summation over all gametes containing the allele A_{ku}. Except in degenerate cases, (7.23) is a system of linear equations for the $\hat{\alpha}_{ku}$ with a unique solution. Standard least-squares theory now shows that the additive gametic variance is

$$2 \sum_{k,u} \{ \Sigma^{(ku)} c_i (m_i - \bar{m}) \hat{\alpha}_{ku} \}. \tag{7.24}$$

In a similar fashion, the additive genetic variance is found by attempting to account for genotypic values as far as possible by the additive effects of alleles. Subject to (7.21) we attempt to minimize

$$2 \sum_i \sum_j c_i c_j \{m_{ij} - \bar{m} - \sum \alpha_{ku}\}^2, \tag{7.25}$$

the inner summation being over all alleles in the genotype (i, j). (If A_{lv} occurs twice in any genotype, the contribution α_{lv} is counted twice.) The minimizing values $\hat{\alpha}_{ku}$ can again be computed, and it is found (Ewens and Thomson, 1977) that these also satisfy (7.23). The sum of squares removed again reduces to (7.24), and thus the additive genetic and additive gametic variances are equal. To find either we thus compute whichever is easier in any given circumstance and note that the identity of the two reinforces the view (Crow and Kimura, 1970) that the expression "genic variance" should be used for both.

It is not, in general, easy to determine the difference between the true additive genetic variance, given by (7.24), and the sum $\sum_k \sigma_A^2(k)$ of the single-locus values, where $\sigma_A^2(k)$ is defined in (7.15). In general the two values are not equal. If, however, all possible two locus linkage disequilibria are zero, so that

$$y_{ku,lv} = x_{ku} x_{lv} \tag{7.26}$$

for all k, l, u, v, then equality does hold. This can be seen immediately from (7.23): when (7.26) holds, the second term on the right-hand side of (7.23) is zero, because of

(7.21), so that $\hat{\alpha}_{ku} = a_{ku}$ and (7.24) reduces immediately to $\Sigma \sigma_A^2(k)$. When two alleles only are possible at each locus and the measurements m_{ij} can be expressed as the sum of single-locus genotypic contributions, Avery and Hill (1978) show that the true additive genetic variance is

$$\sigma_A^2 = \sum_k \sigma_A^2(k) + 4 \sum_{k<l} \sum (a_{k1} - a_{k2})(a_{l1} - a_{l2}) D_{k1,l1}, \qquad (7.27)$$

where $D_{k1,l1}$ is the coefficient of linkage disequilibrium between loci k and l. We check that $D_{k1,l1} = 0$ for all k and l implies identity of the true additive genetic variance and the sum of the single-locus values.

The terms in the second sum in (7.27) can be both positive and negative, and thus considerable cancellation is possible in the summation. However, this need not occur in highly architectural genetic situations when the second term can dominate the first term, and it is of some interest to assess the circumstances under which each of these alternatives occur. Important progress has been made on this point by Bulmer (1976). Bulmer simulated a genetic system in which the value of a given character is determined by the alleles at twelve loci (as well as an independent environmental component), the contributions being additive across loci with $A_{k1}A_{k1}$, $A_{k1}A_{k2}$ and $A_{k2}A_{k2}$ contributing 0, 1 and 2 respectively to the measurement. The genetic contribution to the character thus ranges from 0 to 24, and there are no dominance contributions to the total genetic variance.

We are interested here in the effects in random-mating populations of stabilizing, disruptive and directional selection schemes on the additive variance σ_A^2 and its two components as given in (7.27). If x_{k1} and x_{k2} are both moderate, $\sigma_A^2(k)$ is approximately 0.4 or 0.5 and thus $\Sigma \sigma_A^2(k)$ is approximately 5 or 6. Under both stabilizing and symmetrical disruptive selection $\Sigma \sigma_A^2(k)$ takes values of this order and provides a useful standard against which the value of the second term on the right-hand side of (7.27) can be compared. Under stabilizing selection this term is approximately -1.5 or -2.0 for the simulations of Bulmer: thus linkage disequilibrium lowers the true additive genetic variance somewhat from the value calculated without disequilibrium. However, under disruptive selection this term is extremely large, being ten or eleven times larger than $\Sigma \sigma_A^2(k)$ itself. It is clear why this is so: under strong disruptive selection two gametes — one consisting entirely of A_{k1} genes, ($k = 1, 2, ..., 12$) the other, of A_{k2} genes ($k = 1, 2, ..., 12$) — occur in high frequency, and the genetic system acts very much like a single-locus system with two alleles having values 0, 12 and 24 for the three genotypes. For such a system the additive genetic variance is 72 if the two alleles are of equal frequency, and the contribution to variance of the linkage disequilibrium terms makes up most of the difference between this value and that given by $\Sigma \sigma_A^2(k)$.

The relation between σ_A^2 and $\Sigma \sigma_A^2(k)$ under directional selection is not so easily arrived at intuitively. Bulmer shows that for his simulations $\Sigma \sigma_A^2(k)$ is approximately 25% less than σ_A^2 during the rather small number of generations before fixation of the favoured allele at each locus.

7.5 Particular Models

It is clear from the above that the degree of linkage disequilibrium in a genetic system influences considerably the extent to which multi-locus properties of the system can be determined from a consideration of single-locus properties. It is thus important to assess the extent to which linkage disequilibria will arise in natural populations, particularly at equilibrium, and to make a partial assessment of this we now consider equilibrium properties for certain special fitness models. Note that the only character we consider here is fitness. Since the models considered often have special properties (e.g. of symmetry), some caution is necessary in drawing inferences from them: we gain some generality by considering four different models, noting in particular when the same inference is suggested by all four.

Three of these models require the definition of a "fitness contribution" from each locus given, for the genotype $A_{ku}A_{kv}$ at locus k, by

$$w_k(u, v). \tag{7.28}$$

We assume the w_k (u, v) are such that for a single-locus system with fitness parameters (7.28) there exists a unique internal stable equilibrium point with the frequency of A_{ku} being $\hat{x}_{ku}$.

Assume now the w_k (u, v) are used in some way to define fitnesses for K-locus genotypes. Any equilibrium point of such a system for which the frequency of the gamete $(A_{1u_1}, A_{2u_2}, A_{3u_3}, \ldots)$ is $\hat{x}_{1u_1}\hat{x}_{2u_2}\hat{x}_{3u_3} \ldots$ is called a "product" equilibrium: we will be particularly interested in stability conditions for such equilibria. All coefficients of linkage disequilibrium, of all orders, are zero at such an equilibrium. Note that Roux (1974) and Karlin (1977a) have called such equilibria "Hardy-Weinberg" equilibria, but we prefer here the term "product" to avoid confusion with the slightly different single-locus usage for the term "Hardy-Weinberg".

Suppose first that, in the K-locus system, the fitness of any individual is in the additive form

$$\sum_k w_k(u, v) \tag{7.29}$$

where the sum is taken over his genotypes at all loci. An example of such a scheme (with different notation) is given in (6.21): the notational equivalence is

$$\alpha_1 = w_1(1, 1), \quad \alpha_2 = w_1(1, 2), \quad \alpha_3 = w_1(2, 2),$$
$$\beta_1 = w_2(1, 1), \quad \beta_2 = w_2(1, 2), \quad \beta_3 = w_2(2, 2). \tag{7.30}$$

A generalization of the discussion following (6.21) shows that mean fitness depends on gene frequencies only and is thus non-decreasing from generation to generation: this is probably the only broad model of practical interest for which the Fundamental Theorem holds for an arbitrary value of K. Further (Karlin and Liberman, 1978), the product equilibrium is the only equilibrium of the K-locus system for non-zero recombination rates, and it is globally stable. At this equilibrium the additive genetic variance in any character can be found as the sum of single-locus values, but even though fitnesses are additive over loci, this is not generally true for the character "fitness" for non-equilibrium values.

For the second model it is supposed that fitness is in the multiplicative form

$$\prod_k w_k(u, v). \tag{7.31}$$

This scheme generalizes (6.28), to which it reduces with the identifications (7.30). Considerable speculation on the behavior of real genetic systems has followed from investigation of this model, and we therefore consider its properties in some detail.

Note first that the product equilibrium exists for this model but is not a maximum point of mean fitness. Thus mean fitness can decrease in the neighborhood of the equilibrium, and the Fundamental Theorem fails. We show later that the mean fitness at equilibria other than the product equilibrium is rather higher than at this point.

A completely general analysis is difficult and unrevealing so we consider a simplified case. Suppose the recombination fraction between adjoining loci is R, that there is no interference, and that for all k, u and v

$$w_k(u, u) = 1 - a, \quad w_k(u, v) = 1, \quad (u \neq v) \tag{7.32}$$

where $a > 0$. For $K = 2$ this is a particular case of (6.28), and (6.33) shows that the product equilibrium is stable only if

$$R > a^2/4. \tag{7.33}$$

When (7.33) is violated there exist complementary pairs of stable equilibria with

$$\text{freq} (A_{11}A_{21}) = \text{freq} (A_{12}A_{22}) = \tfrac{1}{4} \pm \tfrac{1}{4}\{1 - 4R/a^2\}^{1/2},$$
$$\text{freq} (A_{11}A_{22}) = \text{freq} (A_{12}A_{21}) = \tfrac{1}{4} \mp \tfrac{1}{4}\{1 - 4R/a^2\}^{1/2}, \tag{7.34}$$

$$D = \tfrac{1}{4}\{1 - 4R/a^2\}^{1/2}. \tag{7.35}$$

For $K = 3$, Feldman et al. (1974) show that the product equilibrium is stable whenever (7.33) holds; when (7.33) does not hold, there exist four stable equilibria analogous to (7.34). Curiously, for a small range of R values ($0.01 < R < 0.0104272$ when $a = 0.2$), in excess of the bound $a^2/4$, these equilibria continue to be stable, but for sufficiently large R ($R > 0.0104272$ when $a = 0.2$) the product equilibrium is the only stable equilibrium.

For $K = 5$, Lewontin (1964a, b) showed by simulation for $a = 0.5$ that whereas the product equilibrium is stable for $R > 0.0625$, the bound given by (7.33), there exists a number of linkage disequilibrium stable equilibria for $0 < R < 0.065$. Note again the small interval of R values for which "$D = 0$" and "$D \neq 0$" stable equilibria exist. For $K = 36$, Franklin and Lewontin (1970) show by simulation that when $a = 0.1$ a large number of stable equilibria exists when $0 < R < 0.0025$; note that

0.0025 is the bound given by (7.33). All of these are in linkage disequilibrium. For 0.0025 $< R <$ 0.01 approximately, these stable equilibria persist with a stable product equilibrium also, while for $R >$ 0.01 approximately, only the product equilibrium is stable. These conclusions jointly suggest that the range of R values for which there exist linkage disequilibrium stable equilibria increases steadily over the bound $a^2/4$ as the number of loci in the system increases. However, a very powerful theorem of Roux (1974) shows that, for *any* multiplicative fitness scheme, the conditions on recombination values that ensure stability of the pairwise product equilibrium for all adjacent loci is sufficient to ensure stability of the product equilibrium in the complete K-locus multiplicative system. In particular, in the present example, (7.33) is sufficient for the stability of the K-locus product equilibrium for all K. This is a most important conclusion and, in conjunction with the simulation conclusions of Franklin and Lewontin, suggests that for very large K a rather wide range of R values will exist for which stable linkage equilibrium and linkage disequilibrium equilibria occur. Note in passing the Roux's theorem requires (7.33) and its generalizations to hold for *all* pairs of adjacent loci: for $K = 3$, $a = 0.2$ (for which (7.33) becomes $R >$ 0.01) Feldman et al. (1974) show that the product equilibrium is not stable if the recombination fraction between loci 1 and loci 2 is 0.0099 and between loci 2 and 3 is 0.0103.

Lewontin (1964a) and subsequently Franklin and Lewontin (1970) noticed two further important properties of stable points of 5- and 36-locus systems respectively. The first is that loci far apart on the chromosome can be held in linkage disequilibrium when the recombination fraction between them is considerably in excess of the limit set by (7.33). This occurs because of loci in the system segregating between these end loci. Thus, for example, when $K = 5$ the recombination fraction between loci 1 and 5 is approximately $4R$ and yet, for $a = 0.5$, these loci can be in linkage disequilibrium for $R = 0.065$ ($4R = 0.26$). Second, the value of D for adjacent interior loci is greater than the value predicted by (7.35). This effect is most marked for large values of R. Thus in the 5 locus case the equilibrium value of D between loci 2 and 3 is about 1.06 times as large as the values predicted by (7.35) for $R = 0.01$ and about 2.91 times as large for $R = 0.06$. In the 36-locus case this effect is even more noticeable. Thus for $a = 0.1$, $R = 0.002$ the value of D from (7.35) is 0.112 whereas Franklin and Lewontin found an average value of $|D|$ for adjacent loci of 0.22. Note that since necessarily $D < 0.25$, this is an extremely large value. For $R >$ 0.0025, two-locus theory does not predict stable linkage disequilibrium equilibria, and yet for $R = 0.004$ the average value of $|D|$ for adjacent loci was found to be as high as 0.185. The average of D^2 for all pairs of loci correspondingly decreases from about 0.05 at $R = 0.0027$ to 0.025 at $R = 0.007$ and essentially zero for $R \geqslant 0.01$.

These latter observations (for $K = 36$) arise because at the equilibria of the system investigated for small R the equilibrium gametic frequencies arise in a highly "architectured" form with two gametes each having frequency of about 0.4 and all 10^9 remaining gametes having total frequency of about 0.2. The two high frequency gametes are "complementary" in that, for the great majority of loci, they carry the alternative forms of the alleles at each locus. (If stochastic loss of alleles at some loci had not occurred, these gametes would be perfectly complementary.) This suggests complex and interesting behavior for equilibria of multiplicative systems for large K, an argument which is taken up in the final section of this chapter.

A further property of highly architectured systems is that the mean fitness is considerably higher than at a product equilibrium. In the present model the mean fitness at a product equilibrium is $(1 - \frac{1}{2}a)^K$, which is 0.158 for $a = 0.1, K = 36$. For equilibria with two almost complementary gametes present in high frequencies, Franklin and Lewontin (1970) found mean fitnesses in excess of 0.4. Note that if only two complementary gametes occur, each with frequency 0.5, each individual is equally likely to be a complete homozygote (fitness = 0.936 = 0.0225) or a complete heterozygote (fitness 1), leading to a mean fitness of 0.511. Of course with recombination between loci, this cannot an equilibrium value of mean fitness, but the equilibrium value will not be much less.

The third model we consider is the "generalized non-epistatic" model of Karlin and Liberman (1978). Here the fitness of any individual is a linear combination of various multiplicative, additive and neutral components, so that this model can be thought of as being intermediate between the two just considered. We consider in detail only the two-locus case in which the fitness of the genotype $A_{1u}A_{1v}A_{2s}A_{2t}$ is

$$b_1 w_1(u, v)w_2(s, t) + b_2 w_1(u, v) + b_3 w_1(s, t) + b_4. \tag{7.36}$$

The requirement that the product equilibrium be stable is that the recombination fraction between the loci exceed

$$\max_{a, b} \frac{b_1 \overline{w}_A \overline{w}_B \lambda_a \mu_b}{b_1 \overline{w}_A \overline{w}_B (1 - \lambda_a)(1 - \mu_b) + b_2 \overline{w}_A (1 - \lambda_a) + b_3 \overline{w}_B (1 - \mu_b) + b_4} \tag{7.37}$$

where the notation is that of (6.36) and (6.37) with

$$a_{uv} = w_1(u, v), \quad b_{uv} = w_2(u, v), \quad A_u = A_{1u}, \quad B_s = A_{2s}$$

and $\lambda_1, \lambda_2, \ldots, \mu_1, \mu_2, \ldots$ are the non-unit eigenvalues of $\{c_{uv}\}, \{d_{uv}\}$ respectively. When $b_2 = b_3 = b_4 = 0$ (so that the model is multiplicative) this requirement reduces to (6.38) while if $b_1 = 0$ (so that the model is additive) it reduces to the known requirement $R > 0$. The condition on the value of R for stability of the product equilibrium is clearly less stringent than the corresponding condition for the multiplicative case with $b_2 = b_3 = b_4 = 0$.

The requirement (7.37) can be extended to an arbitrary number of loci, and an explicit condition (Karlin and Liberman, 1978) can be found for stability of the product equilibrium for general non-epistatic schemes which generalizes the condition (20) of Roux (1974) for purely multiplicative schemes. Although these conditions are not simple, two important conclusions emerge. First, the higher the "mix" of additive components in the fitness, as compared to "multiplicative", the less restrictive are the requirements on recombination for stability of the product equilibrium: this generalizes the discussion below (7.37). Second, stability of the product equilibrium obtains if there is "enough" recombination, whatever the mix may be, and in particular if all loci are unlinked the product equilibrium is stable for any generalized non-epistatic scheme.

We turn finally to a fitness scheme not defined in terms of the $w_k(u, v)$. If, in the two-locus model (6.29), the parameters β and γ are equal, the fitness of any individual depends solely on the number of loci at which he is heterozygous. We consider now the generalization of such a scheme to an arbitrary number K of loci, assuming 2 alleles possible at each locus and that the fitness of an individual heterozygous at k loci is γ_k ($k = 0, 1, 2, ..., K$). All results given below for this model were obtained by Karlin (1977a).

By symmetry, the frequency of each gamete at the product equilibrium is 2^{-K}. If this equilibrium is stable for zero recombination, it is the unique (and globally stable) equilibrium for all recombination patterns. The conditions for stability with no recombination is

$$\sum_{k=0}^{K} \gamma_k \sum_{m=0}^{k} (-1)^m \binom{n}{m} \binom{K-n}{k-m} < 0, \quad n = 1, 2, ..., K \tag{7.38}$$

and for free recombination between all loci is

$$\sum_{k=0}^{K} \gamma_k \left\{ 2 \binom{K-1}{k} - \binom{K}{k} \right\} < 0. \tag{7.39}$$

In line with the discussion above, if (7.38) hold then automatically (7.39) holds. But (7.39) can hold and (7.38) not hold, as the multiplicative fitness scheme shows.

Suppose next that $\gamma_0 < \gamma_1 < ... < \gamma_k < ... < \gamma_K$ so that fitness increases with increasing heterozygosity. Then (7.39) holds so that the product equilibrium is stable for free recombination between loci (and more generally, usually for rather loosely linked loci). However, (7.38) need not necessarily hold, as the multiplicative case again shows. But when the γ_k increase in a *concave* fashion, so that $\gamma_k > \frac{1}{2}(\gamma_{k-1} + \gamma_{k+1})$, (7.38) will hold: this was also noted by Slatkin (1972). In this case the additional fitness component for each additional heterozygous locus decreases with the number of current heterozygous loci, and the product equilibrium is the unique (and globally stable) equilibrium for all recombination patterns. When the γ_i form a *convex* system, $\gamma_k < \frac{1}{2}(\gamma_{k-1} + \gamma_{k+1})$, the product equilibrium is stable only for sufficiently large recombination. Thus (Lewontin, 1964a) if $\gamma_0 : \gamma_1 : \gamma_2 : \gamma_3 : \gamma_4 : \gamma_5$ are in the ratios $2 : 3 : 6 : 11 : 18 : 27$ and the recombination fraction between adjacent loci is R, the product equilibrium is stable only if $R > 0.038$.

Suppose now the product equilibrium is not stable, so that the γ_i do not from a concave sequence and R is sufficiently small. For $R = 0$ and multiplicative fitness there are many stable equilibria, each one consisting of a pair of complementary gametic types. For submultiplicative fitnesses (for example $\gamma_k = (k + 1)^a/(K + 1)^a$, $1 < a < K - 1$), there are stable equilibria with a number of gametic types present, each in moderate frequency. By continuity, for $R \approx 0$ the stable equilibria in the multiplicative model (as the simulations of Franklin and Lewontin, 1970 suggest) are such that two complementary gametes occur in high frequency, with all other gametes at extremely low frequency, while for submultiplicative fitnesses a number of moderate frequency gametes exist at stable equilibrium points.

7.6 The Correlation Between Relatives

In this section we investigate properties of the correlation between relatives for characters that are assumed to depend on a large number of loci. A full discussion of this topic would take into account inbreeding, environmental effects, assortative mating and other factors, and would take us far further into the biometrical aspects of population genetics than we wish to go. Our aim here rather is to restrict attention to the effects of linkage and linkage disequilibrium on standard formulae for these correlations, so here we ignore these complications: a brief verbal discussion of the entire question of the correlation between relatives will be given in Chap. 10.

We first consider characters determined by a single locus and consider a more efficient method of arriving at Eqs. (1.16)–(1.19). In this and in later generalizations to characters determined by many loci, it is convenient to calculate covariances rather than correlations: the latter can always be found from the former by dividing by the total variance σ^2 in the character considered.

Suppose the locus determining the character is locus K. We define $\bar{m}$ and a_{ku} as in (7.9) and (7.13) and write

$$\bar{m}_{ku,kv} = \bar{m} + a_{ku} + a_{kv} + d_{ku,kv},\tag{7.40}$$

so that

$$\sum_u a_{ku} x_{ku} = 0, \quad \sum_u d_{ku,kv} x_{ku} = 0 \quad \text{for all } v.\tag{7.41}$$

The additive and dominance genetic variances are given by (7.15) and (7.16).

Consider now two individuals X and Y, with measurements $m(X)$ and $m(Y)$. The covariance between these measurements is

$$E\{(m(X) - \bar{m})(m(Y) - \bar{m})\}\tag{7.42}$$

and may be determined from (7.40). If the two individuals are unrelated, Eq. (7.41) show that the covariance is zero, as expected. But related individuals can possess genes in common that are "identical by descent" from one or more common ancestors. Thus, for example, the contributions a_{ku} in (7.40) may be identical in both individuals since both possess an A_{ku} gene passed on from a common ancestor. Suppose the genes possessed by X are x_f, x_m, where the suffixes f and m denote the genes passed on from father and mother, respectively, and define y_f, y_m similarly. We use the symbol "$\equiv$" to denote "identical by descent" and define

$$P_{ff} = \text{Prob}\,(x_f \equiv y_f), \quad P_{fm} = \text{Prob}\,(x_f \equiv y_m),$$
$$P_{mf} = \text{Prob}\,(x_m \equiv y_f), \quad P_{mm} = \text{Prob}\,(x_m \equiv y_m).\tag{7.43}$$

Then by inserting (7.40) in (7.42) it is found that

$$\text{covar}\,(X, Y) = \tfrac{1}{2}\,(P_{ff} + P_{fm} + P_{mf} + P_{mm})\sigma_A^2(k) + (P_{ff}P_{mm} + P_{fm}P_{mf})\sigma_D^2(k).\tag{7.44}$$

This formula, due essentially to Malécot (1948), provides a simple and powerful method for deriving covariances for any two related individuals, and we now use it to rederive (1.16)–(1.19).

Consider first the father-son correlation, with X being the father and Y the son. Since the mother and father are assumed unrelated, $P_{mm} = P_{fm} = 0$. Also $P_{ff} = P_{mf} = \frac{1}{2}$, and insertion of these values into (7.44) yields (1.16). If X and Y are full sibs, $P_{ff} = P_{mm} = \frac{1}{2}$, $P_{fm} = P_{mf} = 0$, and insertion of these values in (7.44) gives (1.17). Equations (1.18) and (1.19) may be found equally easily.

We next extend (7.44) to the case where the character in question depends on K loci. Here additive $\times$ additive, additive $\times$ dominance, and further terms enter into the covariances, and it is necessary to develop notation for these. We write $\Sigma \, \sigma^2_{AA}$ for the sum (over all pairs of loci) of additive $\times$ additive interactions, $\Sigma \, \sigma^2_{AD}$ for the sum of all additive $\times$ dominance interactions, and more generally

$$\Sigma \, \sigma^2_{ArDs}, \tag{7.45}$$

for $1 \leqslant r + s \leqslant K$, for the sum of all possible r-wise additive and s-wise dominance interactions.

Consider first the simplest case when all loci involved are unlinked and all pairwise linkage disequilibria are zero. Kempthorne (1954, 1955) has shown that the appropriate generalization of (7.44) is

$$\text{covar}\,(X, Y) = \Sigma'[\,\tfrac{1}{2}(P_{ff} + P_{fm} + P_{mf} + P_{mm})]^r [P_{ff}P_{mm} + P_{mf}P_{fm}]^s \sigma^2_{ArDs} \tag{7.46}$$

where the summation Σ' is over all r and s with $1 \leqslant r + s \leqslant K$. Writing

$$\alpha = \tfrac{1}{2}\,(P_{ff} + P_{fm} + P_{mf} + P_{mm}), \quad \beta = P_{ff}P_{mm} + P_{fm}P_{mf}, \tag{7.47}$$

this becomes for a character depending on two loci

$$\text{covar}\,(X, Y) = \alpha \, \Sigma \, \sigma^2_A(k) + \alpha^2 \sigma^2_{AA} + \beta \, \Sigma \, \sigma^2_D(k) + \beta^2 \sigma^2_{DD} + \alpha\beta\sigma^2_{AD}, \tag{7.48}$$

from which (2.105a) and (2.105b) may be derived immediately. Note also from (7.46) that for a character depending on an arbitrary number of loci,

$$\text{covar}\,(\text{father, son}) = \{\tfrac{1}{2} \Sigma \, \sigma^2_A(k) + \tfrac{1}{4} \Sigma \, \sigma^2_{AA} + \tfrac{1}{8} \Sigma \, \sigma^2_{A3} + \ldots + \tfrac{1}{2^K} \Sigma \, \sigma^2_{AK}\}, \tag{7.49}$$

$$\text{covar}\,(\text{grandfather, grandson}) = \{\tfrac{1}{4} \Sigma \, \sigma^2_A(k) + \tfrac{1}{16} \Sigma \, \sigma^2_{AA} + \ldots + \tfrac{1}{4^K} \Sigma \, \sigma^2_{AK}\}. \tag{7.50}$$

Clearly ancestral line covariances do not contain dominance terms. It is not clear how important the various terms in (7.49) and (7.50) are. While for large r each r-order additive interaction term is probably very small, there are $\binom{K}{r}$ terms in σ^2_{Ar} and, even allowing for the factor 2^{-r}, the total contribution of r-order interaction terms need not be small.

Our next objective is to generalize (7.46) to allow for linkage between loci, continuing however to assume complete linkage equilibrium. This generalization has been considered for two loci by Cockerham (1956) and for K loci by Schnell (1963) and van Aarde (1975). Now the typical term in (7.46) is the sum of expressions of the form

$$\alpha^r \beta^s \sigma^2(k_1, k_2, ..., k_r; l_1, l_2, ..., l_s) \tag{7.51}$$

where $k_1 < k_2 < ... < k_r$, $l_1 < l_2 ... < l_s$ with $k_p \neq l_q$ for all p, q and $\sigma^2(k_1, k_2, ... k_r; l_1, l_2, ..., l_s)$ denotes the contribution to the variance of the interaction of the additive effects at loci $k_1, k_2, ..., k_r$ and the dominance effects at loci $l_1, l_2, ..., l_s$. The coefficient $\alpha^r \beta^s$ is the probability of $r + s$ independent events (the independence arising from the assumptions that the loci are unlinked and that linkage equilibrium obtains) composed of events typified by

$E_{ff}^k = \{$genes derived by X and Y at locus k from their respective fathers are identical by descent$\}$.

For unlinked loci events of the form E_{ff}^k and E_{ff}^l are independent for $k \neq l$ (continuing to assume linkage equilibrium), but this is no longer true for linked loci. Thus, for example, the probability of the compound event

$$E_{fm}^{k1} E_{mf}^{k2} ... E_{ff}^{kr} \tag{7.52}$$

can no longer be obtained by simple multiplication, but will involve the recombination fractions between loci $k_1, k_2, ... k_r$. The appropriate generalization of (7.46) for linked loci is found by replacing the typical term by the sum, for all possible r-wise additive and s-wise dominance contributions, of

$$\left(\tfrac{1}{2}\right)^r \text{Prob}\left[\prod_{p=1}^r \{E_{ff}^{kp} + E_{fm}^{kp} + E_{mf}^{kp} + E_{mm}^{kp}\} \prod_{q=1}^s \{E_{ff}^{lq} E_{mm}^{lq} + E_{fm}^{lq} E_{mf}^{lq}\}\right] \times$$
$$\sigma^2(k_1, k_2, ..., k_r; l_1, l_2, ..., l_s). \tag{7.53}$$

In the calculation of this probability, all product terms are expanded out and the probabilities of sums of compound events of the form (7.52) calculated. This formula, although explicit, yields rather complicated values when more than two loci are involved: note, in particular, that (7.53) does *not* in general provide a contribution to the covariance of the simple form $c \sum \sigma_{ArDs}^2$, for some constant c. We thus exemplify the application of (7.53) for two loci only, assuming the recombination fraction between these loci is R.

Consider first the case of full sibs. Here the events E_{fm} and E_{mf} are impossible (for both loci) and the events E_{ff}, E_{mm} have the same probabilities for both loci. These observations lead to a covariance of

$\frac{1}{2}$ Prob $[E_{ff}^k + E_{mm}^k] \Sigma \sigma_A^2(k) + $ Prob $[E_{ff}^k E_{mm}^k] \Sigma \sigma_D^2(k)$

$+ \frac{1}{4}$ Prob $[(E_{ff}^k + E_{mm}^k)(E_{ff}^l + E_{mm}^l)]\sigma_{AA}^2$

$+ \frac{1}{2}$ Prob $[(E_{ff}^k + E_{mm}^k)E_{ff}^l E_{mm}^l]\sigma_{AD}^2$

$+ $ Prob $[E_{ff}^k E_{mm}^k E_{ff}^l E_{mm}^l]\sigma_{DD}^2$. \hfill (7.54)

Events involving the subscripts *ff* and *mm* are independent. We know that Prob $(E_{ff}^k) =$ Prob $(E_{mm}^k) = \frac{1}{2}$ so to compute (7.54) it is necessary only to compute Prob $(E_{ff}^k E_{ff}^l)$. By considering the various cross-over possibilities it is easy to see that this probability is

$$\frac{1}{2} - R + R^2 \hfill (7.55)$$

and substitution of this value in (7.54), and the same value for Prob $(E_{mm}^k E_{mm}^l)$, leads immediately to the value (2.106).

It is of some interest to compute half-sib and ancestral line covariances. For half-sibs (assuming a common father) Eq. (7.54) still holds (since E_{fm} and E_{mf} are still both impossible events), but now E_{mm} is also an impossible event and (7.54) reduces easily to

$$\text{covar (half-sibs)} = \frac{1}{4} \Sigma \sigma_A^2(k) + \frac{1}{4}(\frac{1}{2} - R + R^2)\sigma_{AA}^2. \hfill (7.56)$$

For ancestral lines there are no dominance components to the covariance but, interestingly, the covariance does depend (except for the father-son value) on R. Thus, for example, (7.49) still applies for all R but, for two loci, (7.50) must be replaced by

$$\text{covar (grandfather, grandson)} = \frac{1}{4} \Sigma \sigma_A^2(k) + \frac{1}{8}(1-R)\sigma_{AA}^2. \hfill (7.57)$$

Note that while (7.56) and (7.57) are identical for unlinked loci, they are no longer equal for $0 < R < \frac{1}{2}$ and that linkage causes a greater change in the grandfather-grandson covariance that it does to the half-sib covariance.

Do simple limiting covariances hold for tightly linked loci? van Aarde (1975) shows that as $R \to 0$,

$$\text{covar }(X, Y) \to \alpha\{\Sigma \sigma_A^2(k) + \frac{1}{2}\sigma_{AA}^2\} + \beta\{\Sigma \sigma_D^2(k) + \frac{1}{2}\sigma_{AA}^2 + \sigma_{AD}^2 + \sigma_{DD}^2\}, \hfill (7.58)$$

where α and β are defined in (7.47). (This formula can also be deduced immediately from Schnell, 1963.) Fisher (1918) asserted that for linked loci the pattern of covariances for traits depending on two loci would not differ significantly from those applying for traits depending on one locus. For tightly linked loci this view is supported by (7.58), which is of the same form as (7.44) with the single-locus additive and dominance variances being replaced by the "generalized" values

$$\Sigma \sigma_A^2(k) + \frac{1}{2}\sigma_{AA}^2, \quad \Sigma \sigma_D^2(k) + \frac{1}{2}\sigma_{AA}^2 + \sigma_{AD}^2 + \sigma_{DD}^2. \hfill (7.59)$$

A parallel remark no doubt holds for a number of closely linked loci. For values of R

not close to 0 or $\frac{1}{2}$, however, slight deviations from the pattern suggested by (7.44) do occur, as may be noted immediately by comparing (7.56) and (7.57). While these covariances are identical at $R = 0$, $R = \frac{1}{2}$, they differ at $R = \frac{1}{4}$ by $\sigma_{AA}^2/64$, a value that is however probably negligible for most traits.

Formulae for covariances when the loci involved are not in linkage equilibrium are extremely complicated. Even for two loci the expressions given by Gallais (1974) and Weir and Cockerham (1977) contain upward of a hundred terms, and there can be no hope of estimating the various components from even the most extensive data. Thus while a useful general formula for correlations appears to be almost impossible to find, some progress can, however, be made by considering special cases. The models considered in Sect. 7.5 suggest that for large recombination fractions equilibrium values of linkage disequilibrium are likely to be small: this is confirmed by simulation for "random" fitness patterns. Thus most interest attaches to the case of small recombination fractions where, as we have noted, both linkage disequilibrium and linkage equilibrium equilibria are both sometimes stable.

For two loci and small R we may compare correlations between relatives by comparing the limiting value (7.58) (for the linkage equilibrium case) with the corresponding limiting value found by direct computation when only two gametes exist (so that there is maximum linkage disequilibrium). It turns out that there is no necessary relation between the two correlations and that they can differ considerably even for simple models. Thus if the matrix of measurements is of the (additive) form

	$A_{21}A_{21}$	$A_{21}A_{22}$	$A_{22}A_{22}$	
$A_{11}A_{11}$	1.1	1.1	0.8	
$A_{11}A_{12}$	1.2	1.2	0.9	
$A_{12}A_{12}$	1.4	1.4	1.1	(7.60)

the correlation between relatives is

$$\text{corr}\,(X, Y) = 0.874\alpha + 0.126\beta \tag{7.61}$$

in the linkage equilibrium case where

$$y_{11,21} = 0.09, \quad y_{11,22} = 0.21, \quad y_{12,21} = 0.21, \quad y_{12,22} = 0.49,$$

[here α and β are defined in (7.47)], and is

$$\text{corr}\,(X, Y) = 0.276\alpha + 0.724\beta \tag{7.62}$$

in the case of extreme linkage disequilibrium where

$$y_{11,21} = 0.3, \quad y_{11,22} = y_{12,21} = 0, \quad y_{12,22} = 0.7.$$

Although this example represents two extreme cases of linkage disequilibrium, it is perhaps disquieting that the correlations (7.61) and (7.62) should be so very different.

7.7 Summary

The two main conclusions reached in this chapter are, first, that opposing views can reasonably be held about the likely extent of linkage disequilibrium in natural populations and, second, that the degree of linkage disequilibrium that does occur will alter, perhaps significantly, the numerical values of several important population genetic parameters and possibly one's view of the likely dynamic and static behavior of genetic populations.

The view that extensive linkage disequilibrium might occur in natural populations has been promoted in particular by Franklin and Lewontin (1970) and Lewontin (1974). This view is supported in some special cases by other authors, for example in the case of disruptive selection by Bulmer (1976). While Franklin and Lewontin show that correlational properties in multi-locus systems can build up linkage disequilibrium values in excess of those predicted by two-locus theory, their analysis applies almost exclusively for multiplicative selective models. Our analysis shows that for many other models rather less linkage disequilibrium can be expected. Even under the multiplicative scheme, stable equilibria with no linkage disequilibrium can exist simultaneously with stable equilibria with high linkage disequilibrium, and it is not certain in general what the domains of attraction of the two types are. For stabilizing selection schemes the numerical calculations of Bulmer (1976) confirm the prediction of Wright (1965b) that the true additive genetic variance is somewhat less than the value computed assuming no disequilibrium, but the decrease is only of order 25% in Bulmer's simulations. We have already recorded in the summary of Chap. 6 the view of Ohta and Kimura (1975) that on the whole linkage disequilibrium is generally small in natural populations and to a first order can be ignored: on the whole this view is also given in Crow and Kimura (1970). It is clear that no uniform view yet exists on this point.

If extensive linkage disequilibrium does occur the genetic properties of a population can be quite complex. The Fundamental Theorem of Natural Selection need not hold although one suspects that a generalization of the concept of quasi-linkage equilibrium ensures that mean fitness "mostly" increases. The mean fitness of a population with strong linkage disequilibrium can be quite high, with only a comparatively small number of possible genotypes represented with non-negligible frequencies. However, such an "architectured" system has less flexibility and thus capacity to cope with altered environmental conditions than does an "unarchitectured" population. The true additive genetic variance may be rather less or considerably more than that calculated without linkage disequilibrium, and in general the properties of the genetic system cannot easily be found by combining single-locus analyses. Finally, the correlation between relatives for measurable characteristics is also affected by the degree of linkage disequilibrium between the loci determining these characteristics.

We conclude with two remarks concerning Chaps. 8 and 9. In Chap. 8 models are considered that recognize the gene as a sequence of nucleotides. The theory of the present chapter is relevant to this analysis with the nucleotide replacing the gene as the fundamental unit. Of course entirely new values for mutation rates, recombination fractions and selective differentials are appropriate at the molecular level, and this will alter the way in which we view certain formulae involving these parameters. Second, in Chap. 9 various tests of the "neutral theory" are described. These tests use the gene frequencies at the locus of interest, and the theory ignores possible effects of linked loci. If extensive linkage disequilibrium does occur gene frequencies reflect more the selective forces acting on segments of the chromosome rather than single loci, and thus much of the testing theory described in Chap. 9 becomes invalid under these circumstances.

8. Molecular Population Genetics

8.1 Introduction

In the theory of the previous chapters the basic genetic unit was taken as the gene, and the basic numerical quantity was the gene frequency. In particular the fundamental unit step in evolution was taken as the replacement of one gene (more strictly, allele) by another in a population, and static genetic polymorphisms were usually described in terms of forces acting on gene frequencies. While in Chaps. 6 and 7 the possibility was raised that such polymorphisms are better viewed through forces acting on sets of genes, it remains true that no genetic unit finer than the gene has yet been considered in this book.

In this comparatively brief chapter we consider the molecular population genetics theory arising from the recognition of the gene as a sequence of nucleotides. The task of placing population genetics theory on a molecular basis was begun by Kimura (1971); see also Nei (1975). To some extent the purely mathematical theory of the previous chapters carries through to the molecular level with the nucleotide frequency replacing the gene frequency as the primary variable, but clearly new models and view-points, as well as new "typical" values for various fundamental genetic parameters, are necessary at the molecular level.

While in the past much of the available molecular genetic data concerned amino acid sequences, recent developments ensure that nucleotide sequences will soon become available for many species and many loci in great profusion. For this reason this chapter considers only the theory relating to such sequences: fortunately, because of problems concerning the genetic code and its redundancy properties, this theory is rather simpler than that for amino acid sequences.

We now consider four points where the mathematical population genetics theory based on nucleotide frequencies differs from the "classical" theory based on gene frequencies. First, the molecular theory is dynamic, in contrast to the often static classical theory. Mutations are usually seen as leading to new genes rather than back to currently (or previously) existing ones, since it is plausible that most nucleotide mutations will lead to sequences not currently existing in the population. The infinite alleles model of Sect. 3.6 was originally proposed by Kimura and Crow (1964) with this view in mind: we discuss this model at greater length in the next section. The dynamic nature of molecular population genetic models has been stressed in particular by Kimura (1971).

Secondly, because of extremely small intracistronic recombination rates (perhaps of order 10^{-5} or less), the assumption that the different sites within one gene evolve independently is particularly questionable. The mathematical theory of Chaps. 6 and 7 (there referring to genes and gametes rather than nucleotides and nucleotide sequences) shows that the evolution of tightly linked systems usually cannot be predicted from independent consideration of the separate loci (or, here, sites). Thus at the nucleotide sequence level various formulae in Chaps. 3, 6 and 7 will be viewed differently than at the gametic level. For example, if (3.92) is used to compute fixation probabilities of two locus gametes, the assumption $NR \gg 1$ may normally be made, in which case the fixation probability (3.92) becomes essentially

$$c_i(0) + \eta_i D(0). \tag{8.1}$$

This is just the product of the probabilities of fixation of the two alleles comprising gamete i so that the fixation processes at the two loci are effectively independent. At the molecular level, on the other hand, it is plausible that NR is small (since we consider two nucleotides within the same cistron) and in this case the fixation probability is close to $c_i(0)$. Thus each two-site "gamete" evolves largely as a unit, and the fixation processes at the two sites are closely associated. Clearly the likely numerical values of the parameter R in the two cases affect the way in which (3.92) can be used to describe nucleotide and gene fixation processes.

Thirdly, while the classical theory concerns the evolution of genes given labels "A_1", "A_2", etc., at the molecular level the actual genetic material is know so that the symbols A, C, G and T refer to specific rather than type entities. The fact that the theory thus concerns ultimate and real entities is of great importance and further reference will be made to it in a moment. It also allows evolutionary inferences not closely associated with classical population genetics theory: for example, the considerable redundancy of the third nucleotide of a triplet in determining amino acids can be used (see Kimura, 1977; Cornish-Bowden and Marson, 1977; Barker et al., 1978; Berger, 1978) for this purpose. We do not pursue these developments here.

Finally, and perhaps most important, molecular considerations often lead to *retrospective* rather than *prospective* evolutionary questions. The great work of Fisher, Haldane and Wright was largely prospective: given reasonable numerical values for various genetic parameters, they showed that evolution as a genetic process could and would occur. In the earlier years of this century such an undertaking was required. It is, however, no longer necessary to do this, and it now appears more useful to attempt to describe, by a retrospective analysis, the course evolution has taken and thus gain empirical insight into evolutionary questions. In such an endeavour knowledge of the actual genetical material is essential, and the entire retrospective analysis must therefore be carried out in the framework of molecular population genetics.

8.2 Infinite Alleles and Infinite Sites Models

It was remarked above that the Kimura-Crow infinite alleles model described in Sect. 3.6 was inspired by the knowledge of the molecular structure of genes. In this section we consider further properties (largely time-dependent properties) of the neutral

Wright-Fisher infinite alleles model, using where necessary various formulae for the stationary case given in Sect. 3.6, where a formal description of the model was also provided. Two selective infinite allele models will be considered in the next chapter in connection with tests of the neutral theory.

We start by considering various properties that can be found immediately from the "two-allele" theory of Chap. 5. That this is possible follows from the fact that, in the neutral infinite allele model, all alleles other than A_1 can be grouped simply as the "allele" "not-A_1" and often this is sufficient to answer certain "infinite alleles" questions. Thus suppose (Li and Nei, 1977) that in the infinite allele model (3.65) the allele A_1 has a current frequency of unity and it is required to find the mean time until A_1 is lost from the population. This mean time can be found immediately from the "two-allele" formula (3.20) with $p = 1$, the mean time (in generations) being

$$T(1) = 4N \sum_{i=1}^{\infty} \{i(i + \theta - 1)\}^{-1}. \tag{8.2}$$

Clearly a similar calculation can be made for any value of p.

An analogous calculation can be made for selective models. Ohta (1974, 1977) has claimed that most gene fixation processes in evolution concern very slightly deleterious alleles. Consider then an infinite alleles model in which a given allele A_1 has frequency unity, and suppose all mutants have heterozygote disadvantage s compared to A_1 (and homozygote disadvantage $2s$). The mean time until one or other deleterious allele fixes must exceed the mean time until loss of A_1: the latter mean time may be found immediately from two-allele theory using a generalization of (3.20) (see Ewens, 1969c, Eq. (5.39), or Li and Nei, 1977, Eq. 1). Writing $\alpha = |2Ns|$ this mean time is, in generations,

$$T(1) = 4N \int_0^1 t(x)dx, \tag{8.3}$$

where

$$t(x) = x^{-1}(1 - x)^{\theta - 1} \exp(2\alpha x) \int_0^x (1 - y)^{-\theta} \exp(-2\alpha y)dy. \tag{8.4}$$

This mean time is calculated by Li and Nei for various θ, α combinations and, as expected, is extremely large even for moderate α values, increasing (for $\theta = 1$) from $40N$ generations for $|\alpha| = 2.5$ to $5 \times 10^6 N$ generations for $|\alpha| = 10$. We conclude that the evolutionary role of such recurrent deleterious mutants is negligible if $|\alpha|$ is 5 or more.

A final formula obtainable from two-allele theory concerns the number of alleles in common in samples from two populations that separated t time units in the past. As a simple model, suppose at $t = 0$ a population of size N duplicates and splits into two identical subpopulations each of size N, which subsequently evolve independently. Suppose the allele A_i has frequency p_i at the time of separation. The density function $f(x_i; p_i, t)$ of the frequency x_i of A_i at time t can be found from two-allele theory (assuming no selection) and is given explicitly by Crow and Kimura (1970, Eq. 8.57). The probability that this allele occurs in a sample of r genes from one subpopulation at time t is thus

$$P(p_i; t) = \int_0^1 \{1 - (1 - x)^r\} f(x, p_i, t) dx, \qquad (8.5)$$

and the mean number of alleles in common to arise in a sample of r taken from each subpopulation is

$$E\{k | p_1, p_2, \ldots\} = \sum_i \{P(p_i; t)\}^2. \qquad (8.6)$$

This probability assumes given values for $p_1, p_2, \ldots$. An unconditional expectation can be found by integrating over the frequency spectrum implicit in (3.83) to obtain

$$E(k) = \theta \int_0^1 p^{-1}(1 - p)^{\theta - 1} \{P(p; t)\}^2 dp, \qquad (8.7)$$

and this can be evaluated numerically for any r, θ and t. Representative values are given by Li and Nei (1977), but we do not here enter into details. Note that while the frequency spectrum $\theta p^{-1} (1 - p)^{\theta - 1}$ relates to the infinite alleles model, its original derivation (cf. the argument leading to (3.82)) was through two-allele theory, and thus the entire argument leading to (8.7) is through this theory.

To arrive at more detailed time-dependent results it is necessary to consider the full infinite alleles model (3.65). This has been done by Griffiths (1979 a, b) who takes the equivalent approach of using K-allele theory and then letting $K \to \infty$. In many respects the various formulae are rather simpler when stated in K-allele terms. Thus while the joint density function of the ordered allele frequencies $x_{(i)}(1 \geqslant x_{(1)} \geqslant x_{(2)} \geqslant \ldots)$ has not yet been given explicitly for arbitrary t (nor indeed, as noted in (5.112)–(5114), in the stationary case), the joint density function of $x_{(1)}, x_{(2)}, \ldots, x_{(K-1)}$ in the K-allele model has been found explicitly by Griffiths as

$$f(x_{(1)}, x_{(2)}, \ldots, x_{(K-1)}) = \frac{K! \Gamma(\theta)}{\{\Gamma(\theta)\}^K} \{x_{(1)} x_{(2)} \ldots x_{(K)}\}^{\epsilon - 1}$$

$$\times \{1 + \sum_{i=2}^{\infty} \rho_i(t) Q_i^*(x, p)\} \qquad (8.8)$$

where

$$\rho_i(t) = \exp \{-\tfrac{1}{2} i(i - 1)t - \tfrac{1}{2} \theta it\}, \qquad (8.9)$$

$\epsilon = \theta/(K - 1)$, and $Q_i^*(x, p)$ is an explicit though complicated polynomial in the $x_{(j)}$ and the initial frequencies $p_{(j)}$ that we do not give in detail here. This expression may be compared with the stationary value (5.11). Note that the leading eigenfunction $Q_1^*(x, p)$ is identically zero and thus omitted from the sum in (8.8): this occurs because the order-statistics process is "unlabelled" (see Ewens and Kirby, 1975, for this concept). The eigenvalue $\rho_i(t)$ is clearly the analogue of the eigenvalue $\lambda_i 2Nt$ defined in (3.81).

For large t terms in $\rho_i(t)$ $(i \geqslant 3)$ may be ignored and then (8.8) become approximately

$$f(x_{(1)}, x_{(2)}, \ldots, x_{(K-1)}) \simeq \text{const exp } \{-\sigma F\} \{x_{(1)} x_{(2)} \ldots x_{(K)}\}^{\epsilon-1}, \tag{8.10}$$

where $F = \Sigma x_{(i)}^2$ and $\sigma = \{E(F) - \Sigma p_{(i)}^2\} \{\text{var } F\}^{-1} \exp \{-t(1+\theta)\}$, and $E(F)$ and Var F, given by (5.118) and (5.119), are the stationary mean and variance of $\Sigma x_{(i)}^2$. This distribution bears a curious resemblance to the selective distribution (5.122), a fact that should be kept in mind in the tests of selective neutrality based on this model considered in the next chapter.

The stationary frequency spectrum $\phi(x) = \theta x^{-1}(1-x)^{\theta-1}$ has already been introduced in Sect. 3.6. Griffiths (1979a) obtains the time-dependent frequency spectrum as

$$\phi(x) = \theta x^{-1}(1-x)^{\theta-1}[1 + \sum_{i=2}^{\infty} \rho_i(t) c_i \psi_i(x) \lambda_i(p_1, p_2, \ldots)] \tag{8.11}$$

where

$$\psi_i(x) = \sum_{j=1}^{i} (-1)^{i-j} \binom{i}{j} \{\Gamma(\theta + i + j - 1)/(j-1)!\} x^j,$$

$$\lambda_i(p_1, p_2, \ldots) = (-1)^i \theta \Gamma(\theta + i - 1) + \sum_j \psi_i(p_j),$$

$$c_i = \{(i-1)(2i + \theta - 1)\} \{i\theta(i + \theta - 1)\Gamma^2(i + \theta - 1)\}^{-1}$$

and the $\rho_i(t)$ are defined by (8.9). The mean number of alleles at time t (as a function of $p_1, p_2, \ldots$) can be found approximately as

$$\int_{(2N)^{-1}}^{1} \phi(x) dx \tag{8.12}$$

and the mean number of alleles in a sample of r genes as

$$\int_0^1 \{1 - (1-x)^r\} \phi(x) dx. \tag{8.13}$$

An explicit expression for this is given by Griffiths (1979b, Eq. (2.10)), who also provides numerical calculations for various r, θ, t, and p_j values. We reproduce some representative calculations in Table 8.1 for two cases: first where there exists initially a single allele in the population and second where there exist initially many alleles of equal frequency. Note that in the former case the approach to the equilibrium value appears rather more rapid than in the latter. It is perhaps of interest to note, in this connection, that the mean value of the homozygosity $F = \Sigma x_i^2$ can be found immediately from (3.66) as

$$E(F_t) = (1+\theta)^{-1} + \{\exp - (1+\theta)t\} \{F_0 - (1+\theta)^{-1}\}. \tag{8.14}$$

Table 8.1. Mean number of alleles observed in a sample of 200 genes for various θ, t values. Unit time = $2N$ generations. Case (i): one initial allele. Case (ii): many initial alleles of equal frequency. From Griffiths (1979b)

				t				
	0.2		0.5		1.0		∞
	(i)	(ii)	(i)	(ii)	(i)	(ii)	(i) & (ii)
0.1	1.31	10.12	1.40	4.62	1.47	2.77	1.57
θ 1.0	4.03	12.39	4.89	7.64	5.49	6.34	5.88
1.5	5.51	13.62	6.74	9.25	7.54	8.18	7.90

Both F and the number of alleles in a sample are measures of genetic variation, and both have been used to estimate θ (see a more complete discussion in the next chapter). A comparison of (8.14) with the value in Table 8.1 may be made to find the extent to which each is affected by non-stationarity, recalling that the initial value F_0 is unity in case (i) and zero in case (ii). Taking the ratio of the time-dependent to the stationary value as a criterion, the mean number of alleles appears in case (i) to approach its stationary value slightly more rapidly than does F for the values of θ considered, but only for sufficiently large θ and t values in case (ii).

A more important result concerns the joint distribution of two samples taken time t apart. Suppose a sample of size r_1 taken at time zero contains k_1 alleles and a sample of size r_2 taken at time t contains k_2 alleles, and that m ($\leqslant \min (k_1, k_2)$) of these alleles are in common. Griffiths (1979b) finds the joint stationary distribution of k_1, k_2 and m (his Eq. (3.7)), and hence the marginal distribution of m. We do not reproduce the rather complex formulae here and note only that for large r_1, r_2, t,

$$\text{Prob}\ (m = 0) \sim 1 - (1 + \theta)\rho_1(t) + 0(\rho_2(t)), \tag{8.15}$$

$$\text{Prob}\ (m = i) = 0(\rho_i(t)) \tag{8.16}$$

and

$$E(m) \sim (1 + \theta) \exp\left(- \tfrac{1}{2}\theta t\right). \tag{8.17}$$

Note the interesting relation between (8.16) and (3.62). Exact values of Prob $(m = i)$ and $E(m)$ are tabled by Griffiths.

An even more interesting and detailed result concerns the joint distributions of allele numbers and frequencies in the two samples, considering also the possibility of alleles in common. Griffiths (1979b) find this joint distribution by introducing a variable $\gamma(i, j)$, defined for all (i, j) other than $(0, 0)$ as the number of alleles having i representative genes in the first sample and j representative genes in the second sample. From the form of this joint distribution several important conclusions follow. For large t, the joint distribution of the $\gamma(i, j)$, given m, can be derived, and it is found that this distribution depends on θ only through $k_1 + k_2$. In other words, given $m, k_1 + k_2$ is a sufficient statistic for θ, and optimal estimation of θ, given m, is carried out using $k_1 + k_2$ only. This conclusion extends that of Sect. 3.6 where (in the notation of that section) k is found to be a sufficient statistic for θ (see (3.78)). More generally a sufficient statistic for (θ, t) is $k_1 + k_2$, and the frequencies $\gamma(i, j)$, for $i, j \geq 1$, while the frequencies $\gamma(i, j)$ for $i, j \geq 1$ are a sufficient statistic for t if θ is known. These conclusions follow from the form of joint distribution of k_1, k_2, m and the $\gamma(i, j)$ at time t.

These considerations are important for various reasons, in particular the construction of phylogenetic trees. The reversibility arguments considered in Chap. 5 show that the stationary joint distribution of two samples taken time t apart is identical to that of two contemporary samples in populations that separated at time $\frac{1}{2} t$ in the past. One method of construction of phylogenetic trees (see Sect. (8.5)) depends on the calculation of a "distance" measure between two populations, based on the gene numbers and frequencies in the two populations. The mean of such a distance is ideally proportional to the time $\frac{1}{2} t$ since the branching occurred, and for efficiency a sufficient statistic for t should presumably be used for the distance measure. Although the above considerations have not yet led to a unique best estimator of t they indicate quite firmly the lines along which such an estimator should be sought. The fact that population genetics theory, which has hitherto had little influence on phylogenetic reconstruction methods, can possible lead to optimal procedures in this undertaking should be noted.

We turn now to the infinite sites model. This was in effect introduced by Kimura (1969) but only named as such by him in 1971 (Kimura, 1971). This model assumes the gene to consist of an effectively infinite sequence of nucleotides. The mutation rate per gene per generation is u (so that the mean number of new mutants per generation in the population is $2Nu$): because an infinite number of sites is available the model assumes that each new mutant arises at a currently non-segregating site. Free recombination is assumed between sites, at least for the moment.

For this model the analogy of the formal mathematics of the segregation process at each site to that of the segregation process at each locus in classical genetics is particularly interesting. Here we may assume, under neutrality, the Wright-Fisher model (1.43) to describe the segregation process at each site, since under our assumptions at most two nucleotide types are possible in the population at any site at any time. From (5.16) the mean number of generations for which the mutant nucleotide assumes a value in $(x, x + \delta x)$ is $2x^{-1}\delta x$, $[(2N)^{-1} \leq x \leq 1]$. Since on average $2Nu$ sites begin segregation in each generation, the mean number of sites at stationarity at which at any time the mutant nucleotide assumes a value in $(x, x + \delta x)$ is

$$4Nux^{-1}\delta x = \theta x^{-1}\delta x. \tag{8.18}$$

The total mean number of segregating sites may be found by integrating (8.18) over $[(2N)^{-1}, 1]$: a slightly more accurate expression [see (5.75)] is

$$\tfrac{1}{2}\theta \{1.355 + 2 \log 2N\}. \tag{8.19}$$

If two nucleotides at a given site segregate with current frequency x, $1 - x$, the probability that a given individual is heterozygous at this site is $2x(1 - x)$. The mean number of heterozygous sites per individual is found by averaging this over the function (8.18) as

$$\theta \int_0^1 x^{-1} \{2x(1 - x)\} dx = \theta. \tag{8.20}$$

Thus for the representative values $N = 10^5$, $u = 10^{-5}$ there will be on the average about 52 sites segregating in the population and, of these, an average of 4 sites segregate in any given individual.

These calculations can be generalized to take into account selection (provided the same fitness structure holds at all sites), if the stochastic processes at the various sites can be assumed independent. Perhaps the most interesting selective scheme is that where the mutant nucleotide is at a slight selective disadvantage $s(< 0)$ over the prevailing type, with no dominance. Here (5.47) shows that (8.18) must be replaced by

$$\theta \{x(1 - x)\}^{-1} \{e^{\alpha(1-x)} - 1\} \{e^{\alpha} - 1\}^{-1} \delta x, \tag{8.21}$$

where $\alpha = |4Ns|$. While no simple explicit form exists for the mean number of segregating alleles in the population, the mean number segregating per individual is readily found from (8.21) as

$$2\theta \{\alpha^{-1} - (e^{\alpha} - 1)^{-1}\}. \tag{8.22}$$

This decreases from θ at $\alpha = 0$ to 0.67θ at $\alpha = 2$ and 0.39θ at $\alpha = 5$ and for larger values of α is essentially $2\theta/\alpha$.

The above computations concern mean values only, and for these the amount of intracistronic recombination is irrelevant. For variance properties it is necessary to make assumptions about recombination as well as the distribution of the "input" mutation process. The latter may reasonably be assumed to be Poisson. In this case if the segregation processes at each site are assumed independent, the number of segregating sites in the population and in each individual will be Poisson variables with (in the neutral case) parameters (8.19) and (8.20) respectively. This would imply, for example, that the probability of no segregation in the population and in any individual would be

$$(2N)^{-\theta} e^{-0.6775\theta} \quad \text{and} \quad e^{-\theta} \tag{8.23}$$

respectively. Clearly, however, the independence assumption is quite unjustified because of the tight intracistronic linkage. If we assume at the other extreme that there is no intracistronic recombination, the present model reduces to the infinite allele model, for which the individual homozygosity probability $(1 + \theta)^{-1}$ [cf. (3.67)] is appropriate. The functions $\exp(-\theta)$ and $(1 + \theta)^{-1}$ are close for small $4Nu$ (as expected since here the degree of intracistronic recombination is less important) but differ increasingly as $4Nu$ increases. Similarly, the former expression in (8.23) can be compared to the value in (5.80), which again assumes no recombination.

Further properties of the Wright-Fisher infinite site model where no recombination is assumed have been found, in the selectively neutral case, by Watterson (1975). Watterson finds in particular the complete distribution of the random variable K_i, defined as the number of segregating sites in a sample of i genes. We have above been considering the means of K_2 and K_{2N} and Prob $(K_2 = 0)$, Prob $(K_{2N} = 0)$. Assuming a Poisson mutation input, Watterson finds approximately

$$\text{Prob } (K_2 = j) = (1 + \theta)^{-1} \{\theta/(1 + \theta)\}^j, \quad j = 0, 1, 2 \ldots \tag{8.24}$$

and

$$E(K_2) = \theta, \quad \text{Var } (K_2) = \theta + \theta^2. \tag{8.25}$$

The value for $j = 0$ in (8.24) agrees with (3.67), and the variance of K_2 exceeds the mean because of correlation between sites caused by linkage.

The distribution of K_{2N} is more complicated. The mean of K_{2N} is given by (8.19), and the variance is approximately

$$\text{Var } (K_{2N}) = E(K_{2N}) + \pi^2 \theta^2 /6. \tag{8.26}$$

Further, for large N, the complete distribution of K_{2N} is approximately Poisson with mean (8.19): this is close to the independent site distribution.

Particularly with reference to the hypothesis testing theory of the next chapter, it is important to ask how satisfactory the sample formulae (3.76)–(3.79) are for an infinite site model with recombination. Unfortunately, in this model the distribution of the number and frequencies of the number of alleles, i.e. distinct nucleotide sequences, appears very difficult to obtain. The view has sometimes been adopted that the generation of a new allele through intracistronic recombination can be regarded for practical purposes as a new "normal" mutation so that (3.76)–(3.79) still apply, to a close approximation, with a new definition of θ embracing the possibility of "mutation" through recombination. Unfortunately this assumption is not necessarily justified as the following analysis, due to Strobeck and Morgan (1978), suggests.

Strobeck and Morgan consider two sites in a gene and suppose that mutation occurs at each site at rate v, all mutations being new. Note that a more realistic model takes into account the fact that there are only three possible "new" mutant nucleotides, but for the small values of v appropriate to nucleotide mutation rates the two models are probably reasonably close. In any event several of the formulae given below are easily amended to the more accurate model. We suppose the recombination rate between sites is $R(R \ll 1)$, the population size is N and a generalized (neutral) Wright-Fisher model is applicable.

Consider four such "two-site" genes, labelled for convenience $(a_1 b_1)$, $(a_2 b_2)$, $(a_3 b_3)$ and $(a_4 b_4)$. Here a_i is the nucleotide at site 1 in gene i and b_i is the nucleotide at site 2 in gene i. (Recall that the model assumes an infinite number of possible nucleotides.) We define the symbol "$\equiv$" to denote identity of nucleotide type and define

$$F_A = \text{Prob } (a_i \equiv a_j), \quad \text{Prob } (b_i \equiv b_j) = F_B, \quad (i \neq j), \tag{8.27}$$

$$F_{AB} = \text{Prob } (a_i \equiv a_j, b_i \equiv b_j), \quad (i \neq j), \tag{8.28}$$

$$G = \text{Prob } (a_i \equiv a_j, b_i \equiv b_k), \quad (i \neq j \neq k), \tag{8.29}$$

$$G^* = \text{Prob } (a_i \equiv a_j, b_k \equiv b_l), \quad (i \neq j \neq k \neq l). \tag{8.30}$$

The formal mathematics of the evolution of the two-site system is now identical to that of the two-locus system considered in Sects. (3.6) and (3.8). In particular, from (3.67), the equilibrium values of F_A and F_B are

$$\hat{F}_A = \hat{F}_B = (1 + \psi)^{-1}, \tag{8.31}$$

where $\psi = 4Nv$. (Note that in the more accurate model allowing four nucleotides only, the equilibrium value from (3.63) with $K = 4$, is $(3 + \psi)/(3 + 4\psi)$).

A recurrence relation analogous to (3.66) can be found for F_{AB}: this was first done (in the context of two-locus models) by Serant (1974). This recurrence relation takes into account the possibilities of no, one or two recombination events between the sites: ignoring terms in N^{-2}, R^2, and v^2 the recurrence relation is

$$F'_{AB} = (1 - 4v)[\,\{1 - 2R\}\{(2N)^{-1} + (1 - (2N)^{-1})F_{AB}\} + RG]. \tag{8.32}$$

Similar recurrence relations hold for G and G^*, and simultaneous equilibrium solutions of all equations may easily be found. The solution depends on the relative order of magnitude assumptions made about R and v: supposing $N \gg 1$ and $v = 0(N^{-1})$ it is found, for example, that

(1) $R \ll v$: $\quad \hat{F}_{AB} \approx (1 + 2\psi)^{-1}, \tag{8.33}$

(2) $R \approx v$: $\quad \hat{F}_{AB} \approx \alpha(\text{see below}), \tag{8.34}$

(3) $R \gg v$: $\quad \hat{F}_{AB} \approx (1 + \psi)^{-2}, \tag{8.35}$

where

$$\alpha = \frac{2\psi^3 + \psi^2\xi + 11\psi^2 + 6\psi\xi + 2\xi^2 + 18\psi + 13\xi + 9}{(1 + \psi)\{4\psi^3 + 64^2\xi + 2\psi\xi^2 + 20\psi^2 + 19\psi\xi + 2\xi^2 + 27\psi + 27\xi + 9\}}$$

and $\xi = 2NR$. It is clear why the values (8.33) and (8.35) arise. In (8.33) the recombination rate is so low that the system can effectively be considered as a one-site system with mutation rate $2v$, while in (8.35) the recombination rate is high enough so that the sites act effectively independently. These conclusions are analogous to the two interpretations of the fixation probability (3.92) for large and small NR considered in Sect. 8.1. Since v is a nucleotide mutation rate (of order 10^{-8} or 10^{-9}) we may expect ψ to be quite small for all populations of size 10^6 or less, in which case the three equilibrium values of F_{AB} are quite close. For larger values of ψ, however, this is not the case: thus for $\psi = 4$, $\hat{F}_{AB}$ decreases from 0.1111 at $R = 0$ to 0.0641 at $R = 4v$ and from 0.0463 at $R = 20v$ to 0.0400 when $R \gg v$. The formulae (8.33)–(8.35) were checked by simulation by Strobeck and Morgan (1978).

These simulations allow a check to be made of the adequacy of Eqs. (3.77) and (3.78) for the distribution of allele number and frequencies in the present model. Watterson (1974b) has shown that if (3.77) and (3.78) hold, the variance in heterozygosity will be given by

$$\text{var}(F) = 2\theta / \{(1 + \theta)^2(2 + \theta)(3 + \theta)\}, \tag{8.36}$$

as in (5.120). This formula can be used for comparison with empirical values of var (F) once an adequate definition of θ can be made. Strobeck and Morgan (1978) do this by defining θ as the solution of the equation

$$\hat{F}_{AB} = (1 + \theta)^{-1}, \tag{8.37}$$

suggested by (5.120), where $\hat{F}_{AB}$ is given by (8.33)–(8.35). (Note that in the next chapter we suggest procedures for estimating θ possibly superior to that using (8.37).) In Table 8.2 we give values of $\hat{F}_{AB}$; θ, computed from (8.37); var (F), computed from (8.36); and empirical values of var F from simulations. The latter differ consistently from the values calculated from (8.36) approximately for $4Nv > 1$ so we conclude, at least for these parameter values, that (3.77) and (3.78) do *not* apply for the two-site model with recombination.

It is difficult to find theoretically properties of the distribution of the number of alleles in this model. Strobeck and Morgan note in their simulations that whereas for $R = 0$ the mean number of alleles somewhat exceeds the variance in the number of

Table 8.2. Values of $\hat{F}_{AB}$ calculated from (8.33)–(8.35), values of θ thus calculated from (8.37), values of var (F) thus calculated from (8.36), and empirical values of var (F) (Strobeck and Morgan, 1978) for $N = 100$ and various values of R and v

	$v = 0.00125$			
R	0	$4v$	$10v$	$20v$
$\hat{F}_{AB}$	0.5000	0.4758	0.4629	0.4552
θ	1.0000	1.1017	1.1603	1.1968
var (F) (theoretical)	0.0417	0.0392	0.0378	0.0370
var (F) (empirical)	0.0410	0.0381	0.0437	0.0391
	$v = 0.005$			
R	0	$4v$	$10v$	$20v$
$\hat{F}_{AB}$	0.2000	0.1477	0.1301	0.1215
θ	4.0000	6.0572	6.6864	7.2305
var (F) (theoretical)	0.0076	0.0037	0.0027	0.0023
var (F) (empirical)	0.0088	0.0050	0.0047	0.0051

alleles, as may be deduced from (3.77), this no longer applies when $R > 0$ so that, for example, for $R = 20v$ the variance is slightly in excess of the mean for $v = 0.00125$ and more than twice the mean for $v = 0.005$. Thus (3.77) cannot hold for such values of R and the conditional distribution (3.78), upon which the tests of neutrality considered in the next chapter are based, is also suspect. These observations confirm those made from consideration of the homozygosity.

It is clearly of great importance to assess realistic values of the scaled parameters $\xi = 2NR$ and $\psi = 4Nv$. Since v is a nucleotide mutation rate we may expect $v \sim 10^{-8}$ or 10^{-9}. Typical values of R are less precise: possibly values of order 10^{-5} may be expected. These values certainly imply $R \gg v$, and the values in Table 8.2 then suggest that (3.77) and (3.78) are in doubt if v is sufficiently large. Unfortunately the simulation values possibly do not cover the (R, v) combinations of most importance, and extrapolation from Table 8.2 is difficult. A conservative argument is to note that the effect of recombination is certainly less for $v = 10^{-8}$ than for $v = 10^{-6}$. But for $v = 10^{-6}$, $R = 10^{-5}$, $N = 125,000$ the values in Table 8.2 suggest that (3.77) and (3.78) might apply to a reasonable approximation, and with this observation we will use (3.78) to carry our hypothesis testing procedures in the next chapter, at least for populations of effective size 100,000 or less. For experimental results relating to the above discussion, see Freeling (1978).

8.3 Genetic Variation Within and Between Populations

In this section we consider how genetic variation at the molecular level can be divided, at least approximately, into "within" and "between" population components by an analysis of variance technique. The approach has points of similarity with that of Lewontin (1973), who uses entropy measures instead of sums of squares, but it is based essentially on the work of Wright (1943, 1951, 1965) and Nei (1973).

Suppose a sample of n genes is taken from each of h populations and that at any chosen nucleotide site only two nucleotides are observed in the entire sample. Define y_{ij} by

$$y_{ij} = \begin{cases} +1 \text{ if the } j\text{th gene in the } i\text{th population contains nucleotide 1,} \\ \\ 0 \text{ if the } j\text{th gene in the } i\text{th population contains nucleotide 2.} \end{cases} \tag{8.38}$$

Then the classical analysis of variances sums of squares

$$\sum \sum (y_{ij} - \bar{y}_i)^2 = \text{within group sum of squares,} \tag{8.39}$$

$$n \sum (\bar{y}_i - \bar{y})^2 = \text{between group sum of squares,}$$

become, with the identification (8.38),

$$n \sum \bar{x}_i (1 - \bar{x}_i) \quad \text{and} \quad n \sum (\bar{x}_i - \bar{\bar{x}})^2 \tag{8.40}$$

respectively, where $\bar{x}_i$ is the frequency of nucleotide 1 in the sample from population i and $\bar{\bar{x}}$ is the average frequency over all samples. If σ_w^2 is the within-group variance in frequency and σ_b^2 the between-group variance, the sum of squares in (8.40) are unbiased estimators of

$$k(n-1)\sigma_w^2 \quad \text{and} \quad (k-1)\sigma_w^2 + n(k-1)\sigma_b^2, \tag{8.41}$$

respectively, so that σ_w^2 and σ_b^2 can be estimated by

$$\hat{\sigma}_w^2 = n\{k(n-1)\}^{-1} \sum \bar{x}_i(1-\bar{x}_i) \tag{8.42}$$

and

$$\hat{\sigma}_b^2 = (k-1)^{-1} \sum (\bar{x}_i - \bar{\bar{x}})^2 - \{k(n-1)\}^{-1} \sum \bar{x}_i(1-\bar{x}_i). \tag{8.43}$$

The estimator (8.43) is necessarily non-negative whereas (8.43) can be negative: if so we conventionally put $\hat{\sigma}_b^2 = 0$.

A measure of within- and between-group variation can now be found by averaging $\hat{\sigma}_w^2$ and $\hat{\sigma}_b^2$ over a number of nucleotide sites: this is in effect the procedure of Lewontin (1973). Note, however, that the ability to allocate individuals to groups with high success on the basis of genetic characteristics is not incompatible with a high $\bar{\hat{\sigma}}_w^2/\bar{\hat{\sigma}}_b^2$ ratio, since such allocation can take advantage of multivariate analysis of variance techniques and does not rely on simple averaging of $\hat{\sigma}_w^2$ and $\hat{\sigma}_b^2$ values.

8.4 Genetic Distance

We now consider how nucleotide frequencies can be used to define a "genetic distance" between two populations, our main purpose being that this distance should provide a rough measure of the time since evolutionary divergence of these populations. It is assumed that the genetic distance is calculated from a sample of individuals from each population and is based on the nucleotides at a specified set of S sites.

Suppose first that there is no genetic variation within populations. The most elementary distance measure is then the number D of sites (out of S) at which the prevailing nucleotides in the two populations differ: this measure was in effect used (although for amino acids rather than nucleotides and with a correction term based on the genetic code) in the fundamental paper of Fitch and Margoliash (1967). Such a distance measure, however, can be improved to allow for the possibility of back or parallel mutation in the following way. Consider first a single population in which, over a given time period, X nucleotide substitution events have occurred. If the time required for each substitution is ignored in comparison with the time between substitutions, the mean number of sites in a set of S for which at least one substitution has occurred is, immediately,

$$S\{1-(1-S^{-1})^X\}, \tag{8.44}$$

provided it is assumed that all sites have equal substitution probabilities. If further any given nucleotide is equally likely to mutate to each of the other three nucleotides, the mean number of sites at which the new nucleotide differs from the original is

$$\tfrac{3}{4} S[1 - \{1 - 4(3S)^{-1}\}^X].$$ (8.45)

Generalizations of (8.45) accomodating generalizations of the above assumptions are possible.

Consider now two populations and suppose X_i ($i = 1,2$) substitutions in population i since the time of branching. The reversibility arguments of Chap. 5 allow us to compare the genetic make-up of these two populations by considering a single evolutionary process in which $X_1 + X_2$ substitutions have occurred. The mean number of sites at which the two populations will differ is then, from (8.45),

$$E(D|X_1, X_2) = \tfrac{3}{4} S[1 - \{1 - 4(3S)^{-1}\}^{X_1 + X_2}].$$ (8.46)

It is also possible to find variance properties, but we do not pursue these here. For any given S and $X_1 + X_2$, $E(D|X_1 + X_2)$ can be calculated from (8.46): representative values for $S = 400$ are given in Table 8.3.

In practice, of course X_1 and X_2 are not known, but Table 8.3 can still be used in the following way. Suppose X_1 and X_2 are independent Poisson random variables with common mean λt, where t is the time since the two populations branched out. This assumption is reasonable if substitutions tend to occur, for whatever reason, at a roughly constant rate in time. Then $X_1 + X_2$ is a Poisson random variable with mean $2\lambda t$ and immediately, from (8.46), the unconditional mean of $E(D|X_1 + X_2)$ is

$$E(D) = \tfrac{3}{4} S \{1 - e^{-8\lambda t/3S}\}.$$ (8.47)

For any given S and observed value D, λt can be estimated from this equation by substituting D for $E(D)$: with minor manipulations the values in Table 8.3 (or its analog for other values of S) can be used if desired for this purpose.

The above procedure assumes a constant value of λ from site to site. If, as is undoubtedly the case in practice, λ varies from site to site, (8.47) gives an underestimate of $\bar{\lambda}t$ (Nei, 1975, pp. 225–226), the extent of which will depend on the variance in λ from site to site. Nevertheless (8.47) provides a first-order correction to D as a distance measure between two populations, with λt replacing D as the distance measure.

Table 8.3. Values of $E(D|X_1 + X_2)$ [see (8.46] for $S = 400$ and various values of $X_1 + X_2$

$X_1 + X_2$	0	10	25	50	100	200	400	∞
$E(D|X_1 + X_2)$	0.0	9.8	24.0	46.1	85.0	146.0	220.9	300

A more general definition of distance must allow for nucleotide segregation within populations. Here various measures (Sokal and Sneath, 1963; Rogers, 1972; Hedrick, 1971; Cavalli-Sforza and Edwards, 1967; Nei, 1972) have been proposed for different purposes. For evolutionary considerations we require a measure for which, if substitutions occur at a constant rate, is proportional to the time t between the splitting of the two populations considering. Nei (1976) showed by computer simulation that the expected value of his genetic distance measure D_N, defined for the infinite alleles model by

$$D_N = -\log\left[\left\{\sum_{loci}\sum_i (x_i y_i)\right\} / \left\{\sum_{loci}\sum_i (\sum x_i^2 \sum y_i^2)^{1/2}\right\}\right] \tag{8.48}$$

(where x_i and y_i denote the frequencies of the allele A_i at a typical locus in the two populations) increases almost linearly with time and that this property is not shared by the other distance measures above.

For nucleotide sequence data the infinite allele model is no longer necessarily appropriate, and it does not necessarily follow that D_N has the desired property. Nevertheless it seems useful to examine properties of D_N (where x_i and y_i are now nucleotide frequencies) for a nucleotide model, and here the theory of Chaps. 3 and 5 gives information of immediate use, at least for neutral models.

Suppose at a given site the frequencies of the four nucleotides are p_1, p_2, p_3 and p_4 at time 0 and x_1, x_2, x_3 and x_4 at time $2t$. Assuming a symmetrical neutral mutation Wright-Fisher model, the theory of Chaps. 3 and 5 shows that

$$E(x_i|p_i) = \tfrac{1}{4}(1 - e^{-8vt/3}) + p_i e^{-8vt/3}, \tag{8.49}$$

where v is the mutation rate. Thus

$$E\sum x_i p_i = \tfrac{1}{4}(1 - e^{-8vt/3}) + e^{-8vt/3}E\sum p_i^2$$

$$= \tfrac{1}{4}(1 - e^{-8vt/3}) + \{(4Nv + 3)/(16Nv + 3)\}e^{-8vt/3}$$

by using (3.63) for the mean value of $\sum p_i^2$. If summation is carried out over many neutral nucleotide sites with the same value of v, reversibility arguments show that for two contemporary populations that branched out time t in the past,

$$D_N \sim -\log\left[\{(16Nv + 3)/(16Nv + 12)\}\{1 - e^{-8vt/3}\} + e^{-8vt/3}\right]. \tag{8.50}$$

This is essentially a linear function of t only when $vt < 0.1$ approximately: because of the very low value of v, however, this may be a very long time. Problems of variation in v from site to site, preferred mutation rates and selective substitutions have not yet been considered so that, while D_N remains the most reasonable distance measure, its full properties (and those of any possible superior estimator) are currently unknown.

8.5 Constructing Phylogenetic Trees

One of the most useful applications of a genetic distance measure is in the construction of phylogenetic trees of evolution. We therefore suppose now that some distance measure has been decided upon and discuss how it can be used for this purpose.

Consider data from s species or populations from which $\frac{1}{2} s(s - 1)$ distance measures have been constructed. Various algorithms are possible to construct phylogenetic trees from these distances: we outline here that of Fitch and Margoliash (1967). For $s = 3$ we may suppose $d(1, 2) < d(1, 3), d(2, 3)$ where $d(i, j)$ is the distance between species i and j and construct a tree as in Fig. 8.1, with the branch lengths a, b and c being calculated from

$$a + b = d(1, 2), \quad a + c = d(1, 3), \quad b + c = d(2, 3). \tag{8.50}$$

These equations lead to unique values for a, b and c, and if it is assumed that the length of the branch from the initial bifurcation to species 3 is the average of the lengths to species 1 and 2, the lengths e and f may be calculated also.

For $s > 3$ the algorithm proceeds by a sequence of steps analogous to the above, where collections of species are temporarily grouped to give average distances. Thus, supposing the smallest of the original $\frac{1}{2} s(s - 1)$ distance measures is between species 1 and 2, the initial step requires calculation of the average distance $d(1, \cdot)$ from species 1 to all species other than 1 and 2, with a similar calculation for $d(2, \cdot)$. We can then draw a tree as in Fig. 8.1 with "3" representing all species other than 1 and 2, and calculate a and b from

$$a + b = d(1, 2), \quad a + c = d(1, \cdot), \quad b + c = d(2, \cdot). \tag{8.52}$$

The values for a and b will be used in the final tree. Species 1 and 2 are, from now on until the final tree is drawn, grouped as the amalgamation (1, 2), and the distance of

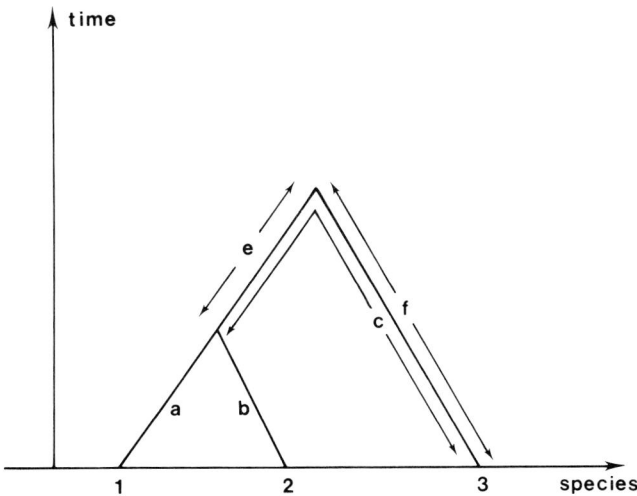

Fig. 8.1

any other species i from this amalgamation is $\frac{1}{2}\{d(1, i) + d(2, i)\}$. The smallest of the $\frac{1}{2}(s - 1)(s - 2)$ distances remaining is now taken and the above procedure repeated: this continues until the complete tree is drawn. Suppose, for example, that the original distance matrix is

		species					
		1	2	3	4	5	
	1	0					
	2	4	0				
species	3	12	10	0			(8.53)
	4	22	26	24	0		
	5	26	27	28	6	0	

Note that species 1 and 2 are the two closest species so we calculate

$$d(1, \cdot) = \tfrac{1}{3}(12 + 22 + 26) = 20, \quad d(2, \cdot) = \tfrac{1}{3}(10 + 26 + 27) = 21. \tag{8.54}$$

Using Fig. 8.1, this gives

$$a + b = 4, \quad a + c = 20, \quad b + c = 21 \tag{8.55}$$

from which $a = 2.5$, $b = 1.5$. Species 1 and 2 are now grouped and the new distance matrix is

	(1, 2)	3	4	5
(1, 2)	0			
3	11	0		
4	24	24	0	
5	26.5	28	6	0

The smallest distance is now between species 4 and 5. We find

$$d(4, \cdot) = 24, \quad d(5, \cdot) = 27.25, \tag{8.56}$$

and these values, together with $d(4, 5) = 6$, may be used to compute branch lengths (analogous to a and b in Fig. 8.1) leading to species 4 and 5 of 1.375, 4.265. Species 4 and 5 are now grouped and the final distance matrix is

	(1, 2)	3	(4, 5)
(1, 2)	0		
3	11	0	
(4, 5)	25.25	26	0

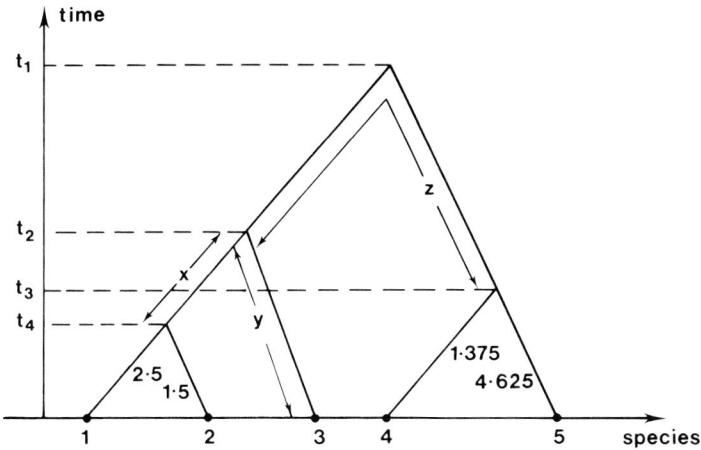

Fig. 8.2

The smallest distance is now between (1, 2) and 3. This shows that the shape of the tree is as in Fig. 8.2, with lengths already calculated marked in. The values for x, y and z are found (by averaging) from

$$2 + x + y = 11, \quad 2 + x + z + 3 = 25.25, \quad y + z + 3 = 26,$$

so that x = 3.125, y = 5.875, z = 17.125, and the entire tree has been determined. The position of the initial bifurcation can again be calculated if it is assumed that the average distance along the lines leading to species 1, 2 and 3 is equal to the average along the lines leading to species 4 and 5.

The above procedure has several possible defects. First, there will usually be some discrepancy between the actual distance between two species and the distance computed in the final tree; thus $d(3, 4) = 24$ and the distance computed from the final tree is 5.875 + 17.125 + 1.375 = 24.375. It is not necessarily true that the algorithm leads to a "closest-fitting" tree. Fitch and Margoliash (1967) suggest various rearrangements to try to achieve closer fits of observed and calculated distances. Second, there is no guarantee that all distances calculated will be positive, and in the reconstruction of real phylogenies negative distances have been found. Despite these and lesser problems the above algorithm, clearly well suited to electronic computers, is possibly still the most valuable method available for estimating phylogenies from species distance measures.

We mention only briefly a second general approach to phylogeny construction, namely the maximum parsimony approach of Fitch (1971). Here the general shape of the tree is assumed known (for example from palaeontological evidence), and the principal aim is to find the most parsimonious (i.e. minimum mutation) evolutionary process leading to observed nucleotides. One important by-product of this approach is the set of ancestral nucleotides consistent with the most parsimonious tree.

The method proceeds on a site-by-site basis. Given contemporary nucleotide observations, the algorithm initially proceeds "backwards" in time to find candidate nucleotides at the various speciation points, up to the earliest one. The second phase of the algorithm then works "forwards" in time modifying these candidate values in the light of "earlier" possible values. A final phase leads to a tree which accounts for the current observations with a minimal number of mutational changes. Details of the algorithm are complicated and are not pursued here.

It is clear that tree construction can proceed using various criteria for optimality. Those discussed above are reasonable but are largely independent of statistical optimization theory and population genetics theory. It would be desirable to introduce these two into tree construction methods. Thus Felsenstein (1973) shows that under certain restrictions the maximum likelihood tree (which may be expected to have optimal statistical properties) is also the most parsimonious tree. But the integration of population genetics theory (in particular as considered in Sect. 8.2) into tree construction has hardly begun, and it is perhaps in this area that the most valuable developments of population genetics will soon arise.

9. The Neutral Theory

9.1 Introduction

The theory considered in the preceding chapters is largely deductive: assuming a certain model for the evolutionary behavior of a population and the consequent implications for the behavior of gene frequencies we can predict, assuming certain values for the genetic parameters, the evolution (or in stochastic models the likely evolution) of a genetical system. Of course any such model will reflect the real world only imperfectly, and this fact must be taken into account when discussing the relevance of the predictions of the model to the evolution of real biological systems.

In the later sections of this chapter we will consider, by contrast, what has become the leading problem of induction in theoretical population genetics. Here certain aspects of the evolution of a real genetical system are observed (in particular the current gene frequencies at various loci in a population and differences of gene frequencies and structure between populations), and we ask whether these observations conform reasonably to a particular hypothesis of genetic evolution. To answer any such inductive question we must first consider deductively what observations might be expected to arise under such a hypothesis: this chapter therefore contains both deductive and inductive considerations.

The hypothesis in which we are interested is the so-called "non-Darwinian" theory of Kimura (1968). Under this theory it is claimed that, whereas the gene substitutions responsible for obviously adaptive and progressive phenomena are clearly selective, there exists a further class of gene substitutions (perhaps in number far exceeding those directed by selection) which have occurred purely by chance through stochastic processes acting in finite populations and for which the replacing gene has no selective advantage over (and which for all practical purposes may be considered to be equivalent to) the replaced gene. A better name for the theory would thus be the "extra-Darwinian" theory, although here we adhere to the standard expression given above.

In a broader sense the theory asserts that a large fraction of currently observed genetic variation between and within populations is non-selective. In this more extreme sense the theory has been described as the "neutral alleles" theory, although this term and the term "non-Darwinian" have been used interchangeably in the literature and will be so used here.

This theory is, not surprisingly, controversial, not only among theoreticians but also among practical geneticists, and the question of whether certain specific substitutions have been neutral has been argued strongly (cf. King and Jukes, 1975; Blundell and Wood, 1975; Jukes, 1976). We do not consider particular cases here. Rather our aim is, first, to discuss (in Sects. 9.2 and 9.3) those mathematically based arguments that first led to the neutral theory, and second (in Sects. 9.4–9.10) to consider various deductive properties of neutral models, and then to consider various formal and less formal inductive procedures that have been put forward on the basis of these properties to test the neutral theory.

9.2 Arguments Leading to the Neutral Theory: Loads

The prime argument (Kimura, 1968) that led to the neutral theory was the view that the amount of genetic substitution estimated to have taken place in evolution (in particular in mammalian evolution) could not be explained by selective processes because of the claimed unbearable (substitutional) genetic load that selective substitutions would imply. This argument was amplified by Kimura and Ohta (1971a) with calculations for the substitutional load more explicit than those initially given. These authors estimated that in the evolutionary history of mammals approximately six substitutions have been completed per generation in any evolutionary line. In the notation of Sect. 2.10, this requires us to put $n = 1/6$, and if this value is used in (2.112) a substitutional load of $e^{180} \approx 10^{78}$ is arrived at. Thus (Kimura and Ohta 1971a)

> "to carry out mutant substitution at the above rate, each parent must leave e^{180}
> $\approx 10^{78}$ offspring for only one of the offspring to survive. This was the main reason why random fixation of selectively neutral mutants was first proposed by one of us as the main factor in molecular evolution."

It is thus necessary to assess the assumptions, both explicit and implicit, that lead to this overwhelming value of load and to clarify in more detail what this calculation really refers to.

We note immediately that the formula (2.112) refers not to the offspring requirement of every individual (as the above quotation implies) but to the requirement of individuals of the maximum possible fitness. To the extent that selection occurs through differential fertility rather than viability, this is not a calculation of any relevance to the average individual. This point is of some importance, and since the calculation (2.112) on which this observation is based uses the perhaps obscure "total load" L, we now analyse the above argument by rederiving (2.112) without any reference to L in order to clarify the calculations as much as possible.

We assume (as is implicit in substitutional load arguments: the validity of the assumption is discussed below) a sequence of loci substituting because of selective differences of the form (1.24b) at each locus with $h = \frac{1}{2}$. Fitnesses are assumed multiplicative over loci, and complete linkage equilibrium also obtains. The value of s is also assumed the same at all loci. The population mean fitness contribution for any locus at which the frequency of the favored gene is x is $1 + sx$. Multi-locus theory shows that under the above assumptions the population mean fitness is

$$\bar{w} = \prod_i (1 + sx_i),$$
(9.1)

where x_i is the frequency of the favored allele at the ith locus substituting. As a first approximation, if there are J loci substituting at any time and since the x_i take values symmetrically around 0.5, we may make the approximation

$$\bar{w} \approx (1 + \tfrac{1}{2} s)^J. \tag{9.2}$$

If each substitution takes T generations and n substitutions start per generation, then $J = T/n$ and thus

$$\bar{w} \approx (1 + \tfrac{1}{2} s)^{T/n} \approx \exp (\tfrac{1}{2} sT/n). \tag{9.3}$$

The fitness of an individual homozygous for the favored allele at all loci substituting is

$$w_{max} = (1 + s)^{T/n} \approx \exp (sT/n), \tag{9.4}$$

and thus if fitnesses are rescaled so that $\bar{w} = 1$ (as is assumed for these purposes to apply for a population of stable size, cf. Wright, 1977, p. 480), the offspring requirement for the most fit genotype is

$$l = \exp (\tfrac{1}{2} sT/n). \tag{9.5}$$

To calculate this we must calculate T, whose value will depend on the values assumed for the initial and final frequencies of the substituting allele. We take (with Kimura and Ohta and Haldane, 1957) these values to be 0.0001, 0.9999: substitution in (1.27) then gives

$$T = \int_{0.0001}^{0.9999} \{\tfrac{1}{2} sx(1 - x)\}^{-1} dx = 36.8/s \tag{9.6}$$

generations. Substitution in (9.5) gives $l = \exp (18.4/n)$. Replacing, as in Sect. 2.10, the value 18.4 by the "representative value" 30 to allow for cases involving dominance, we get

$$l \approx \exp (30/n), \tag{9.7}$$

and substitution of $n = 1/6$ leads to the value $\exp (180)$ for the output requirement of the optimal genotype, as previously calculated. This calculation can be made more precise by using (9.1) rather than the approximation (9.2), but the substance of the conclusion still applies.

Note that in reaching (9.7) we have made several crucial assumptions. The first is that fitnesses are fixed constants (and not, for example, frequency-dependent); the second is that fitness is "multiplicative over loci"; and the third is that total linkage equilibrium always obtains.

We have seen in Chap. 2 that frequency-dependent selection can reduce the segregational load to zero. Such a reduction is less likely for the substitutional load since with a change in gene frequencies due to selection, some selective differentials are necessary and hence some load. Little information is available on the extent to which frequency-dependent selection can reduce substitutional load. As far as linkage disequilibrium is concerned, we have seen in Chap. 6 that if it occurs, it will increase equilibrium mean fitness for multiplicative schemes and hence reduce load. The extent to which this occurs in dynamical situations is unknown, but some reduction is clearly possible: note, from (9.6), that if $s = 0.01$ there will be 22,080 substitutions occurring at any one time, so that many of the loci involved must be closely linked.

By far the most important assumption made in the above analysis is that fitnesses are multiplicative over loci, since the large offspring requirement of the most fit individual is a direct consequence of this assumption. It is certainly true in nature that epistasis occurs, and if this is so there will be a considerable reduction to the load from that calculated formally by using marginal fitnesses and multiplicativity. Thus, for example, with the epistatic scheme (2.75), the true genetic load at the equilibrium point (2.110) is 0.0233, and yet for tight linkage, the sum of the marginal one-locus loads is 0.0422. This arises because of the strong linkage disequilibrium at the point (2.110). A greater discrepancy still between the true load and the formally computed multiplicative load arises when many loci are involved. This point has been made strongly by Lewontin (1974, pp. 289–290) in a slightly different context in discussing the effect of linkage disequilibrium on mean fitness.

Apart from the major difficulty with load calculations just noted through the implicit assumption of multiplicativity, there is a second and perhaps more important reason why we should regard the load calculations given earlier as being of dubious value. As we have noted, the calculation refers to the offspring requirement of an individual of optimal genotype, i.e. an individual homozygous for the favored allele at all loci substituting. It is extremely unlikely that such an individual ever exists. If we take the parameter values $s = 0.01$, $h = 0.5$, $n = 1/6$, initial frequency $= 0.0001$, final frequency $= 0.9999$, there will be 22,080 loci substituting at any one time and the favored allele will take a variety of frequencies in $(0, 1)$ at these loci. In particular at those loci where the substitution has only recently started, the frequency of the favored allele will be quite low. By calculating the frequencies $x_1, x_2, \ldots$ of the favored allele at the loci substituting through (1.27) it is found that the probability that an individual taken at random is of this optimal genotype is $10^{-23,200}$. This value is so extremely small that a theory basing its numerical computations on the offspring requirement of such an individual must demand reconsideration. This point has been stressed in particular by Wright (1977, p. 481).

Some progress on this point may be made by using the statistics of extreme values in a population of given finite size (Kimura, 1969; Ewens, 1970). It is convenient, for purposes of illustration only, to maintain the multiplicativity assumption here so as to discuss the point at issue. We have just shown how, from the frequencies $x_1, x_2, \ldots$ of the favored allele at the loci substituting, the probability can be found that an individual taken at random is of the optimal genotype. In fact, it is possible to find various characteristics of the distribution of the fitness of an individual taken at random from the population and in particular to find the variance of this distribution if the mean is scaled to unity. The resulting variance value is s/n (see Ewens, 1970, Eq. 16; or Crow and Kimura, 1970, p. 252). For $s = 0.01$, $n = 1/6$, this is a variance of 0.06, and hence a standard deviation of approximately 0.245. The rather low value for this standard deviation arises because it is most unlikely that any individual will have a genetic constitution which differs markedly, in terms of the number of favored genes carried, from the average.

If s is extremely small we may suppose, to a first approximation, that the distribution of fitness is a normal distribution. The statistics of extreme values (see Pearson and Hartley, 1958, Table 28) show that, for example in a population of size 10^5, the most fit individual that is likely to occur will have a fitness approximately four standard deviations in excess of the mean. In the present case this implies a fitness of $1 + 4 (0.245) = 1.98$. On average then the most fit existing individual is required to produce only about two offspring in order to effect the gene substitutions observed. For viability selection all individuals must, on average, produce this number of individuals with differential survival, with survival probability proportional to $(1 + s)^a (1 + \frac{1}{2} s)^b$ for an individual homozygous for the favoured allele at a loci and heterozygous at b loci, normalized so that the total population size remains unchanged. Clearly such a task is well within the realms of possibility.

The essence of the above argument is that in a finite population only a minute proportion of all theoretically possible genotypes are realized and that those that are realized are not normally very "extreme". In particular the fitness of the most fit existing genotype is not extreme, and substitutions at the required rate can easily be achieved through each individual's producing as many offspring as this most fit existing genotype with consequent differential viability effecting the required substitutions. To the extent that fertility selection occurs, the offspring requirement is correspondingly lowered.

If we take into account, then, the unreasonable multiplicative fitness requirement implicit in load calculations and the unreasonable concentration on the fitness requirement of essentially impossible genotypes, together with the possibility of linkage disequilibria, frequency-dependent fitnesses and ecological arguments concerning the real nature of selective processes, it appears that there is no compelling reason for the neutral theory on the basis of load arguments.

9.3 Arguments Leading to the Neutral Theory: Substitution and Mutation Rates

A second argument leading to the neutral theory concerns equality of substitution rates in phylogenetic trees constructed from contemporary protein sequences from various species. Consider a given nucleotide site in a particular gene in a population of fixed size N. If the mutation rate per nucleotide site is u (perhaps of order 10^{-9}) the mean number of new mutations per generation at this site is $2Nu$. For sufficiently small u and assuming selective neutrality, the probability that a designated one of these mutants increases in frequency eventually to unity is essentially $(2N)^{-1}$, as was found in Chap. 1. Thus the mean number of mutants to arise per generation and ultimately reach fixation is $(2Nu)(2N)^{-1} = u$, which is independent of the population size N. If the mutation rate can be assumed the same from one population to another, we can then expect approximately equal rates of evolution under neutrality in populations of quite different sizes. Note that no statement can be made, under the neutral (or indeed a selective) theory, about the comparative rates of evolution from one protein to another, since different values of u are known to apply for different proteins.

We shall consider in Sect. 9.10 a test of the neutral theory based on this argument, using the phylogenetic trees of Chap. 8, since it is claimed by neutralists that observed trees do exhibit this equality. Two problems with this claim may now be noted. First, if the observed equality occurs, it is on a "clock" basis, whereas the neutral theory predicts equality on a "per generation" basis. When the data are put, as closely as possible, on a "per generation" basis, a picture rather less favorable to the neutralist position emerges. Second, the above line of argument uses the survival probability, $(2N)^{-1}$, of a new (neutral) mutant. It thus assumes that a substitution starts by a new neutral mutant's arising and ultimately reaching fixation. It is, however, possible that a pool of low-frequency deleterious alleles exists, and one of them, through a change in evironment, becomes neutral and increases in frequency. If this is so a rather different calculation for the substitution rate will emerge. The contrasting views of Wright and Fisher on the nature of evolutionary processes, as outlined in Chap. 1, are relevant to this point.

As far as mutation rate is concerned, it has been argued that the assumption that most substitutions observed in evolution are directed by selection would require an unrealistically high mutation rate to favourably new alleles. If a species is of size N, the rate of substitution of $1/n$ per generation, the average fitness differential between favored and unfavored alleles of s and the mutation rate per gamete to favorable new alleles of U, then (since the survival probability of a favorable new mutant is approximately $2s$)

$$n^{-1} = 4NUs. \tag{9.8}$$

Using the values $n^{-1} = 6$ (see Sect. 9.2), $N = 10^5$ and $s = 0.001$, we arrive from (9.8) at the value $U = 0.015$, or a mutation rate of about 3×10^{-7} per locus in a diploid individual containing 10^5 loci. This rate is unacceptably high, since it implies that the fraction of favorable new mutants among all mutations is of order 10%. Again, difficulties arise with this argument. First, the determination of U and of the per locus mutation rate involve estimation of three other parameters, none of which can be known with much certainty. Large order-of-magnitude differences can arise by the adoption of one reasonable set of parameter values instead of another. Second, the argument (as with that of substitution rates) uses fixation probabilities for new mutants and if, as noted above, fixations occur because environmental changes cause previously deleterious low-frequency alleles to become selectively favored, new calculations are required.

It will be clear that neither load, substitution rate, nor mutation rate arguments provide firm support for the neutral hypothesis, even though they were used as the first theoretical supports for it. Perhaps because of the vast accumulation of information during the last decade or more on the nature and extent of genetic variation in populations, theoretical analyses of the neutral theory have largely moved to an assessment of what patterns of genetic variation might be expected under neutrality and to methods of testing whether observed patterns agree reasonably with neutral predictions. We now turn to a discussion of this line of approach.

9.4 Neutral Models

The data presently available on gene frequency patterns in natural population arise from a large number of loci, many different species, various geographical areas, and several different experimental techniques. The very abundance of these data poses difficulties: should all the data be treated by one overall statistical analysis with an overall assessment of neutrality being given, or is a collection of tests of smaller scope preferable? The arguments against the former procedure are obvious enough: selection surely acts at some loci, and yet other loci are surely almost selectively neutral. At the very least a percentage allocation between selective and neutral cases seems desirable. However, such an allocation can presumably come about only through some form of locus-by-locus, population-by-population testing, and the problems in this are obvious enough also.

In the testing procedures considered in this chapter no account will be taken of geographical subdivision, so that panmixia in all populations is implicitly assumed. (In Sect. 10.3 a test using geographical subdivision as an essential starting point is mentioned.) The extent to which subdivision must be taken into account in testing neutrality is not clear. We shall see in Chap. 10 that if migration is isotropic, only a small amount of migration is necessary for a subdivided population to act effectively as one large random-mating population. However, if migration is not isotropic (e.g. if migration occurs only between adjoining subpopulations), the subdivisional structure is more important. No firm conclusion is available on the desirability of ignoring population subdivision for purposes of testing neutrality.

Perhaps the main way in which data used for testing neutrality can be differentiated is through the experimental techniques by which they are obtained. Here we outline three different mathematical models of genetic variation (and the tests of neutrality derived from them), each inspired by one or other real or imagined experimental technique. The first model is the "charge-state" model of Ohta and Kimura (1973), inspired by the technique of electrophoresis, from which most present information on genetic variation has been found. The second is the "infinite alleles" model of Kimura and Crow (1964): this model assumes an experimental technique that can indicate whether two genes are of the same or different allelic types. In practice, such a situation might approximately arise through the combination of sequential electrophoresis and other laboratory techniques. The third model is that of "infinite sites" (Kimura, 1969). This model assumes a technique yielding the nucleotide sequence of any gene, so that the ultimate level of genetic variation is finally reached. Techniques yielding this information are now increasingly available.

Unfortunately the matching between mathematical model and experimental technique is not completely resolved, a matter discussed further in Sect. 9.9. Thus some caution must be exercised in applying any of the tests described to a given set of data. At the same time there is sufficient evidence of reasonable agreement between model and technique so that arbitrary matching of data and model cannot be allowed, and at least some thought must be given to the question of which mathematical model is most appropriate to the data at hand.

9.5 Charge-State Models

The charge-state model was inspired by the technique of electrophoresis. A single quantity (namely the electric charge) on any protein is measured: it follows immediately, since two proteins differing by one or more amino acids may have the same charge, that not all genetic variation can be uncovered by this technique. It is estimated, in fact, that only about one-third of all genetic variation can be so resolved.

The charge-state model contains two main ingredients. The first is that the permissible charge levels are discrete and equally spaced units (which we label ..., $-2, -1, 0, 1, 2, ...$), while the second is that mutation causes a change of $+1$, 0 or -1 in the charge level (sometimes $+2$ and -2 are also allowed). The electrophorectic technique measures the charge level but no further characteristic.

Clearly several different evolutionary models may be introduced to describe charge-level behavior. Here we consider only the generalized Wright-Fisher model, defined as follows. Consider a gene locus in a diploid population of fixed size N. Suppose in any generation there are exactly N_i proteins of charge level i ($i = 0, \pm1, \pm2, ...,$ $\Sigma N_i = 2N$). It is assumed that (by mutation) the charge level of a "daughter" protein differs from that of its "parent" by j units ($j = 0, \pm1, \pm2, ...$) with probability u_j. Note that u_j is assumed independent of the charge level of the parent protein. The probability π_i that a daughter protein taken at random is of charge level i is thus

$$\pi_i = \{... + u_2 N_{i-2} + u_1 N_{i-1} + u_0 N_i + u_{-1} N_{i+1} + u_{-2} N_{i+2}...\}/(2N). \qquad (9.9)$$

The model now assumes that the probability that there are M_i ($i = 0, \pm1, \pm2, ...$) proteins of charge level i in the daughter generation is

$$\frac{(2N)!}{\prod_i M_i!} \prod_i \pi_i^{M_i}. \qquad (9.10)$$

After sufficient time has passed the joint distribution of the number of proteins in the various charge states will settle down to a "quasi-stationary" form. No strict stationary distribution is possible in this model because any such distribution would clearly be invariant under translation ($i \to i + k$), and this implies infinite probability mass. However, a well-defined concept of stationarity of the *relative* positions of the occupied charge states exists (Moran, 1975; Kingman, 1976). There will therefore exist well-defined stationary distributions for sample characteristics that refer solely to relative, rather than absolute, positions. Two such characteristics are the "covariance measure"

$$C_k = \sum_i N_i N_{i+k}/(2N)^2 \qquad (9.11)$$

and the variance-like quantity

$$V = \Sigma i^2 N_i - (\Sigma i N_i)^2. \qquad (9.12)$$

We consider some of the stochastic properties of these characteristics. Suppose first that $u_j = 0$ for $|j| \geqslant 2$ so that mutation can change the charge state by at most one position. Put

$$\theta = 4N(u_{+1} + u_{-1}).$$ (9.13)

Then (Ohta and Kimura, 1973) the mean value of C_k, when the form of stationarity described above has been reached and when selective neutrality is assumed, is

$$E(C_k) = z^{|k|}/\{1 + 2\theta\}^{1/2},$$ (9.14)

where

$$z = \theta \{1 + \theta - (1 - 2\theta)^{1/2}\}.$$ (9.15)

In particular

$$E(F) = \{1 + 2\theta\}^{-1/2},$$ (9.16)

where F is the random population "homozygosity" measure $\Sigma N_i^2/(2N)^2$. The variance of F can also be found (Moran, 1976): while the exact formula is complicated, a satisfactory approximation is

$$\text{Var}\,(F) = \theta \{3 + 11.25\theta + 13\theta^2 + 1.7\theta^3\}^{-1}.$$ (9.17)

Suppose now two-step mutations are also possible and put $\psi = 4N(u_{+2} + u_{-2})$. Then the polynomial equation

$$\psi z^4 + \theta z^3 - (2 + 2\theta + 2\psi)z^2 + \theta z + \psi = 0$$ (9.18)

clearly has the property that if z_1 is a solution of the equation, then so is z_1^{-1}. Let z_1 and z_2 be the two solutions of (9.18) within the unit circle and put $w_1 = z_1 + z_1^{-1}$, $w_2 = z_2 + z_2^{-1}$. Then assuming selective neutrality, the mean value of C_k at configuration equilibrium is

$$E(C_k) = z_1^{|k|} \{\tfrac{1}{2}\,\theta + \psi w_1(w_1^2 - 4)^{1/2}\}^{-1} - z_2^{|k|} \{\tfrac{1}{2}\,\theta + \psi w_2(w_2^2 - 4)^{1/2}\}^{-1}.$$ (9.19)

Time-dependent expectations $E(C_k(t))$ for both models have been found by Li (1976) and Wehrhahn (1975), but we do not pursue the details here.

We turn next to the quantity V defined by (9.12). Writing the value of V at time t as $V(t)$, Moran (1975) has shown that under selective neutrality

$$E\{V(t)\} = \{1 - (2N)^{-1}\}E\{V(t-1)\} + 2(u_{+1} + u_{-1}) + 8(u_{+2} + u_{-2}),$$ (9.20)

where expectations are conditional on the configuration of charge states at time zero. Taking limits as $t \to \infty$,

$$\sigma^2 = E\{V(\infty)\} = \theta + 4\psi,$$ (9.21)

and, further, the rate at which configuration equilibrium of $E\{V(t)\}$ is reached is, from (9.20),

$$\lambda_1 = 1 - (2N)^{-1}.$$ (9.22)

Despite the simplicity of this value, the complete eigenstructure of the charge-state model is complex. Maruyama (1977) has shown that there exists a continuum of eigen-values as contrasted, for example, with the discrete set (3.17); but this continuum arises because the model assumes an infinity of possible charge levels and thus on a practical level may not be of much significance.

How may tests of selective neutrality be based on the above conclusions? Equations (9.14)–(9.22) all assume neutrality, and thus an approach to testing neutrality can be made by comparing observed data with the predictions these equations imply. There are several problems with any such procedure. First, stationarity must be as-sumed: (9.22) shows that stationarity is reached slowly for these models, a point we take up again later. Second, the equations given refer to population random variables. In practice the data are sample random variables which will have somewhat different distributions (because of the extra component of variation caused by the sampling pro-cess) than the corresponding population variables. Essentially no theory on sample variables is known for the charge-state model, and testing proceeds under the assump-tion that, to a sufficient accuracy, population variable theory will apply to sample variables. Third, it is usually not known whether a one-step or two-step mutation model is appropriate, yet (9.14) and (9.19) show that different formulae apply in the two cases. Finally, the above theory is sometimes used with data from many loci under the assumption that a common value of θ (and ψ) may be taken at all loci considered. Such an assumption, if made, is rarely justified by objective tests, and attempts to cir-cumvent it by additional assumptions concerning locus-to-locus variation in θ are not usually successful.

The conclusion of any test of neutrality based on the charge-state model must be assessed with these difficulties in mind. We now discuss two (not completely distinct) testing methods that have been put favored in the literature on the basis of the above theory, indicating in particular where the problems just listed affect these tests.

Suppose a sample of r proteins is subjected to electrophoresis and it is found that r_i proteins occur at charge level $i(i = 0, \pm 1, \pm 2, ..., \Sigma r_i = r)$. We form the sample analog C_k^* of C_k from

$$C_k^* = \Sigma r_i^2/r^2.$$ (9.23)

It is assumed under both approaches described below that $E(C_k^*) = E(C_k)$: this as-sumption is probably accurate if terms of order r^{-1} are ignored. The two procedures we consider, by Weir et al. (1975) and Wehrhahn (1975), both use C_k^* to test neutral-ity, but in slightly different ways.

The test of Weir et al. proceeds as follows. Note that $C_k^* \equiv C_{-k}^*$, so that $E(C_k^*) = E(C_{-k}^*)$. Define then $D_k^* = 2C_k^* \; (k \neq 0)$, $D_k^* = C_k^* \; (k = 0)$. A measure of the extent to which the observed D_k^* values differ from their means is provided by the chi-square-like statistic

$$\chi^2 = r \, \Sigma \, \{D_k^* - \mathrm{E}(D_k^*)\}^2 / \mathrm{E}(D_k^*). \tag{9.24}$$

The values of $\mathrm{E}(D_k^*)$ are found via (9.19) and depend on θ and ψ. It is possible to minimize χ^2 by suitable choice of θ and ψ, yielding a minimizing value $\chi_{\min}^2$. The larger the value of $\chi_{\min}^2$ the less well the D_k^* values fit their neutral theory expectations and thus the less acceptable is the neutral theory. To find significance points the neutral theory distribution of $\chi_{\min}^2$ must be found: once this distribution is found an objective assessment can be made of the acceptability of the neutral theory.

Wehrhahn's approach is slightly different. If a particular value is assumed for the ratio ψ/θ, an estimate of θ can be made from (9.21) by equating the sample estimate of V to $\theta + 4\psi$. The observed values of C_k^* are then compared to the theoretical values, using (9.19) with the estimates of θ and ψ found through V. An essentially identical procedure is to note that for any symmetrical probability distribution, $p_i (i = 0, \pm 1, \pm 2, \ldots, p_i = p_{-i})$, the quantities γ_k, determined by

$$\gamma_k = \Sigma \, p_i p_{i+k}, \tag{9.25}$$

define the p_i uniquely. Assuming the distribution of charge is symmetrical, the probability distribution can then be estimated by replacing γ_k by C_k^*, and a visual comparison then made between observed and expected distributions. This comparison will usually require a location adjustment because the sample mean position of electrophoretic charge values will usually not be an integer.

Both the above approaches carry out the test of neutrality on a locus-by-locus basis and thus do not require a common estimate of θ and ψ for several loci. Although they are incomplete to the extent that an objective statistical test is not fully arrived at, they without doubt offer the best lines of approach currently available to testing the neutral theory from electrophoretic profile data.

9.6 Infinite Allele Models

We turn next to a model whose application assumes an apparatus capable of indicating whether two genes are of the same or different allelic types. This is the infinite allele model described in some detail in Sects. 3.6 and 8.2. This model possesses the added attraction that a sampling theory is available for it: details of sampling distribution formulae are given in Eqs. (3.76)–(3.79). In this section we describe the test of "strict" neutrality deriving from these formulae and in the following section consider the extension to testing "generalized" neutrality.

Before doing this we digress to consider the problem of the estimation of θ, given (in the notation of Sect. 3.6) a sample vector $\{k, r_1, r_2, \ldots, r_k\}$ with neutral theory distribution (3.76). We note from (3.78) that the conditional distribution of $r_1, \ldots, r_k$, given k, is independent of θ. Thus, once k is given, the observations $r_1, \ldots, r_k$

provide no further information concerning θ beyond that already provided by k. In statistical language k is a sufficient (or exhaustive) parameter for θ. Statistical theory and common sense alike show that inferences concerning θ are optimally carried out using k (and k only): to introduce r_1, ..., r_k into the inference procedure only introduces useless noise that must make the inference procedure less efficient. In particular optimal estimation of θ (or of any function of θ) is carried out by using k only. The maximum likelihood estimator $\hat{\theta}$ of θ is, from (3.77), found as the solution of the equation

$$k = \frac{\hat{\theta}}{\theta} + \frac{\hat{\theta}}{\hat{\theta} + 1} + \frac{\hat{\theta}}{\hat{\theta} + 2} + \ldots + \frac{\hat{\theta}}{\hat{\theta} + r - 1} ; \tag{9.26}$$

this bears an interesting resemblance to (3.79). Similarly optimal estimation of $(1 + \theta)^{-1}$ is found by using k only: while there exists a function of k that estimates $(1 + \theta)^{-1}$ unbiasedly, it is probably more convenient to use instead the estimator $(1 + \hat{\theta})^{-1}$, where $\hat{\theta}$ is given in (9.26), even though this estimator is slightly biased.

Curiously, it is sometimes preferred to estimate $(1 + \theta)^{-1}$ by $\hat{F}$, defined by

$$\hat{F} = \Sigma r_i^2 / r^2 . \tag{9.27}$$

This is a poor estimator in that it uses precisely that part of the data that is *least* informative about $(1 + \theta)^{-1}$. The estimator of θ derived from $\hat{F}$, namely

$$\hat{\theta}_F = \hat{F}^{-1} - 1 , \tag{9.28}$$

has been shown (Ewens and Gillespie, 1974) to be strongly biased and to have mean square error approximately six or eight times larger than that of $\hat{\theta}$. In the context of strict neutrality there is no excuse for using (9.28) as an estimator of θ: in the next section we consider the extent to which this is true in the context of generalized neutrality.

We turn next to tests of the hypothesis of "strict" neutrality, i.e. that the model (3.65) holds. The sufficiency of k for θ, implying that the conditional distribution (3.78) is free of all unknown parameters and thus can be used for an objective test of the neutrality hypothesis, is of central importance to this procedure. The broad aim of the hypothesis testing method is to test whether the observed values of r_1, ..., r_k conform reasonably to what is expected from the distribution (3.78) for the given values of r and k for any particular sample.

In the standard Neyman–Pearson theory of testing statistical hypotheses, such a test is carried out by computing a test statistic (i.e. a single function of the observations r_1, ..., r_k) and determining whether the observed value of this test statistic is an extreme one relative to its neutrality distribution. This procedure require a well-defined selective alternative hypothesis (or class of hypotheses), an objective choice of test statistic and a calculation of its neutrality distribution. The choice of test statistic is comparatively straightforward: it is determined by the ratio of the likelihood of the observations under selection to the likelihood under neutrality. Since presumably selective schemes can be contrived for which the conditional distribution of r_1, ..., r_k, given k, is identical to the neutral form (3.78), we must limit attention to a specific class of selective alternative hypotheses for which this does not occur. We choose here the selective scheme defined either by the heterotic model deriving (as $K \to \infty$) from (5.124) or the deleterious alleles model deriving from (5.125). For these selective schemes, Watterson (1977a) shows that for small selective values the probability distribution of the sample $\{k; r_1, \ldots, r_k\}$ is

$$\text{Prob}\{k; r_1, \ldots, r_k\} = \frac{r! \theta^k}{1^{\alpha_1} 2^{\alpha_2} \ldots r^{\alpha_r} \alpha_1! \alpha_2! \ldots \alpha_r! S_r(\theta)} [1 + A\beta + 0(\beta^2)] \qquad (9.29)$$

where α_i and $S_r(\theta)$ are as defined after (3.76),

$$A = r\{(1 + \theta)^{-1} - r\hat{F}(r + \theta)^{-1}\}\{r + 1 + \theta\}^{-1} \qquad (9.30)$$

and $\beta = 2Ns$ in the heterotic selection scheme, $\beta = -2(2Ns)^2 \gamma(1 - \gamma)$ in the deleterious alleles scheme. Note that the ratio of (9.29) to the neutral value (3.76) depends on the observations, for small β at least, only through $\hat{F}$, and we therefore choose $\hat{F}$ as our test statistic. The subjectively chosen statistic

$$I = -\Sigma(r_i/r) \log(r_i/r) \qquad (9.31)$$

first used (Ewens, 1972) for testing purposes is less efficient (Anderson, 1978) than $\hat{F}$, and its use should be discontinued.

After $\hat{F}$ is selected as test statistic, it is necessary to establish what values of $\hat{F}$ will reject the neutral theory. Clearly $\hat{F}$ will tend to be smaller under heterotic selection than under the neutrality since this form of selection will tend to equalize allele frequencies compared to the neutral case, thus decreasing $\hat{F}$. Under the deleterious allele model $\hat{F}$ will tend to exceed its neutral theory value. These observations are confirmed by noting (Watterson, 1978) that for small selective differences,

$$E(\hat{F}|k) = E(\hat{F}|k, \text{neutrality}) + \beta[2\theta/\{(1 + \theta)^2(2 + \theta)(3 + \theta)\}] + 0(\beta^2, r^{-1}) \qquad (9.32)$$

where $\beta = -\alpha$ in the heterotic model deriving from (5.122) and $\beta = 2\alpha^2 \gamma(1 - \gamma)$ in the deleterious alleles model deriving from (5.125). Thus neutrality tends to be rejected in favor of the heterotic scheme if $\hat{F}$ is "too small" and in favor of the deleterious alleles scheme if $\hat{F}$ is "too large".

To determine how large or small $\hat{F}$ must be before neutrality is rejected it is necessary to find its neutral theory probability distribution. This may be found in principle from (3.78). Unfortunately, in practice, difficulties arise with the calculations because of the form of the distribution (3.78), and approximate procedures are needed.

One approach is to use a computer to draw random samples from the distribution (3.78): efficient ways of doing this are given by Watterson (1978). If a large number of such samples is drawn, an empirical estimate can be made of various significance level points. This has been done by Watterson (1978, Table 1), and his table, expanded by further simulations of Anderson (1978), is given in App. C. Use of the table in

App. C, with interpolation for k, r values not listed, gives probably the most direct and useful test of selective neutrality using $\hat{F}$.

A second approach centres on a given data set with a given k, r. Here the probability that $\hat{F}$ is more extreme than its observed value can be calculated either by simulating a large number of samples, using (3.78), with this k, r value, or exactly by summing (3.78) over all those $r_1, r_2, \dots, r_k$ combinations that lead to a more extreme $\hat{F}$. The latter procedure, while exact, may be impossible in practice if an extremely large number of sample points is involved. Examples of computing significance levels by both methods are given in Watterson (1978, Table 4).

A third approach involves a perhaps less satisfactory approximation. It is clear that $\hat{F} \geq k^{-1}$ and, to a sufficient approximation, $\hat{F} \leq 1$. Then the variable

$$\hat{G} = (k\hat{F} - 1)/(k - 1) \tag{9.33}$$

lies in $(0, 1)$, and its mean and variance may be found from those of $\hat{F}$. Suppose, now, as an approximation, that $\hat{G}$ has the beta distribution

$$f(\hat{G}) = \frac{\Gamma(p + q)}{\Gamma(p)\Gamma(q)} \; \hat{G}^{p-1}(1 - \hat{G})^{q-1}. \tag{9.34}$$

The mean and variance of this distribution are $p/(p + q), pq/\{(p + q)^2(p + q + 1)\}$ respectively, and thus p and q may be found by solving the equations

$$\frac{p}{p + q} = \frac{k E(\hat{F}|k) - 1}{(k - 1)}, \tag{9.35}$$

$$\frac{pq}{(p + q)^2(p + q + 1)} = \frac{k^2 \, \text{Var} \, (\hat{F}|k)}{(k - 1)^2}. \tag{9.36}$$

These equations give

$$p = \frac{[k E(\hat{F}|k) - 1][\,\{1 - E(\hat{F}|k)\}\,\{k E(\hat{F}|k) - 1\} - k \, \text{Var} \, (\hat{F}|k)]}{(k - 1)k \, \text{Var}(\hat{F}|k)} \tag{9.37}$$

$$q = \frac{k[1 - E(\hat{F}|k)]p}{k E(\hat{F}|k) - 1}. \tag{9.38}$$

If $\hat{G}$ does have the beta distribution (9.16) then the random variable

$$q\hat{G}/\{p(1 - \hat{G})\} \tag{9.39}$$

has a variance-ratio distribution with $2p, 2q$ degrees of freedom. (We do not use the standard terminology "F distribution" because of possible confusion with the variable $\hat{F}$.) For any given r and k, p and q may be computed from (9.19) and (9.20), and hence the observed value of $q\hat{G}/\{p(1 - \hat{G})\}$ may be compared to significance points of the various ratio with $2p$, $2q$ degrees of freedom.

Tables of $E(\hat{F}|k)$ and Var $(\hat{F}|k)$ for various k and r values are given in App. D. We do not give further details of this approximation procedure since its interest is mainly historical and because the tables in App. C provide a more accurate method for testing neutrality. The extent to which the approximation through the beta distribution is inaccurate is indicated in numerical examples below.

A final approach to testing neutrality in the present model does not make use of any formal testing procedure. Let $a(i)$ be the number of alleles in the sample that are represented by exactly i genes. For given k and r the mean value of $a(i)$ may be found directly from (3.78): we find

$$E\{a(i)|k,r\} = \frac{r!}{i(r - i)!} \frac{|S_{k-1}^{r-i}|}{|S_k^r|} \tag{9.40}$$

where the S_j^i are values of Stirling numbers of the first kind as discussed after (3.77). It is not difficult to compute (9.40) from recurrence relations for Stirling numbers (see Ewens, 1973). The acceptability of the neutral theory may now be assessed by a visual comparison of data with the theoretical mean $a(i)$ values.

We now illustrate the above methods of testing neutrality by applying them to particular data. The data concern numbers and frequencies of different alleles at the Esterase-2 locus in various *Drosophila* species and are quoted by Ewens (1974) and Watterson (1977a). Since the data are obtained by electrophoresis it is quite possible that the infinite allele model is not appropriate for them so that the calculations given below are for illustrative purposes only.

Table 9.1. *Drosophila* sample data

Species	r	k	r_1	r_2	r_3	r_4	r_5	r_6	r_7
willistoni	582	7	559	11	7	2	1	1	1
tropicalis	298	7	234	52	4	4	2	1	1
equinoxalis	376	5	361	5	4	3	3		
simulans	308	7	91	76	70	57	12	1	1

The data are displayed in Table 9.1. For each set of data we compute $\hat{F}$, the observed homozygosity. Then the exact probability P (given in Table 9.2) that the homozygosity is more extreme than its observed value may be calculated (except for the *simulans* case where the computations are prohibitive). The simulated probabilities P_{sim} are also given in Table 9.2: these are in reasonable agreement with the exact values, so that some confidence can be placed in the values listed in App. C, which were found by simulation. The probabilities computed from the beta distribution approximation are clearly less reliable.

Table 9.2. Sample statistics, means and variances and probabilities for the data of Table 9.1

Species	$\hat{F}$	E($\hat{F}$)	Var($\hat{F}$)	P	P_{sim}	P_{beta}
willistoni	0.9230	0.4777	0.0295	0.007	0.009	0.002
tropicalis	0.6475	0.4434	0.0253	0.130	0.134	0.120
equinoxalis	0.9222	0.5654	0.0343	0.036	0.044	0.021
simulans	0.2356	0.4452	0.0255		0.044	0.084

We finally provide an illustration of the technique of assessing neutrality by a comparison of observed and expected $a(i)$ values. Coyne (1976) provides data with $r = 21, k = 10$, such that

$$r_1 = r_2 = \ldots = r_9 = 1, \quad r_{10} = 12.$$

Equation (9.22) shows that

$$E\{a(i)|k,r\} = \frac{21!}{i(21-i)!} \frac{S_9^{21-i}}{S_{10}^{21}}, \tag{9.41}$$

and this may be evaluated for $i = 1, 2, \ldots, 12$ (the only possible values in this case). A comparison of the observed $a(i)$ values and the expected values calculated from (9.23) is given in Table 9.3. It appears very difficult to maintain the neutral theory in the light of this comparison.

Table 9.3. Comparison of observed and expected data of Coyne (See text for details)

						i						
$a(i)$	1	2	3	4	5	6	7	8	9	10	11	12
expected	5.2	2.1	1.1	0.7	0.4	0.2	0.1	0.1	0.0	0.0	0.0	0.0
observed	9	0	0	0	0	0	0	0	0	0	0	1

If data from several different locations or species are available, it may be judged inappropriate to carry out a formal testing procedure for each data set as has been done above. In such a case it might be preferred to calculate, for each data set, the index

$$I = \frac{\hat{F} - E(\hat{F}|k,r)}{\sqrt{\text{Var}(\hat{F}|k,r)}}, \tag{9.42}$$

which measures the deviation of $\hat{F}$ from its neutral theory mean in standard deviation units. A visual comparison of I for all the data at hand might provide useful evidence on the neutrality question. One problem with this procedure is that the distribution of

$\hat{F}$ is not close to the normal distribution (see, for example, Fig. 1 in Watterson, 1978) so that the usual two standard deviation limits, arrived at ultimately from the normal distribution, may not be of much value. The values of I for the four species of Table 9.1 are 2.59, 1.28, 1.93 and −1.31 respectively. These values agree reasonably with the probability levels in Table 9.2 except for the last one: it is clear that values of $\hat{F}$ falling short of the mean are significant at a smaller number of standard deviations than those in excess of the mean. This is because of the skewness of the distribution of $\hat{F}$ to the right and may be noted by comparing the $2\frac{1}{2}\%$ and $97\frac{1}{2}\%$ significance curves in Fig. 9.1 below.

9.7 Infinite Allele Model: Generalized Neutrality

The testing procedures described in the previous section are a test of strict neutrality: under this hypothesis no selective forces of any kind are assumed to arise. Sometimes this requirement is too severe: recurrent deleterious mutations are of little evolutionary significance, and thus for evolutionary purposes a generalized neutrality hypothesis, allowing such deleterious alleles, might be preferred. (However, if interest centers on the genetic polymorphism itself, maintained by any form of selection, then the strict neutrality test is appropriate.) In this section we consider testing the generalized neutrality hypothesis.

No progress has been made in constructing a formal statistical test in which the deleterious model is taken as the null hypothesis. Such a model contains at least two parameters (the heterozygous and homozygous fitness disadvantages of the deleterious alleles) beyond those in the test of strict neutrality, and it is most unlikely that sufficient statistics for these parameters can be found. A parameter-free distribution (analogous to 3.78) available for testing purposes almost certainly does not exist. Here we take a different line and consider how well the formal machinery arrived at in the previous section for testing strict neutrality works in the wider context of generalized neutrality.

We begin by viewing the procedure of the previous section from a different angle. Recall that for the neutral model (and using the same notation as in the previous section),

$$E(k) = \frac{\theta}{\theta} + \frac{\theta}{\theta + 1} + \frac{\theta}{\theta + 2} + \dots + \frac{\theta}{\theta + r - 1} , \qquad (9.43)$$

$$E(F) = \frac{1}{1 + \theta} . \qquad (9.44)$$

Equations (9.43) and (9.44) determine a parametric relation between $E(F)$ and $E(k)$. As the parameter θ varies a curve of $E(k)$ against $E(F)$ is traced out for any given value of r: such a curve, for $r = 200$, is given by the solid (middle) curve in Fig. 9.1. Essentially the same curve arises if $E(F)$ is replaced by $E(\hat{F})$, where F is the sample homozygosity (9.27).

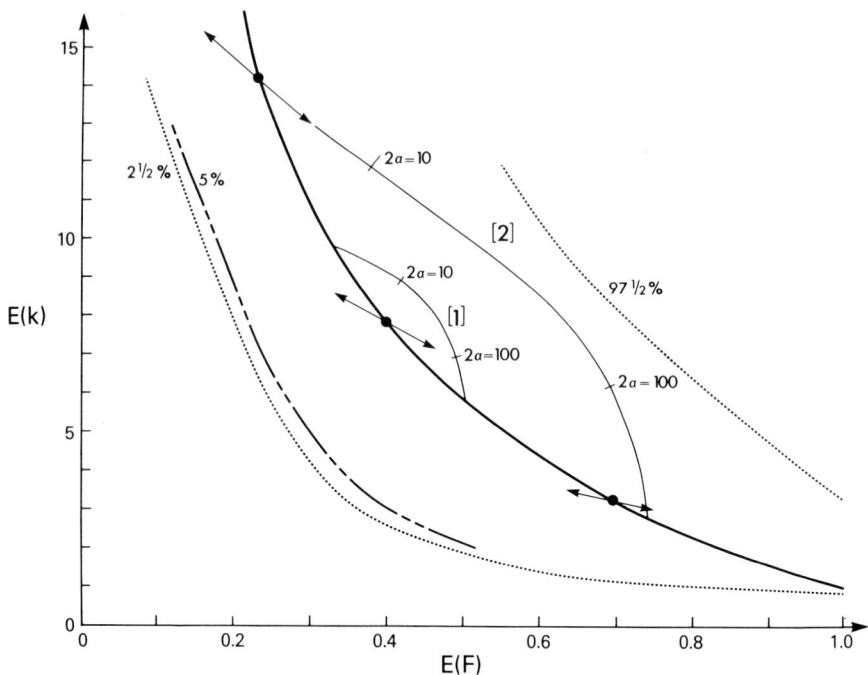

Fig. 9.1. Values of $E(k)$ and $E(F)$ (*thick central curve*) for a neutral sample of 200 genes and various θ. Also shown are confidence limits for F for given k and curves of $E(k)$ against $E(F)$ for various selective models

Suppose now a sample of gene frequencies is given which yields an $(\hat{F}, k)$ pair precisely on this curve (for example at $\hat{F} = 0.440$, $k = 7$). This sample then yields very strong support for the (strict) neutral theory: indeed a perfect fit to the neutrality prediction has been found. The more $\hat{F}$ differs from the value on the curve for the given value of k, the less likely one is to accept neutrality. For values of $\hat{F}$ sufficiently far from the curve (to the left or the right) neutrality will be rejected. Critical values of $\hat{F}$ are given in App. 3 and, for $r = 200$, are sketched (dashed curves) in Fig. 9.1, with the appropriate significance level indicated. For example, if $k = 7$, a two-sided test rejects the hypothesis of neutrality for $\hat{F}$ outside $(0.22, 0.78)$, and a one-sided test (with the alternative of heterotic selection) rejects neutrality if $\hat{F} < 0.23$.

All the above assumes strict neutrality. If some alleles are deleterious, the observed value of k will tend to decrease and the observed value of $\hat{F}$ will tend to increase. Thus if a fraction γ of new mutants are deleterious and individuals with i deleterious genes have fitness $1 - is$, Eqs. (5.136) and (5.135) show that, for $\alpha = 2Ns$ small,

$$E(k) = E_{neut}(k) - 2\theta\alpha^2\gamma(1 - \gamma)(1 + \theta)^{-2} + 0(r^{-1}),\tag{9.45}$$

$$E(\hat{F}) = E_{neut}(\hat{F}) + 4\theta\alpha^2\gamma(1 - \gamma)\{(1 + \theta)^2(2 + \theta)(3 + \theta)\}^{-1} + 0(r^{-1}).\tag{9.46}$$

For a heterotic model the same formulas apply if $2\alpha^2\gamma(1 - \gamma)$ is replaced by $-\alpha$ (where s is now the selective advantage of the heterozygote). Note that the ratio of the changes in $E(k)$ and $E(\hat{F})$ depend only on θ (and not on γ and α): more exactly,

$$\frac{\delta E(k)}{\delta E(\hat{F})} = -\frac{1}{2}(2 + \theta)(3 + \theta). \tag{9.47}$$

For any given θ, (9.47) determines the mean line along which, for small selective differentials, the values of k and $\hat{F}$ under selection tend to lie. In particular, for the deleterious mutations case, the pair $(\hat{F}, k)$ will tend to lie down and to the right, in the direction indicated by the arrows, from the neutral value on the curve.

Before discussing the importance of this observation we note that the calculations on which it relies assume that α is small. In practical circumstances α (being proportional to population size) might be quite large, possibly of the order of $10-10,000$. It is therefore necessary to establish the properties of $E(k)$ and $E(\hat{F})$ (or, effectively equivalently, $E(F)$) for larger values of α. This can be done immediately once an expression for the frequency spectrum $\phi(x)$ (generalizing (5.130) to larger values of α) can be obtained. This frequency spectrum has been found by Li (1977) for the deleterious allele case. Li finds, for arbitrary α, that for $(2N)^{-1} \le x \le 1$,

$$\phi(x) = \theta x^{-1}(1 - x)^{\theta - 1} \sum_{i=0}^{\infty} \frac{(1-x)^i}{i!\Gamma(i + \theta)}[c_1(1 - \gamma)(-2\alpha)^i\Gamma(i + \gamma\theta)$$

$$+ c_2\gamma(2\alpha)^i\Gamma(i +(1 - \gamma)\theta)] \tag{9.48}$$

where $\Gamma(j)$ is the gamma function and

$$c_1^{-1} = \sum_{i=0}^{\infty} (-2\alpha)^i\Gamma(i + \gamma\theta)/\{i!\Gamma(i + \theta)\},$$

$$c_2^{-1} = \sum_{i=0}^{\infty} (2\alpha)^i\Gamma(i +(1 - \gamma)\theta)/\{i!\Gamma(i + \theta)\}. \tag{9.49}$$

It is easy to see that for $\alpha = 0$ (i.e. no selection), $\phi(x)$ reduces to its neutral theory value (5.129). Using (9.48) in (5.127) and (5.128) it is found that

$$E(\hat{F}) = 1 - \theta \sum_{i=0}^{\infty} \left[\frac{c_1(1 - \gamma)(-2\alpha)^i\Gamma(i + \gamma\theta) + c_2\gamma(2\alpha)^i\Gamma(i + (1 - \gamma)\theta)}{i!(i + \theta + 1)\Gamma(i + \theta)}\right], \tag{9.50}$$

$$E(k) = \sum_{i=0}^{\infty} \left[\left\{\frac{c_1(1 - \gamma)(-2\alpha)^i\Gamma(i + \gamma\theta) + c_2\gamma(2\alpha)^i\Gamma(i + (1 - \gamma)\theta)}{i!\Gamma(i + \theta)}\right\}\right.$$

$$\left. \times \left\{\sum_{j=0}^{r-1} (i + j + \theta)^{-1}\right\}\right] \tag{9.51}$$

These expressions reduce to $(1 + \theta)^{-1}$ and $\theta \sum\limits_{j=0}^{r-1} (\theta + j)^{-1}$ when $\alpha = 0$, as we expect from (9.44) and (9.43), while for small α, they confirm the expressions (9.46) and (9.45).

It is possible (Ewens and Li, unpublished) to represent $\phi(x)$, $E(\hat{F})$ and $E(k)$ in terms of the confluent hypergeometric function $M(a, b, z)$, defined (for suitable a and b) by the series

$$M(a, b, z) = 1 + \frac{az}{b} + \frac{a(a + 1)z^2}{b(b + 1)2!} + \frac{a(a + 1)(a + 2)z^3}{b(b + 1)(b + 2)3!} + \dots . \tag{9.52}$$

$$= \frac{\Gamma(b)}{\Gamma(a)} \sum_{n=0}^{\infty} \frac{\Gamma(a + n)z^n}{\Gamma(b + n)n!} .$$

This function has been tabulated by Abramowitz and Stegun (1965) who also list various of its properties, some of which we now note. When $0 < a < b$ we can represent $M(a, b, z)$ in integral form as

$$M(a, b, z) = \frac{\Gamma(b)}{\Gamma(b - a)\Gamma(a)} \int\limits_0^1 e^{zt} t^{a-1} (1 - t)^{b-1} dt. \tag{9.53}$$

From this representation it is easy to see that $M(a, b, z)$ satisfies

$$M(a, b, z) = e^z M(b - a, b, -z), \tag{9.54}$$

and further, if $|z|$ is large, that

$$M(a, b, z) \sim \Gamma(b) e^z z^{a-b} / \Gamma(a), \quad z \gg 0, \tag{9.55}$$

$$M(a, b, z) \sim \Gamma(b) |z|^{-a} / \Gamma(b - a), \quad z \ll 0. \tag{9.56}$$

The representation (9.52) can be used directly to show that

$$\int\limits_0^1 (1 - x)^{b-1} M(a, b, c(1 - x)) = b^{-1} M(a, b + 1, c) \tag{9.57}$$

and

$$(1 + a - b)M(a, b, z) = aM(a + 1, b, z) - (b - 1)M(a, b - 1, z). \tag{9.58}$$

We use (9.53)–(9.58) below in developing properties of $\phi(x)$, $E(\hat{F})$ and $E(k)$. Perhaps the most important observation is to note that $\phi(x)$ can be represented in various ways using the confluent hypergeometric function. Thus from (9.48) and (9.52),

$$\phi(x) = \theta x^{-1}(1-x)^{\theta-1}$$

$$\left[\frac{(1-\gamma)M(\gamma\theta, \theta, -2\alpha(1-x))}{M(\gamma\theta, \theta, -2\alpha)} + \frac{\gamma M((1-\gamma)\theta, \theta, 2\alpha(1-x))}{M((1-\gamma)\theta, \theta, 2\alpha)}\right], \quad (9.59)$$

and (9.54) can now be used to show that the representations

$$\phi(x) = \theta x^{-1}(1-x)^{(1-\gamma)\theta-1}$$

$$\times \{(1-\gamma) + \gamma e^{-2\alpha}\}\left\{e^{2\alpha x}(1-x)^{\gamma\theta}\frac{M((1-\gamma)\theta, \theta, 2\alpha(1-x))}{M((1-\gamma)\theta, \theta, 2\alpha)}\right\} \quad (9.60)$$

and

$$\phi(x) = \theta x^{-1}(1-x)^{(1-\gamma)\theta-1}$$

$$\times \{(1-\gamma) + \gamma e^{-2\alpha}\}\left\{(1-x)^{\gamma\theta}\frac{M(\gamma\theta, \theta, -2\alpha(1-x))}{M(\gamma\theta, \theta, -2\alpha)}\right\} \quad (9.61)$$

are also possible. For different purposes, one or other of (9.59)–(9.61) will be the most useful representation. Thus the normalization requirement

$$\int_0^1 x\phi(x)dx = 1$$

follows by noting that, with the representation (9.59),

$$\int_0^1 x\phi(x)dx = (1-\gamma)\frac{M(\gamma\theta, \theta+1, -2\alpha)}{M(\gamma\theta, \theta, -2\alpha)} + \gamma\frac{M(1-\gamma)\theta,\theta+1, 2\alpha)}{M((1-\gamma)\theta, \theta, 2\alpha)} \quad (9.62)$$

$$= \frac{(1-\gamma)M(\theta(1-\gamma)+1, \theta+1, 2\alpha) + \gamma M((1-\gamma)\theta, \theta+1, 2\alpha)}{M((1-\gamma)\theta, \theta, 2\alpha)} = 1.$$

The three steps in this derivation use (9.57), (9.54) and (9.58) respectively. Further, the first term in (9.62) is the probability that a gene taken at random from the population is of non-deleterious allelic type and is thus asymptotically (i.e. for large α),

$$1 - \gamma\theta/(2\alpha) = 1 - \gamma u/s, \quad (9.63)$$

from (9.56). This is what we expect from (1.35) and (5.72). The function $\xi(x)$, defined by

$$\xi(x) = e^{2\alpha x}(1-x)^{\gamma\theta}\frac{M((1-\gamma)\theta, \theta, 2\alpha(1-x))}{M((1-\gamma)\theta, \theta, 2\alpha)}$$

which enters (9.60) and, in effect, (9.61), has the interesting property that $\xi(x) < 1$ if $(1 - \gamma)\theta > 1$ and, from (9.55), for any fixed x, $\xi(x) \sim 1$ as $\alpha \to \infty$. We use this latter property later when considering asymptotic formulae.

We turn now to $E(\hat{F})$ and $E(k)$ which, from (9.50) and (9.51), can also be written conveniently in terms of the confluent hypergeometric function. After some manipulation we find

$$E(\hat{F}) = \frac{(1 - \gamma)M((1 - \gamma)\theta + 2, \theta + 2, 2\alpha) + \gamma M((1 - \gamma)\theta, \theta + 2, 2\alpha}{(\theta + 1)M(1 - \gamma)\theta, \theta, 2\alpha)}$$

$$= \frac{1}{1 + (1 - \gamma)\theta} + \frac{\gamma\theta}{\alpha(1 + (1 - \gamma)\theta)} + 0(\alpha^{-2}). \tag{9.64}$$

Note that as $\alpha \to \infty$, $E(\hat{F})$ approaches $\{1 + (1 - \gamma)\theta\}^{-1}$ as we might expect, since the large selective disadvantage of the deleterious alleles now has the effect that these alleles occur with negligible frequency and the population behaves, for practical purposes, as a neutral population with θ replaced by $\theta(1 - \gamma)$.

It is convenient, in calculating $E(k)$, to calculate separately the mean number $E(k_1)$ of normal alleles (in a sample of r genes) and the mean number $E(k_2)$ of deleterious alleles in the sample. It can be shown from (9.51), (Ewens and Li, unpublished), that for large α,

$$E(k_1) \sim 1 + \frac{(1 - \gamma)\theta}{(1 - \gamma)\theta + 1} + \frac{(1 - \gamma)\theta}{(1 - \gamma)\theta + 2} + \ldots + \frac{(1 - \gamma)\theta}{(1 - \gamma)\theta + r - 1}, \tag{9.65}$$

$$E(k_2) \approx \gamma\theta \int_0^1 \{1 - (1 - x)^r\}x^{-1}(1 - x)^{(1 - \gamma)\theta - 1}e^{-2\alpha x}\xi(x)dx.$$

The factor $e^{-2\alpha x}$ in the integral ensures that only small values of x contribute significantly to the integral and examination of $\xi(x)$ shows that it is precisely for these values that the approximation $\xi(x) \approx 1$ is most accurate. Thus to a high level of accuracy

$$E(k_2) \approx \gamma\theta \int_0^1 \{1 - (1 - x)^r\}x^{-1}(1 - x)^{(1 - \gamma)\theta - 1}e^{-2\alpha x}dx$$

$$\approx \gamma\theta \sum_{j=0}^{r-1} \int_0^1 (1 - x)^{j + (1 - \gamma)\theta - 1}e^{-2\alpha x}dx$$

$$\approx \gamma\theta \sum_{j=0}^{r-1} \int_0^\infty e^{-(j + (1 - \gamma)\theta + 2\alpha - 1)x}dx$$

$$\approx \gamma\theta \log \frac{\gamma\theta + 2\alpha + r - 1.5}{\gamma\theta + 2\alpha - 1.5}. \tag{9.66}$$

An upper bound for $E(k_1)$ and $E(k_2)$ can be found, when $(1 - \gamma)\theta > 1$, from using the upper bound $\xi(x) \leqslant 1$. We find that the right-hand side in (9.65) is an upper bound for $E(k_1)$, while $E(k_2)$ is bounded above by

$$\gamma\theta[\{\gamma\theta + 2\alpha - 1\}^{-1} + \{\gamma\theta + 2\alpha\}^{-1} + \ldots + \{\gamma\theta + 2\alpha + r - 2\}^{-1}].$$

When $(1 - \alpha)\theta < 1$ both these upper bounds must be multiplied by ξ_{max}, where ξ_{max} is the maximum of $\xi(x)$, $0 \le x \le 1$, and is easily computed in any particular case. A lower bound can be found, by a similar bounding procedure, to be

$$E(k_1) > (1 - \gamma)\theta[\theta^{-1} + (1 + \theta)^{-1} + \ldots + (r - 1 + \theta)^{-1}],$$

$$E(k_2) > \gamma\theta[\theta + 2\alpha\}^{-1} + \{\theta + 2\alpha + 1\}^{-1} + \ldots + \{\theta + 2\alpha + r - 1\}^{-1}].$$

As a check on the accuracy of the approximations (9.65) and (9.66), we may compare the values given by these formulae with exact values computed directly from (9.51) by Li (1977 and personal communication). Consider first the case $\theta = 2$, $\gamma = \frac{1}{2}$, $r = 200$, $\alpha = 15$. Here (9.65) and (9.66) yield respectively $E(k_1) = 5.88$, $E(k_2) = 2.05$ so that $E(k) \approx 7.93$. This value is in very close agreement with the exact value $E(k) = 7.89$ calculated by Li (1977). We may thus use (9.65) and (9.66) with some confidence when $\alpha > 15$ and in Table 9.4 we list values of $E(\hat{F})$ and $E(k)$ for various α, calculated exactly by Li for $\alpha \le 15$ and using (9.64)–(9.66) for $\alpha > 15$.

Our interest in this curve is to assess the effect of deleterious mutations on the one-sided test for heterotic selection. For a fixed value of k, the existence of deleterious alleles clearly moves the mean value of $\hat{F}$ to the right of the "strict" neutral expectation. However for no value of α is this movement in excess of about 0.04 and we may conclude that; for the parameter values considered, formal application of the "strict neutrality" test against heterosis will, in the presence of deleterious alleles, cause a slight but not important tendency towards a conservative test.

This calculation assumes $r = 200$. To a first order of approximation the above analysis shows that the changes in $E(\hat{F})$ and $E(k)$ brought about by deleterious alleles

Table 9.4. Values of $E(\hat{F})$ and $E(k)$ computed from (9.34), (9.36), and (9.44) for various values of 2α (Note: $r = 200$)

2α	$E(\hat{F})$	$E(k)$
0.0	0.3333	9.766
1.0	0.3361	9.740
5.0	0.3773	9.321
10.0	0.4199	8.812
30.0	0.4689	7.89
50.0	0.4808	7.48
100.0	0.4902	6.97
200.0	0.4951	6.57
400.0	0.4975	6.28
1,000.0	0.4990	6.06
∞	0.5000	5.88

Table 9.5. Values of $E(\hat{F})$ and $E(k)$ for various values of 2α for $\theta = 3.36$, $\gamma = 25/2, r = 200$ (cf. Li, 1977)

2α	$E(F)$	$E(k)$
0	0.229	14.30
1.0	0.230	14.30
5.0	0.262	13.85
10.0	0.386	12.26
30.0	0.598	9.08
50.0	0.647	7.69
100.0	0.691	6.20
200.0	0.713	5.01
400.0	0.727	3.95
1,000.0	0.731	3.49
∞	0.735	2.94

compared to the neutral theory values are independent of r: this is confirmed by the numerical values in Table 1 of Li (1977). This conclusion thus obtains, to a first approximation, for any value of r.

It is perhaps more realistic to suppose that γ is close to unity and we thus consider the calculations for the case $\theta = 3.36$, $\gamma = 25/28, r = 200$ and various α. Values of $E(\hat{F})$ and $E(k)$ for $\alpha \leqslant 15$ have been computed exactly by Li (1977 and personal communication) and for $\alpha > 15$ (9.65) and (9.66) can again be used to a sufficient approximation. Values for various α are tabulated in Table 9.5 and are plotted as curve [2] in Fig. 9.1. From this curve we note that $E(\hat{F})$ decreases by more than 0.05 from its neutral theory value when $20 \leqslant 2\alpha \leqslant 500$ or $5N^{-1} \leqslant s \leqslant 125 N^{-1}$. Decreases of 0.10 or more may be expected when $35 \leqslant 2\alpha \leqslant 250$ or $9N^{-1} \leqslant s \leqslant 63N^{-1}$. For $N = 100,000$ these ranges are $(0.00005, 0.00125)$ and $(0.00009, 0.00063)$. Clearly even moderately detrimental alleles (say $s > 0.001$) have very small effect on the test for neutrality against heterotic selection for populations of this size. In this connection it is of interest to note that Simmons and Crow (1977) find a mean heterozygote selective disadvantage of deleterious mutants of about 0.01. Nor do the very slightly deleterious mutants (with values of s of order 10^{-5}, so that $\alpha \sim 1$) in a population of 10^5 considered for example by Ohta (1976) appear to be of significant evolutionary importance. If deleterious mutants tend to have either very small or comparatively large selective disadvantages, with moderate values occuring comparatively rarely, then we may expect only an unimportant bias in using the "strict neutrality" test against heterosis if in fact deleterious alleles occur.

We conclude this section by considering two questions. The first concerns testing for deleterious, rather than heterotic, alleles. The arrows in Fig. 9.1, when extended indefinitely, do not cross the 95% significance curve, and we conclude from this and from the nature of curve [1] and [2] that it is not easy to reject neutrality in favor of deleterious selection of the form we consider. The *D. willistoni* and *D. equinoxalis* samples of Table 9.2 are possibly significant because the model we use does not conform to the biological mechanism leading to these samples. There is probably essentially no recurrent neutral mutation in these cases, and the high-frequency allele is at a strong selective advantage to the remaining alleles.

Secondly, we consider estimation of θ in the presence of deleterious alleles. In the previous section it was shown that, under strict neutrality, estimation of θ using k only is far superior than estimation using $\hat{F}$: the latter estimator has high bias and large variance. In the presence of deleterious alleles the question immediately arises whether the "total" value θ or the "neutral only" value $\theta(1 - \gamma)$ is to be estimated. In the former case Fig. 9.1 shows that the superiority of k as an estimator is maintained in that the bias introduced by the existence of deleterious alleles is less using k than using $\hat{F}$. Thus, for example, if $k = 9$, $F = 0.61$ (which lies on curve [2]), the estimate of θ is 1.86 using k and 0.639 using $\hat{F}$ (the correct value is 3.36). Thus if estimation of the "total" θ is required, k is always a superior estimator. Estimation of $(1 - \gamma)\theta$ can be performed only if the value of 2α is known to be large. If this is the case it might appear that $\hat{F}$ is a superior estimator to k, but this conclusion is not necessarily correct. For very large α values the system behaves, as has been noted, essentially as a neutral system, and in this case the superiority of estimation through k rather than $\hat{F}$ will be maintained. For a range of large, but not too large, α values it is possible that estimation of $\theta(1 - \gamma)$ through $\hat{F}$ will lead to more satisfactory values than estimation through k, but since the value of α is unknown in practice it is difficult to know when to take advantage of this possibility.

9.8 Infinite Site Models

It was remarked in Chap. 8 that the complete nucleotide sequences of genes will soon become available in large number: such data represent an ultimate state of knowledge of the gene, and it is therefore important to assess how data of this form can be used for testing neutrality. The most useful model on which to build any such test is the infinite site model of Kimura (1969), described in Sect. 8.2: we now see how this model may be used for testing neutrality.

We recall the assumption that segregation at any site starts as a result of a unique mutational event (with a mean number $2Nu$ new mutations per gene per generation), so that at most two nucleotides will segregate at any site, and such segregation continues until the new mutant nucleotide is either lost or fixed. Consider then a sample of r genes. If two nucleotides are segregating in the population at a given site and have current frequencies x, $1 - x$, the probability that both are observed in the sample is

$$1 - x^r - (1 - x)^r. \tag{9.67}$$

From (1.51), the probability that the population frequency of a segregating mutant nucleotide is in $(x, x + \delta x)$ is proportional to $x^{-1}\delta x (2N)^{-1} \leqslant x \leqslant 1 - (2N)^{-1})$. Averaging (9.67) over this function it is found that the probability that both nucleotides are observed in the sample is, to a very close approximation,

$$\text{Pr (sample segregation)} = S_{r-1}/S_{2N}, \tag{9.68}$$

where $S_i = 1 + 1/2 + 1/3 + \ldots + 1/i$. Further the probability that, if both nucleotides are observed in the sample, the mutant nucleotide is observed in exactly j of the r genes is

$$S_{r-1}^{-1} \binom{r}{j} \int_0^1 x^{-1} \{x^j(1-x)^{r-j}\}dx = j^{-1} \{S_{r-1}\}^{-1} \tag{9.69}$$

for $1 \leqslant j \leqslant r - 1$. Note also that the mean time that segregation continues at any site in the population is, from (1.51), approximately $2S_{2N-1}$, and this implies that the mean number of segregating sites in the population at any time is approximately

$$\theta S_{2N}. \tag{9.70}$$

The mean number of segregating sites observed in the sample is thus, from (9.68) and (9.70), to a close approximation,

$$\theta S_{r-1}. \tag{9.71}$$

This expression is correct irrespective of intracistronic recombination: if there were free recombination the number of segregating sites observed in the sample has, to a close approximation, a Poisson distribution with mean (9.71). In this case the variance is also (9.71). Intracistronic linkage implies that the Poisson distribution does not strictly apply, and the variance of the observed number of segregating sites is greater than (9.71). At the upper limit of complete linkage, Watterson (1975) has shown that the variance of the number of segregating sites in a sample of r genes is

$$\theta S_{r-1} + \theta^2 \{1 + \tfrac{1}{4} + \tfrac{1}{9} + \ldots + (r-1)^{-2}\}. \tag{9.72}$$

Suppose we wish to use the data provided by the r genes to estimate θ. It is clear that the actual frequencies of the two nucleotides segregating do not depend on θ, so that these frequencies should not be used in the estimation procedure. All information concerning θ resides simply in the number k^* of segregating sites. Thus estimation of θ should be carried out using k^* only (just as, for the infinite alleles model, the number of observed alleles, rather than their frequencies, is used to estimate θ). From (9.71) an unbiased estimator of θ, irrespective of the extent of intracistronic linkage, is

$$\hat{\theta} = k^*/S_{r-1}. \tag{9.73}$$

The variance of this estimator will be somewhere between the values applying for completely free recombination and no recombination, viz.

$$\theta S_{r-1}^{-1} \quad \text{and} \quad \theta S_{r-1}^{-1} + \theta^2 \{1 + \tfrac{1}{4} + \ldots + (r-1)^{-2}\} S_{r-1}^{-2}. \tag{9.74}$$

The bounds are close for small θ and large r values, as is expected. The standard deviation of the estimate of θ approaches zero at the very slow rate $(\log r)^{-1/2}$.

Before turning to consider tests of neutrality we recall that the infinite sites model without recombination is identical to the infinite alleles model. Thus the mean number of different alleles (i.e. sequences of nucleotides) in a sample of r genes is (9.43), and the probability that two different genes taken at random have identical nucleotide sequences is (9.44). The sampling distribution (3.76)–(3.78) also holds true. We return to this observation later when considering tests of neutrality.

Having estimated θ by using k^* we turn, as with the infinite allele model, to observed frequencies to test for neutrality. Unfortunately it is not clear, for the infinite sites model, how this is best done since the correlation of the site-to-site frequencies is unknown because of the unknown intracistronic recombination pattern. A site-by-site test may be carried out in the following way. Passing to a continuous approximation we find from (9.69) that the frequency $x = j/r$ of the mutant nucleotide has probability distribution

$$f(x) = x^{-1} \{\log (r - 1)\}^{-1}, \quad r^{-1} \leqslant x \leqslant 1 - r^{-1}. \tag{9.75}$$

Since in practice it will usually be unknown which of two segregating nucleotides is the mutant, it is preferable to use as test statistic some function which remains unaltered if x is replaced by $1 - x$. The most convenient such statistic is

$$w = |\log \{(1 - x)/x\}|/\log (r - 1). \tag{9.76}$$

This statistic has a uniform distribution on $(0, 1)$ under the hypothesis of selective neutrality. Alternatively, under neutrality,

$$y = - 2 \log w \tag{9.77}$$

has a chi-square distribution with 2 degrees of freedom. If we are interested in heterotic selection we reject the neutrality hypothesis for significantly large values of y.

The real problem with testing neutrality is in aggregating the test results for the various sites. If segregation at the sites were independent, a suitable test statistic of neutrality (against heterosis) would be $\Sigma\, y_i$ which under neutrality has a chi-square distribution with $2k^*$ degrees of freedom, k^* being the number of segregating sites. Unfortunately independence of segregation at the various sites is unlikely, and this approach cannot be used. At the other extreme, when complete linkage obtains, the fact that the infinite sites model reduces to the infinite allele model implies that the theory of Sect. 9.6 can be applied, where distinct alleles are now distinct nucleotide sequences. In reality a situation between these two extremes must obtain, and at present the best approach to testing neutrality seems to be to carry out the "free recombination" test (using $\Sigma\, y_i$) and the "no recombination" test, using Sect. 9.6, and make a judgement for or against neutrality on the basis of the results of the two tests and the known or estimated recombination structure.

9.9 Relations Between Models

It was remarked earlier that neither the charge-state nor the infinite alleles model probably describes faithfully the behavior of a presently observed real system so that in some cases some form of "mixture" of these models is appropriate. We consider in this section some relationships between all the models considered above, concentrating in particular on the way in which models can be defined to bridge the gaps between those we have considered.

The first such bridging model is that of Kingman (1977a). This is a multi-dimensional model generalizing, to an arbitrary number of dimensions, the one-dimensional charge-state model. Suppose that d characteristics of a gene can be measured: $d = 1$ for the (protein) charge-state model. Each characteristic is assumed to take one or other of the values $\{\dots -2, -1, 0, 1, 2 \dots\}$. If a parent gene has measurements $(y_1, y_2, \dots, y_d)$ we suppose a daughter gene has measurements $(y_1 + \delta_1, y_2 + \delta_2, \dots, y_d + \delta_d)$ with probability $u\, p(\delta_1, \delta_2, \dots, \delta_d)$, provided that not all the δ_i are zero. Here u is the total mutation rate so that $\Sigma\, p(\delta_1, \delta_2, \dots \delta_d) = 1$, the sum being taken over possible δ_i values except $\delta_1 = \delta_2 = \dots = \delta_d = 0$. Introduce the symmetrized value

$$\tilde{p}(\delta_1, \dots, \delta_d) = \tfrac{1}{2}\, [p(\delta_1, \dots, \delta_d) + p(-\delta_1, \dots, -\delta_d)]. \tag{9.78}$$

Consider now the random walk (in d dimensions) such that the probability of moving in one step from $(a_1, a_2, \dots a_d)$ to $(a_1 + \delta_1, a_2 + \delta_2, \dots, a_d + \delta_d)$ is $\tilde{p}(\delta_1, \dots, \delta_d)$. Let $\tilde{p}_n(\delta_1, \delta_2, \dots, \delta_d)$ be the n-fold convolution probability for this random walk, i.e. the probability of moving from $(a_1, a_2, \dots, a_d)$ to $(a_1 + \delta_1, a_2 + \delta_2, \dots, a_d + \delta_d)$ at the end of exactly n steps. Then Kingman (1977a) has shown that in a Wright Fisher model generalizing (9.10) the stationary probability $E(F)$ that the d measurements of two genes taken at random from the population are all identical is

$$E(F) = (1 + \theta)^{-1} + \sum_{n=1}^{\infty} \theta^n\, \tilde{p}_n(0, 0, \dots 0)(1 + \theta)^{-n-1}. \tag{9.79}$$

Here, as usual, $\theta = 4Nu$, N being the population size and u the total mutation rate. Equation (9.79) is useful in describing the way in which the homozygosity F depends on the dimension d of the process. Thus if, in one dimension,

$$\tilde{p}(-1) = \tilde{p}(+1) = \tfrac{1}{2},$$

then

$$\tilde{p}_n(0) = 0 \text{ for } n \text{ odd},$$

$$\tilde{p}_n(0) = \binom{n}{\tfrac{1}{2}n} (\tfrac{1}{2})^n \text{ for } n \text{ even}.$$

Thus, from (9.79),

$$E(F) = (1 + \theta)^{-1} + \sum_{n \text{ even}} \binom{n}{\tfrac{1}{2}n} (\tfrac{1}{2}\theta)^n (1 + \theta)^{-n-1},$$

and standard formulae show that this becomes

$$E(F) = (1 + 2\theta)^{-1/2}$$

in agreement with (9.16). In the limiting case $d \to \infty$, $p_n(0, 0, \dots 0) \to 0$ for all n and thus $E(F) \to (1 + \theta)^{-1}$ in agreement with (9.44). Unfortunately closed expressions for $E(F)$ cannot be found for general values of d, but Kingman shows that in the case where the possible values for $(\delta_1, \delta_2, \dots \delta_d)$ are $(+1, 0, 0, \dots 0), (-1, 0, 0, \dots, 0), \dots,$ $(0. 0, \dots, 0, -1)$, each with probability $(2d)^{-1}$, the bounds

$$\left[1 + \theta - \frac{(1+\theta)\theta^2}{2d(1+\theta)^2 + \theta^2}\right]^{-1} \leqslant E(F) \leqslant \left[1 + \theta - \frac{(1+\theta)\theta^2}{2d(1+2\theta) + \theta^2}\right]^{-1} \quad (9.80)$$

may be obtained. These inequalities show that $E(F)$ rapidly approaches the "$d = \infty$" value $(1 + \theta)^{-1}$ as d increases. It is reasonable to assume, then, that if $d = 5$ or more, the infinite allele theory of Sect. 9.6 and 9.7 can be used to a reasonable approximation to test for selective neutrality in the d-dimensional model, although clearly the geometrical structure of the latter model implies that some loss of information is involved in doing this.

A second model, leading to a similar conclusion, is that of Li (1976). In this model mutation can be "stepwise" or "non-stepwise", the former class following a charge-state model and the latter the infinite alleles model. There are three mutation probabilities: u_{+1} and u_{-1} (for stepwise mutations of +1 or −1) and v (for a unique non-stepwise mutation of the infinite allele type). Put

$$\theta = 4N(u_{+1} + u_{-1} + v), \quad \alpha = 4N(u_{+1} + u_{-1}), \quad \beta = 4Nv. \quad (9.81)$$

Then at equilibrium the probability that two genes drawn at random are identical (i.e. occupy the same state in the stepwise model and are identical by descent with respect to the non-stepwise mutations) is

$$E(F) = \{(1 + \theta)^2 - \alpha^2\}^{-1/2}. \quad (9.82)$$

For $\alpha = 0$, this reduces to (9.44), and for $\alpha = \theta$ it reduces to (9.16). Clearly, unless β is very small, $E(F)$ is much closer to $(1 + \theta)^{-1}$ than to $(1 + 2\theta)^{-1/2}$ so that, as with the d-dimensional process considered above, the infinite allele model provides the best approximation upon which to base a test of neutrality.

We next compare the genetic variation in the charge-state and the infinite sites models. Consider the infinite sites model with no recombination. A mutation at any site is assumed to change the charge level by +1, 0, or −1 with probabilities β, $1 - 2\beta$, β, respectively. These probabilities are the same for all sites and the mutations arise independently at the various sites according to a Poisson process. Let the mutation rate be u and put (as usual) $\theta = 4Nu$, where N is the population size. Suppose at stationarity two genes are taken at random from the population. These two genes will differ at i sites and will have charge level difference k ($i \geqslant k$). Then, for given k, the mean value of i is (Chakraborty and Nei, 1976)

$$\mu_k = \frac{\theta(1 - 2\beta + 2\beta\theta)}{1 + 4\beta\theta} + \frac{|k|(1 + \theta)}{(1 + 4\beta\theta)^{1/2}} \cdot \tag{9.83}$$

In particular, two electrophoretically identical genes will differ on average at

$$\mu_0 = \frac{\theta(1 - 2\beta + 2\beta\theta)}{1 + 4\beta\theta} \tag{9.84}$$

sites which, in the particular case $\beta = \frac{1}{2}$, reduces to

$$\mu_0 = \theta^2/(1 + 2\theta). \tag{9.85}$$

The variance of the number of different sites, and indeed the complete distribution, can also be obtained (Chakraborty and Nei, 1976). In particular the probability that two electrophoretically identical genes are identical at all sites is

$$(1 + 4\beta\theta)^{1/2}/(1 + \theta),$$

which, for $\beta = \frac{1}{2}$, becomes

$$(1 + 2\theta)^{1/2}/(1 + \theta). \tag{9.86}$$

This formula can be arrived at immediately from (9.16) and (9.44). With no intracistronic recombination (as we assume) the infinite site model acts, as is noted above, as an infinite allele model. Standard conditional probability formulae then yield (9.86) immediately.

It was observed in the previous section (and has been noted again in the preceding paragraph), that with no intracistronic recombination the infinite sites model behaves as an infinite alleles model. To the extent that intracistronic recombination is very rare, the testing theory of Sects. 9.6 and 9.7 can then be used for nucleotide sequence data: this will certainly give a better test than that assuming free recombination between sites and proceeding via Σy_i (see Eq. 9.77). Since the generalized charge-state model, for five or more dimensions, seems also to converge rapidly on the infinite alleles model, the theory of Sects. 9.6 and 9.7 appears to apply in a very wide variety of practical circumstances: this is why considerable attention has been paid to this theory above. It is thus perhaps fortunate that a more complete testing theory exists for the infinite alleles model than for any other model.

Because of this it is important to discuss the formula (3.78) upon which the infinite allele test ultimately depends. This formula was arrived at using a particular model, namely the generalized Wright-Fisher model (3.65). If (3.78) applies for this model only, it is not of much use for testing real data, since the population process generating the data might not even approximate (3.65). Fortunately this is not the case: Kingman (1977b) has shown that (3.78) applies for essentially any neutral model of the infinite alleles type. This implies that (3.78), and any test deriving from it, may be used with confidence in assessing whether a set of allele frequencies conforms reasonably to neutral theory predictions.

One problem concerning testing neutrality, whichever model is used, concerns non-stationarity. All models considered have as leading non-unit eigenvalue a quantity of the form $1 - CN^{-1}$, where C is of order unity and N is the population size. This implies a very slow rate of approach to stationarity so that many of the above formulae, which have been derived under a stationarity assumption, must be viewed with some reservation. A number of analyses consider rates of approach to stationarity of various sample and population characteristics but so far these have not been useful in deriving tests that work under non-stationary conditions.

9.10 Other Tests of Neutrality

A variety of less formal tests, more or less based on the theory given in this chapter, has been put forward to test the neutral theory. These include comparison of observed and expected frequency spectra (Ewens, 1969d; Maruyama and Yamazaki, 1974), tests based on estimated and observed values of θ (Ayala et al. 1972), tests correlating various genetic observations with environmental phenomena (Mitton and Koehn, 1973; McNaughton, 1974; Schaeffer and Johnson, 1974; Christiansen and Frydenberg, 1974). Such informal tests are often to be preferred to the more formal procedures given above, especially in ecological circumstances where a useful mathematical theory is unlikely. We do not present details of these tests here and close this chapter by outlining a test of selective neutrality based on phylogenetic trees (such as that of Fig. 8.2) constructed by a comparison of contemporary amimo acid sequences between species.

The neutral theory predicts "equal" rates of evolution: more specifically (see Sect. 9.3) it predicts that for any protein, if the neutral mutation rate is the same for all species, the rate of evolution will be the same for all species, irrespective of size. Of course this equality is on a "per generation" basis rather than a "clock" basis but, for a tree such as that in Fig. 8.2, this can be accommodated by a redefinition of the "nodal" times $t_1, t_2, \dots$.

This implies that the number of substitutions along any arm of the tree (in Fig. 8.2) is, under neutrality, a Poisson random variable with mean determined by the length of the arm and the mutation rate for the protein in question. (Gillespie and Langley, 1979, have shown that the Poisson assumption is not entirely accurate, but we nevertheless use it here.) Suppose we have data on the amino acid sequences of p proteins and assume that these proteins evolve at respective rates $\lambda_1, \lambda_2, \dots, \lambda_p$ ($\Sigma \lambda_i = 1$). For any arm on the tree joining nodes at times t_j, t_k, the mean number of substitutions in the mth protein is thus proportional to $\lambda_m(t_j - t_k)$: we may take this proportionality to be an equality by suitable redefinition of the t_j. If this is labelled the ith arm of the tree, the probability under neutrality that there will exist $x_{m,i}$ substitutions in the mth protein along this arm is then

$$L_0(m, i) = \exp\{-\lambda_m(t_j - t_k)\}\{\lambda_m(t_j - t_k)\}^{x_{m,i}}/x_{m,i}! \tag{9.87}$$

The neutral theory probability for the entire phylogeny is then

$$L_0 = \prod_m \prod_i L_0(m, i). \tag{9.88}$$

Assuming the $x_{m,i}$ values have been estimated (for example by the parsimony principle described in Chap. 8), the function L_0 may be maximized with respect to the λ_m and t_j to obtain maximum likelihood esimates $\hat{\lambda}_1, \ldots, \hat{\lambda}_p, \hat{t}_1, \hat{t}_2, \ldots$ of $\lambda_1, \ldots, \lambda_p, t_1, t_2, \ldots$. From these we compute the expected number $e_{m,i}$ of substitutions in the mth protein along the ith arm by the formula $e_{m,i} = \hat{\lambda}_m(\hat{t}_j - \hat{t}_k)$. The test of neutrality now reduces to a test of the compatibility of the "observed" values $x_{m,i}$ with the expected values $e_{m,i}$.

We do not enter into the details of this testing procedure (see Langley and Fitch, 1974). As it happens, available data reject the neutral theory at a probability level of order 10^{-5}. Various subhypotheses may also be tested, and adjustments to the test made because of the bias in estimating the $x_{m,i}$ as a result of the parsimony principle: in all cases neutrality is still rejected. Despite this, the vast amount of information currently appearing on nucleotide sequences can be used to give more efficient testing procedures than these, and a final judgement on the neutral theory, at least from the point of view of reconstructed phylogenetic trees, should be suspended until such tests have been carried out.

10. Generalizations and Conclusions

10.1 Introduction

This comparatively brief chapter has two main aims. The first is to recast and redevelop some of the arguments raised in Chap. 1 in a more modern mathematical form and to give a more up-to-date account of the purely mathematical aspects of various evolutionary questions treated by Fisher and Wright. This will be done in Sect. 10.2. Second, in Sects. 10.3–10.7, various important special topics treated only sketchily (or not at all) in previous chapters will be discussed. Even here our account will be incomplete (to a great extent because some of these topics require extensive verbal as well as mathematical discussion): each of these topics would require a monograph in itself for a proper presentation. Finally, in Sect. 10.8, a brief summary of this book will be given.

10.2 Mathematical Extensions

In this section we discuss, at greater length than was done in Chap. 1, aspects of the mathematical theory of fitness, the evolution of mutation rates, branching processes, sex-ratio and the correlation between relatives.

In most of this book "fitness" has been taken to mean viability fitness, i.e. a measure of the capacity of an individual of a given genotype to survive from the time of conception to the age of reproduction. In Sect. 2.6 fertility selection was discussed briefly, but nowhere have we considered the fitness component relating to mating success. Fitness properly embraces all three aspects and, to be considered in greater detail, should involve age distribution, ecological and other factors, very few of which are mentioned in the literature in any detail. Indeed the concept of fitness is possibly too complex to allow of a useful mathematical development: nevertheless, since it enters fundamentally into many population genetics considerations, it is remarkable how little attention has been paid to it.

The most complete discussion of the concept of fitness is that of Kempthorne and Pollak (1970), who emphasize several points worth repeating. First, while it is uniformly agreed that fitness is a property of an entire individual (or genome), it is also apparently agreed, with Wright (1931), that to a first approximation, for a short time, a constant net selective value for any allele may usefully be defined. However, evolutionary arguments require consideration of very long time periods, and it is not certain that this approximation is adequate for them. The analysis of the interactive effects of genes at several loci given in Chaps. 6 and 7 is relevant to this point. Second, the obscurity of the Fisherian definition of fitness through Malthusian parameters is to be stressed, particularly since this definition does not usually fit with well-formed mathematical evolutionary models. Finally the very concept of fitness (as mentioned above), in particular when fecundity parameters are not in the multiplicative form (2.27), appears elusive: at best the fitness of any genotype must be frequency-dependent, and it then appears impossible to secure an analog of the Fundamental Theorem under these circumstances. The concept of fitness is not well defined in any other than very simple models.

We turn next to the evolution of mutation rates. It is known that these are subject to evolutionary adjustment, and we have noted the qualitative argument (in Sects. 1.4 and 1.5) that mutation rates have reached presently observed values by natural selection as a compromise between higher rates (favoring adaptability) and lower rates (favoring present adaptedness). This argument has been discussed quantitatively by Kimura (1967) and Leigh (1973). Both authors consider the introduction of genes modifying mutation rates and note that the optimality theory invokes intergroup selection arguments. In Sect. 10.5 we discuss evolution in fluctuating environments: Leigh (1973) discusses the possible advantage of an increased mutation rate for very rapid fluctuations and finds little theoretical support for such an advantage, concluding that the problem is best treated theoretically by comparing the cost of introducing mutation rate suppressor genes with the benefit of such suppression. No theoretical work has been done along this interesting line.

We consider next Fisher's branching process discussion outlined in Sect. 1.4 and culminating in Eqs. (1.81) and (1.84). These conclusions may be found as particular cases of the general theory of branching processes, a field inspired in part by Fisher's early work in population genetics theory. The appropriate generalization of (1.81) is due to Kesten et al. (1966): if the offspring distribution in a branching process has mean unity, finite variance σ^2, and is otherwise arbitrary, and if Z_n is the number of individuals in the process in generation n, then

$$E[Z_n | Z_n > 0] \sim \tfrac{1}{2} n \sigma^2 \tag{10.1}$$

and

$$\text{Prob}\,[2Z_n / n\sigma^2 \leqslant x | Z_n > 0] \to 1 - e^{-x}, \quad x > 0. \tag{10.2}$$

If the mean offspring distribution is slightly in excess of unity, Fahady et al. (1971) have shown that, under certain mild restrictions,

$$\text{Prob}\,[Z_n / \alpha_n \leqslant x | Z_n > 0] \sim 1 - e^{-x}. \tag{10.3}$$

where $\alpha_n = E[Z_n | Z_n > 0]$. This is the required generalization of (1.84).

The problem of the evolution of sex-ratio was discussed in Sect. 1.5, and the "non-genetic" argument of Fisher concerning this evolution was noted. A number of genetically based arguments for sex-ratio adjustment have been put forward since

Fisher's time: see in particular Shaw and Mohler (1953), Shaw (1958) and Eshel (1975). For theoretical analyses of Fisher's theory see Kolman (1960), Bodmer and Edwards (1960), Edwards (1961) and Verner (1965). We briefly describe the most recent approach to this problem (Uyenoyama and Bengtsson, 1979) which is of particular interest since the analysis is genetically based and is close in spirit to Fisher's original argument.

Consider, in a diploid population, an autosomal locus admitting alleles A_1 and A_2 and hence genotypes A_1A_1, A_1A_2 and A_2A_2 (which we call genotypes 1, 2, 3 respectively). The frequency of males (females) of genotype i is $m_i(f_i)$ and $\Sigma m_i = \Sigma f_i = 1$. We make two assumptions concerning the female genotypes. The first is that different genotypes have different brood sizes: suppose females of genotype i have brood size proportional to σ_i. Second we suppose that the sex-ratio among offspring depends on the maternal genotype: let females of genotype i produce a fraction α_i of male offspring and $1 - \alpha_i$ of female offspring.

The values α_i and σ_i specify the recurrence relations of the f_i and m_i. We are particularly interested in the equilibria of these recurrence relations and will consider these only in the particular case where the condition

$$\sigma_1\sigma_2(\alpha_1 - \alpha_2) + \sigma_2\sigma_3(\alpha_2 - \alpha_3) + \sigma_3\sigma_1(\alpha_3 - \alpha_1) = 0 \tag{10.4}$$

holds; we show later that an argument using the parental expenditure concept leads to (10.4). The recurrence relations have three equilibria, one of which is symmetric ($m_i = f_i$) and the others asymmetric ($m_i \neq f_i$). We focus attention here on the asymmetric equilibria.

The frequencies M of males and F of females in the population are

$$M = \Sigma f_i\sigma_i\alpha_i/\Sigma f_i\sigma_i, \quad F = \Sigma f_i\sigma_i(1 - \alpha_i)/\Sigma f_i\sigma_i. \tag{10.5}$$

Shaw and Mohler (1953) and others define the mean fitness W for this model as

$$W = \Sigma f_i\sigma_i\{\alpha_iM + (1 - \alpha_i)F\}. \tag{10.6}$$

Subject to $\Sigma f_i = 1$, W may be maximized with respect to the f_i, and, assuming that (10.4) holds, the maximizing values are found to occur precisely at the asymmetric equilibria of the system. In other words, if the system evolves towards such an equilibrium, it is evolving in such a way as to optimize the sex-ratio (as measured by W). The optimizing value is

$$F/M = \{\sigma_1(1 - \alpha_1) - \sigma_2(1 - \alpha_2)\}/\{\sigma_2\alpha_2 - \sigma_1\alpha_1\}. \tag{10.7}$$

It remains for us to justify that (10.4) will be true under the parental expenditure concept. Suppose the ratio of the parental expenditure required to raise a female offspring to maturity compared to that to raise a male offspring is $\phi:1$: the mean expenditure per offspring of a female of genotype i is then proportional to $\alpha_i + \phi(1 - \alpha_i)$. If females of all genotypes make the same total expenditure for their entire brood, then the brood size σ_i must statisfy

$$\sigma_i = K \{\alpha_i + \phi(1 - \alpha_i)\}^{-1} \tag{10.8}$$

for some constant K. Values of σ_i satisfying (10.8) automatically satisfy (10.4) and furthermore lead to an F/M value of ϕ^{-1}, as might be expected: the optimal sex-ratio is determined entirely by the relative expenditure in rearing male and female offspring.

The final topic considered in this section is the correlation between relatives for a metrical trait determined by many loci and also by the environment. In Chaps. 2 and 7 the purely genetic aspects of such correlations were considered under the assumption of random mating with respect to the trait of interest. For practical applications it is important to discuss the consequences of assortative mating and environmental effects on these correlations.

Almost all authors, when considering assortative mating and environmental effects, have assumed (usually implicitly) that the loci controlling the trait are unlinked, that complete linkage equilibrium holds, and that no epistatic effects arise. In this case (7.46) simplifies to

$$\text{Covar}\ (X, Y) = \alpha \sum \sigma_A^2(k) + \beta \sum \sigma_D^2(k) \tag{10.9}$$

where α and β are determined in (7.47). We rewrite this as

$$\text{corr}\ (X, Y) = \alpha H + \beta D, \tag{10.10}$$

where $H = \sum \sigma_A^2(k)/\sigma^2$, $D = \sum \sigma_D^2(k)/\sigma^2$, a notation more in line with that of biometrical genetics and one we follow for the remainder of this section. We emphasize the strong assumptions that have been made in replacing complex equations such as (7.53), (which is itself far less complex than the corresponding equation when linkage equilibrium is not assumed) by the simple Eq. (10.10). (Note that in the biometrical literature, e.g. Eaves et al. 1977, it is sometimes implied that absence of assortative mating ensures linkage equilibrium, but the theory of Chap. 7 shows that this implication is incorrect.)

The case of assortative mating (without environmental effects) was considered by Fisher (1918) and many subsequent authors. We assume a correlation r_{pp} between mating individuals for the trait in question: $r_{pp} = 0$ leads to equations like (10.10). When there is no dominance, Nagylaki (1978a) arrives by very simple arguments at formulae for the correlation between relatives while, when dominance does exist, Wright (1921), Crow and Felsenstein (1968) and Wilson (1973) arrive, with rather greater labor, at extensions of these formulae. Some important examples of these correlations are the following:

corr (parent-offspring) $\quad = r_{po} = \frac{1}{2}(1 + r_{pp})H,$

corr (full sibs) $\quad = r_{ss} = \frac{1}{2}(1 + r_{pp})H + \frac{1}{4}D,$

corr (nth generation grandparent) $= r_{po}\ \{\frac{1}{2}(1 + r_{pp}H)\}^{n-1},$ (10.11)

corr (uncle, nephew) $\quad = r_{un} = \{\frac{1}{2}(1 + r_{pp}H)\}^2 H + \frac{1}{8}r_{pp}DH,$

corr (first cousins) $\quad = r_{fc} = \{\frac{1}{2}(1 + r_{pp}H)\}^3 H + D\{\frac{1}{4}r_{pp}H\}^2.$

More complex formulae for models where selection acts against individuals with extreme values of the character have been given by Wilson (1973, 1976).

We turn now to environmental effects. These were already considered by Fisher (1918) but only for the simplified model where the phenotypic value P of the trait can be written as $P = G + E$, G and E being genetic and environmental contributions, and for which

$$\sigma_p^2 = \text{Var}(P) = \text{Var}(G) + \text{Var}(E). \tag{10.12}$$

The equation $P = G + E$ implies no interaction between genotype and environment, while (10.12) further implies no covariance. While these assumptions may be reasonable as approximations in various controlled breeding experiments, they are severe assumptions for a trait such as IQ in humans. The entire work of Burt (1971) on this trait, for example, makes precisely these assumptions. [Burt's approach is to define $E = \text{Var}(E)/\sigma_p^2$ and to add E to each correlation displayed in (10.11), where now H and D are redefined using σ_p^2 instead of σ^2 as divisor. Burt also uses a variance term for assortative mating but this appears unjustified and is not followed by other authors. If any two correlations can be estimated from data, these formulas then give H, D and E (note that $H + D + E = 1$): Burt uses the three correlations r_{po}, r_{pp} and the correlation $(1 + E)^{-1}$ of identical twins, since under his model a further correlation is required to estimate the "assortive mating" component of variance.]

A more realistic model clearly allows potential genotype-environment interaction and covariance and also distinguishes the almost certainly different enironmental correlations for individuals of different degrees of relatedness. Various models, of increasing complexity, have been put forward (largely in the psychological literature) to this end. Jinks and Eaves (1974) assume an added environmental correlation only for those individuals in the same family (i.e. parents and offspring living together): this applies for the first two correlations in (10.11). A revised model adds a correlation only for sibs raised together. Eaves et al. (1977) present a model with two environmental correlations, E_1 for "within families" and E_2 for "between families": here E_1 is added to the first two correlations in (10.11) and E_2 to the last three.

With empirical values for correlations between a variety of relatives (perhaps 10–15 relationships) the parameters of such a model can be estimated by least squares and the goodness of fit of the model tested by chi-square, the argument in this procedure being that if a fit is not rejected by chi-square the model can be accepted and a more complex model is not needed. Perhaps the major difficulty with this procedure is the low power of the goodness-of-fit procedure. Thus Last (1976) shows that for reasonable parameter values sample sizes in excess of 5,000 are required to be fairly certain of detecting genotype-environment interactions of some magnitude. The question of the adequacy of various models and of the above fitting procedure has led to much acrimonious discussion in the literature: see Eaves et al. (1977), Mather and Jinks (1977) and Goldberger (1978a, b) for summarizing (and contrasting) views. Our interest here is more in the genetic aspects, and we conclude by emphasizing that essentially all models used in the biometrical literature, on all sides, have used simple formulae, such as (10.9), for the purely genetic components of correlation and that in practice far more complex formulae are possibly required. While simplifying assumptions must be made in these analyses at some level, it is not yet certain that the level chosen, i.e. leading to (10.9), is a satisfactory one.

10.3 Geographical Structure

The importance of geographical structure to the evolutionary theories of Wright (1931, 1969a, b, 1977, 1978) has been noted in Chap. 1. The effects of such structure have been considered at some length in the literature and here we refer, very briefly, to several features of the analysis.

The two main types of model of geographical structure considered in the literature are an "island" model of distinct subpopulations, or demes, and a continuous cline model (in one or two dimensions). The most elementary fact about an island model is that, if random mating obtains within any island, the fraction of individuals who are heterozygotes (at a single locus admitting two alleles) is necessarily less than (or equal to) the fraction for a large random-mating population with the same mean allelic frequency. If the frequency of A_1 in island i is x_i, and island i comprises a fraction f_i of the entire population, this (Wahlund) inequality can best be demonstrated through the equation

$$2\bar{x}(1 - \bar{x}) = 2 \sum f_i x_i (1 - x_i) + 2 \sum f_i (x_i - \bar{x})^2, \tag{10.13}$$

where $\bar{x} = \sum f_i x_i$. We use this fact in a moment when considering mean fixation times.

One interesting class of problems concerns quantities that are not affected by geographical structure. For a finite population, Maruyama (1970, 1971, 1974) has found two such quantities, at least for selectively neutral loci and certain genetic models. The first of these is the probability of fixation of a given allele, and the second is the mean total number of heterozygotes to appear as a result of a single new mutation. It then follows automatically from (10.13) that the mean time to fixation in the subdivided case is larger than that in the undivided case, and this has been confirmed (Maruyama, 1971) by simulation. In the former case, on average, a smaller number of heterozygotes appears per generation, but for a greater number of generations, than in the latter case.

The eigenvalues in a genetic model involving geographical subdivision were given in (3.114), and the consequent effective population size was noted in (3.115). Note that except for very small migration rates, this does not differ much from the actual population size. We conclude that in this model the effect of subdivision is not important and that for many purposes the population can be taken as one large random-mating population. To this extent the evolutionary theory of Wright, depending to some extent on population subdivision, must be questioned. However, in the model leading to (3.114), the migration is isotropic, and a much less extreme conclusion holds for structured populations where migration is most likely to occur to (and from) neighbouring subpopulations. Further eigenvalue questions have been discussed by Maruyama (1970, 1971, 1972), Nagylaki (1974b, 1976, 1977c), and Kimura and

Maruyama (1971). Kimura and Maruyama also note one further important result: even in the selectively neutral case clines of gene frequency can occur, the propensity for this depending on the subpopulation sizes and migration rates. Nagylaki (1978b) has obtained mean gene frequencies at any point in the habitat and covariances between frequencies at two points in a stochastic model incorporating selection as well as migration and random genetic drift in a cline. Slatkin and Maruyama (1975) discuss the effect of stochastic gene frequency fluctuations on the slope of gene frequencies in a cline and show that the slope is decreased through such fluctuations.

A second form of stochastic fluctuation occurs in infinite populations where the migration rate and the gene frequencies of immigrants into any subpopulation are random variables. This model has points of similarity with those discussed in Sect. 10.5, and we do not discuss it in detail here (see Nagylaki, 1978c).

A further question we discuss concerns the possibility of using observed gene frequencies in various subpopulations as a test for selective neutrality, perhaps of an informal nature such as those discussed in Sect. 9.10. Two such tests have been proposed by Lewontin and Krakauer (1973), and we now describe briefly the model behind these tests and the nature of the tests themselves. Consider n subpopulations, of respective sizes $N_1, N_2, ..., N_n$, each exchanging migrants from a "mainland" population on which the frequency of some allele A_1 is constant at x^*. The stochastic theory shows easily that the frequency x_i of A_i in subpopulation i has, at stationarity, the distribution

$$f(x_i) = \text{const } x_i^{4N_imx^*-1}(1 - x_i)^{4N_im(1-x^*)-1} \tag{10.14}$$

where m is the migration rate. This distribution has mean and variance

$$x^* \quad \text{and} \quad x^*(1 - x^*)/(1 + 4N_im), \tag{10.15}$$

respectively. If $\bar{x} = \Sigma N_i x_i / \Sigma N_i$, it follows that to a first approximation the mean value of F, defined by

$$F = \Sigma (x_i - \bar{x})^2 / \{(n - 1)\bar{x}(1 - \bar{x})\} \tag{10.16}$$

is independent of x^*. Lewontin and Krakauer then compute values $F_1, F_2, ..., F_K$ of F at K different loci and from these derive the standardized variables

$$G_j = (n - 1)F_j/\bar{F}, \quad j = 1, 2, ..., K, \tag{10.17}$$

where $\bar{F} = K^{-1}(F_1 + F_2 + ... + F_K)$. Their first test for neutrality consists of a comparison of the empirical distribution of the G_j's to a chi-square distribution with $n - 1$ degrees of freedom, the neutral theory being accepted if the fit is satisfactory. However, G_j has a chi-square distribution in effect only if the subpopulation gene frequencies have a normal distribution, and this is not implied by selective neutrality. Indeed the selective stationary distribution (5.70) can be normal, and thus this test has no real foundation in statistical theory.

The second testing procedure uses as test statistic the quantity k, defined by

$$k = (n - 1)S_F^2/\bar{F}^2,$$ (10.18)

where

$$S_F^2 = \Sigma (F_i - \bar{F})^2/(K - 1).$$ (10.19)

Neutrality is rejected if k significantly exceeds 2. Now if the x distribution has mean μ, variance σ^2, and fourth central moment μ_4, to a first approximation the mean value of k is

$$2 + \gamma_2,$$ (10.20)

where $\gamma_2 = (\mu_4/\sigma^4) - 3$, the kurtosis of the x distribution. However, there is no reason why the kurtosis of a neutral stationary distribution should not exceed zero: thus for the neutral distribution (5.69) with $2\beta_1 = 0.1$, $2\beta_2 = 0.9$, the mean of k is about 7.61, well in excess of the proposed neutral limit of 2. We conclude again that this testing procedure has little basis in statistical theory. For further remarks see Nei and Maruyama (1975), Robertson (1975) and Ewens and Feldman (1976).

A considerable literature also exists on the deterministic theory of geographically structured populations, originating with the remarkable pioneering paper of Fisher (1937). Of particular interest, in view of the theory of Chap. 6, is the demonstration by Li and Nei (1974) that linkage disequilibrium can be generated by geographical subdivision, even in the absence of epistasis. We consider here, however, an example of two questions specific to geographic subdivision, namely whether selection can maintain a cline of gene frequencies from west to east if A_1 is favored in the west and A_2 in the east, and second whether the frequency of A_1 can be sustained at positive values when A_1 is favoured only in a finite interval of the entire east-west line. These questions have been considered in particular by Nagylaki (1975), and we follow his analysis of them closely.

Consider the line $[L, R]$, where possibly $L = -\infty$ or $R = +\infty$, and suppose at the point x on this line that the fitnesses of A_1A_1, A_1A_2 and A_2A_2 are $1 + sg(x), 1 + hsg(x)$ and $1 - sg(x)$. (Note that in this notation $h = 0$ corresponds to no dominance.) The small parameter s is a measure of the strength of selection. Each individual is supposed to migrate, from the time of birth to the time of reproduction, by a random amount y, where y has a normal distribution with mean 0, variance σ^2. Migration of different individuals is independent. The frequency $p = p(x; t)$ of A_1 at the point x at time t then satisfies the partial differential equation

$$\frac{\partial p}{\partial t} = \tfrac{1}{2}\sigma^2 \frac{\partial^2 p}{\partial x^2} + sg(x)p(1 - p)\{1 + h - 2hp\},$$ (10.21)

together with the boundary conditions $\partial p/\partial x = 0$ at $x = L, x = R$. This is a generalization of the formula of Fisher (1937). If a stationary cline exists there must be a solution of (10.21) with $\partial p/\partial t = 0$. In the important case $h = 0$ the equilibrium cline equation is thus

$$\frac{d^2 p}{dx^2} + \frac{2s}{\sigma^2} g(x)p(1-p) = 0, \tag{10.22}$$

subject to the boundary condition $dp/dx = 0$ at $x = L, x = R$. This equation always has the trivial solutions $p(x) \equiv 0, p(x) \equiv 1$, and our aim is to find conditions for non-trivial solutions. These will clearly depend on the numerical value of the parameter $2s/\sigma^2$ as well as the nature of $g(x)$. In the particular case where $L = 0, R = +\infty, s > 0$, and

$$g(x) = \begin{cases} 1, & 0 \leqslant x \leqslant a \\ \\ -\alpha^2, a < x < \infty \end{cases} \tag{10.23}$$

(so that A_1 is favored when $x \leqslant a$ and A_2 when $x > a$), Naylaki shows that the necessary and sufficient condition for a unique non-trivial solution of (10.22) is

$$a\sqrt{2s} > \sigma \arctan \alpha. \tag{10.24}$$

Clearly (10.24) will apply for sufficiently large values of a and s and will not apply for sufficiently large σ and, to a much smaller extent, α. Note that the right-hand side in (10.24) increases only from 0.79 to 1.59 as α increases from 1 to ∞, but increases linearly with the migration distribution standard deviation σ. Clearly, while intuitively an inequality of the general from (10.24) is to be expected, the particular form of (10.24) is perhaps surprising and indicates explicitly the relevance of the various parameters involved to the maintenance of A_1.

This analysis can be taken over immediately to the case of a "pocket" in which A_1 is favoured. By reflecting the interval $(0, a)$ about $x = 0$, we find that if

$$g(x) = \begin{cases} 1, & -a \leqslant x \leqslant a \\ \\ -\alpha^2, |x| > a \end{cases} \tag{10.25}$$

(so that A_1 is favoured in the "pocket" $(-a, a)$ but not elsewhere), the condition that A_1 can be maintained in the population is again (10.24).

Further analyses can be made for other functional forms for $g(x)$, but we do not pursue the details. Analyses of this sort are most relevant to Wright's theory of evolution and, further, to important questions concerning theories of allopatric and sympatric modes of speciation, as discussed most recently, for example, in White, 1978.

10.4 Age Structure

So far all our analyses have ignored age structure: perhaps curiously, the effect of age structure has been considered far more in mathematical ecology than in mathematical population genetics. To take account of age structure one must specify age-specific reproductive and survival schedules for all genotypes of both sexes and must also make assumptions concerning the sex-ratio and the mating process. While Norton (1928) and Haldane (1927) considered age-structured population many years ago, it is only recently that further attention has been paid to them in any detail (see, for example, Demetrius, 1971, 1974, 1975, 1976, 1977; Charlesworth, 1970, 1971, 1972, 1973, 1974). An excellent summary of the topic is given by Charlesworth (1976). Perhaps the most important aim, for age-structured populations, is to establish natural definitions of fitness that allow much of the classical theory to be applied. In this direction, Charlesworth (1976) has shown that under certain conditions a natural definition of the fitness of any genotype exists, and that, with this definition, natural selection leads to equilibria of the form (1.30) in the overdominant case and to dynamical equations of the form (1.25) if one allele is becoming fixed in the population. Similarly, Demetrius (1974) has given an analog of the Fundamental Theorem for age-structured populations. It thus appears likely that age structure does not introduce radically new behavior in populations compared to that expected from classical analyses. For this reason we do not consider it in any further detail in this book.

10.5 Random Environments

Two models involving random temporal changes in selective parameters in infinite populations were introduced and considered in Sect. 5.8. The theory of such models is analysed further in this section, both finite and infinite populations being considered. The case of spatial as well as temporal variation is discussed, and the section closes with remarks concerning the usefulness of these models in explaining the nature and extent of observed genetic polymorphisms.

We consider first an infinite diploid population where in generation n, the fitnesses of the various genotypes are

$$
\begin{array}{ccc}
A_1A_1 & A_1A_2 & A_2A_2 \\
1 + \eta_n & 1 & 1 + \epsilon_n.
\end{array}
\tag{10.26}
$$

(Note that since only ratios of fitnesses are relevant we do not lose generality by fixing the fitness of A_1A_2 at unity.) The fitnesses η_n and ϵ_n are the realized values (in generation n) of random variables η and ϵ. These variables are possibly correlated for a given n (for example, in one model mentioned below, $\eta_n \equiv -\epsilon_n$), but we assume (η_n, ϵ_n) to be independent of (η_m, ϵ_m), $(n \neq m)$. Throughout we assume

$$E(\eta) = \mu_1 \delta, \quad Var(\eta) = \sigma_1^2 \delta, \quad E(\epsilon) = \mu_2 \delta, \quad Var(\epsilon) = \sigma_2^2 \delta \qquad (10.27)$$

for some small parameter δ, while higher moments are $O(\delta^2)$ or less.

Let x_n be the frequency of A_1 in generation n. Our objective is to examine properties of the random sequence $x_1, x_2, x_3, \ldots$ as determined by the distribution of (η, ϵ), and the starting point for this is clearly the recurrence relationship

$$x_{n+1} = x_n(1 + \eta_n x_n)/\{1 + \eta_n x_n^2 + \epsilon_n(1 - x_n)^2\}. \qquad (10.28)$$

Further progress is made by introducing the concept of the stochastic local stability of an equilibrium. An equilibrium x^* is said to be stochastically locally stable (SLS) (Karlin and Liberman, 1974) if for any $\gamma > 0$ there exists a neighborhood $x^* - \xi, x^* + \xi$ such that for any initial frequency x_1 in this neighborhood,

$$\text{Prob } \{\lim_{n \to \infty} x_n = x^*\} > 1 - \gamma. \qquad (10.29)$$

We have already encountered this concept for the model (5.81): here $x = 1$ is SLS when $E \log (1 - s) < 0$ and $x = 0$ is SLS if $E \log (1 - s) > 0$. The selective parameters (5.81) are equivalent to $(1 - s)^{-1}$, and $1 - s$. This conclusion is a particular case of a more general result, namely that in the notation (10.26) $x = 0$ is SLS if $E \log (1 + \epsilon) > 0$ and $x = 1$ is SLS if $E \log (1 + \eta) > 0$. For a parallel conclusion for diffusion processes, see Levikson and Karlin (1975).

These conclusions emphasize two important features of stochastic environment models. The first is that it is the geometric mean fitness (the geometric mean of a positive random variable y defined as $\exp (E \log y)$) that determines stability behavior, rather than the arithmetic mean. The second feature, which arises from the first, is that the variance in the distribution of a selective parameter is just as important as the mean in determining evolutionary behavior. This point has been stressed by Gillespie (1974a, b) and forms a central theme of his theory of stochastically varying environments considered later. Thus since

$$E \log (1 + \eta) = (\mu_1 - \tfrac{1}{2}\sigma_1^2)\delta + O(\delta^2), \qquad (10.30)$$

the condition that $x = 1$ be SLS is $\mu_1 - \tfrac{1}{2}\sigma_1^2 > 0$, showing the importance of the variance in determining SLS behavior at $x = 1$. A parallel remark applies for $x = 0$.

Consider now the model (5.91) with $\alpha = 1$. Here the equilibrium at $x = 0.5$ is SLS if $E \log \{(2 + s)/(2 + 2s)\} > 0$: for small s this is equivalent to $E(s) - \tfrac{1}{2}Var (s) < 0$. The case $\alpha = -1$ is of greater interest since then fitnesses are additive. For fixed fitnesses A_1 will become fixed (when $s > 0$) or A_2 will become fixed (when $s < 0$). When s is a random variable, however, neither $x = 0$ nor $x = 1$ is necessarily SLS: the conditions that $x = 0$ and $x = 1$ are SLS are, for small s,

$$-E(s) - \tfrac{1}{2}Var(s) > 0, \quad E(s) - \tfrac{1}{2}Var (s) > 0, \qquad (10.31)$$

respectively. Clearly if $Var(s)$ is sufficiently large compared to $E(s)$, neither $x = 0$ nor $x = 1$ is SLS.

We turn now to finite populations. Here stochastic fluctuations in gene frequency occur for two reasons, random sampling effects and stochastic variation in fitness. We consider here only diploid models with fitness scheme (10.26). If $E(\eta) = E(\epsilon)$, $Var(\eta) = Var(\epsilon) = \sigma^2$, corr $(\eta, \epsilon) = r$ and A_1 mutates to A_2 at rate u with reverse mutation also at rate u, there will exist a symmetric (about $x = 0.5$) stationary distribution of the frequency of A_1 for any finite population size N. This distribution can be found explicitly (Avery, 1977) and increasingly concentrates near $x = \frac{1}{2}$ as σ^2 increases. In other words, increasing the variance of the homozygote fitnesses increases the degree of heterozygosity in the population. The degree of heterozygosity also increases with r. These interesting and important conclusions differ qualitatively from those found by incorrect analyses of Kimura (1954) and Ohta (1972).

If there is no mutation, interest centers on fixation probabilities and mean fixation times. Avery (1977) shows that the probability of fixation of a new mutant increases substantially as σ^2 increases but is almost invariant to the value of r. The conditional mean times to fixation and to loss also increase substantially with σ^2 and with r: again the earlier conclusions of Kimura and Ohta are qualitatively incorrect. Note the general observation that in all cases, increasing the variance of homozygote fitness tends to increase genetic variation in populations, often by substantial amounts. This observation is clearly of some importance in possibly explaining at least in part the large genetic variation actually observed in natural populations.

We turn now to spatial variation. Although normally this is considered when temporal variation also occurs, we consider first a model (due to Levene, 1953) where fixed fitness regimes occur in each of M habitats, the fitnesses in habitat i being $1 + \eta^{(i)}$, 1, $1 + \epsilon^{(i)}$. In this "fixed Levene model" the entire population mates at random in a common area and then disperses to the various habitats, a fraction c_i entering habitat i. The recurrence relation for the frequency x of A_1 is

$$x_{n+1} = \sum_i c_i[\{x_n + \eta^{(i)} x_n^2\}/\{1 + \eta^{(i)} x_n^2 + \epsilon^{(i)}(1 - x_n)^2\}]. \qquad (10.32)$$

Of particular interest are the equilibrium solutions of this system. Karlin (1977b) has proved the remarkable result that in general at most three internal equilibria of (10.32) can exist and that in certain cases at most one can exist. In particular, when

$$\eta^{(i)} + \epsilon^{(i)} + \eta^{(i)} \epsilon^{(i)} \leqslant 0 \qquad (10.33)$$

for $i = 1, 2, 3, ..., M$, at most one internal equilibrium exists, and if one does exist it is globally stable. If no such equilibrium exists there is a unique globally stable fixation point.

One case where (10.33) plainly applies is where $\eta^{(i)} < 0$, $\epsilon^{(i)} < 0$ for all i. Here there always exists a (unique globally stable) internal equilibrium, analogous to (1.30)). A second case where (10.33) holds is the linear scheme $\eta^{(i)} = -\epsilon^{(i)}$. Whether or not an internal equilibrium exists depends in a complicated way on the $\eta^{(i)}$ and c_i values. A parallel remark holds when fitnesses are of the multiplicative form where $1 + \epsilon_i = (1 + \eta_i)^{-1}$. The location of the equilibrium must be found numerically from (10.32).

As might be expected, if $1 + \eta^{(i)} > 1 > 1 + \epsilon^{(i)}$ for all i, A_1 becomes fixed in the population, with a corresponding conclusion for A_2 when $1 + \eta^{(i)} < 1 < 1 + \epsilon^{(i)}$.

We consider finally models involving both spatial and temporal variation. Gillespie (1974a, b, 1976a, b, 1977a, b, 1978) and Gillespie and Langley (1974, 1976) have analysed an increasingly complex series of such models, culminating in a "stochastically additive scale, concave fitness function" (SAS-CFF) model that reaches broad predictions concerning natural populations that fit well with observed genetic patterns. While various other spatial-temporal models also have to be considered, we concentrate here entirely on the Gillespie-Langley model and its practical implications.

Consider first a single population with fitnesses of the additive form

A_1A_1	A_1A_2	A_2A_2
$1+s$	$1+\frac{1}{2}(s+t)$	$1+t$

$$(10.34)$$

Here the values of s and t vary from generation to generation according to an independent process for which

$$\mathrm{E}(s)=\mu_s, \quad \mathrm{Var}(s)=\sigma_s^2, \quad \mathrm{E}(t)=\mu_t, \quad \mathrm{Var}(t)=\sigma_t^2, \quad \mathrm{corr}(s,t)=\rho. \qquad (10.35)$$

By suitably amending (10.31) or by direct argument, Gillespie (1974a) proves that a polymorphic stationarity distribution exists if $|\alpha|<1$, where

$$\alpha = 4|\mu_s-\mu_t+\tfrac{1}{2}(\sigma_t^2-\sigma_s^2)|/\{\sigma_s^2+\sigma_t^2-2\rho\sigma_s\sigma_t\}, \qquad (10.36)$$

and that when this is so the stationary distribution of the frequency x of A_1 is

$$f(x)=\mathrm{const}\, x^\alpha\,(1-x)^{-\alpha}. \qquad (10.37)$$

Consider now a random Levene model which is identical to the fixed Levene model considered above, except that the fitnesses in any subpopulation varies through time. Thus at time n, the fitnesses in the ith subpopulation are $1+s_n^{(i)}, 1+\frac{1}{2}(s_n^{(i)}+t_n^{(i)}), 1+t_n^{(i)}$, respectively. These fitnesses are assumed independent over time but not necessarily from one subpopulation to another at any given time. Then (Gillespie, 1974a) the appropriate generalization of the above result is that a stationary polymorphic distribution for the frequency x of A_1 exists if and only if

$$4|\mu_s-\mu_t+\tfrac{1}{2}(\sigma_t^2-\sigma_s^2)|/\{(\sigma_s^2+\sigma_t^2-2\rho\sigma_s\sigma_t)(1+\pi(1-k))\}<1, \qquad (10.38)$$

where

$$\pi = 2\sum_{i>j} c_i c_j, \quad k=(\sigma_{ss}+\sigma_{st}-2\sigma_{sstt})/(\sigma_s^2+\sigma_t^2-2\rho\sigma_s\sigma_t),$$

$\sigma_{ss}(\sigma_{tt})$ = covariance of the $s(t)$ values in two subpopulations at the same time,
σ_{sstt} = covariance of the s value from one subpopulation with the t value from another subpopulation at the same time.

The stationary distribution is still (10.37). Note that increased patchiness (as measured by increased π) increases the chances of polymorphism, as does a decrease in the spatial correlations: for completely correlated environments $k = 1$, in which case (10.38) reduces to $|\alpha| < 1$, while for $k < 1$ the requirement (10.38) is easier to satisfy than the condition $|\alpha| < 1$. Note that the means and variances of s and t are both important determinants of whether or not a stationary polymorphic distribution exists.

All the above assumes additive fitnesses. Gillespie (1976a, b, 1978) and Gillespie and Langley (1974, 1976) argue more generally that it is more likely that the enzymatic activities of the various genotypes are additive and that the fitnesses of any genotype are concave functions of these activities. Under this argument the activities of the three genotypes are of the form z_1, $\frac{1}{2}(z_1 + z_2)$, z_2 and the fitnesses are $\phi(z_1)$, $\phi(\frac{1}{2}(z_1 + z_2))$, $\phi(z_2)$, where $\phi(z)$ is a concave function satisfying

$$\phi'(z) > 0, \quad \phi''(z) < 0, \quad \lim_{z \to \infty} \phi(z) = K < \infty. \tag{10.39}$$

Perhaps the simplest function having these properties is

$$\phi(z) = (1 + c)z/(c + z). \tag{10.40}$$

It is clear that the chances of polymorphism in this (SAS-CFF) model are increased compared to the cases when the fitnesses themselves are additive, since $\phi(\frac{1}{2}(z_1 + z_2)) > \frac{1}{2}\phi(z_1) + \frac{1}{2}\phi(z_2)$. If z_1 and z_2 have means μ_1 and μ_2, common variance σ^2, correlation ρ the polymorphism requirement generalizing (10.38) is

$$2|\mu_1 - \mu_2|/[\sigma^2(1 - \rho)\{\phi'(1)(1 + \pi(1 - k)) - \phi''(1)/\phi'(1)\}] < 1, \tag{10.41}$$

where k and π are as defined above, the covariances σ_{ss}, σ_{tt} and σ_{sstt} referring to the distribution of the z's. The stationary distribution, when it exists, is a beta distribution generalizing (10.37).

Gillespie (see in particular 1977a, pp. 305–311) shows that the predictions of this model, especially for $\phi(z)$ of the form (10.40), fit in remarkably well with a large series of observations of natural polymorphisms. We do not pursue these comparisons here (nor the generalizations of the model to the multi-locus case where the expected nature of observed linkage disequilibria are also discussed), since they go beyond the purely mathematical analyses that are our primary concern.

10.6 Ecological Considerations

There is now a vast literature on the mathematical theory of ecological processes (see May, 1975, 1976; Pielou, 1974 for summaries), including static and dynamic theories of the growth of a number of interacting populations. In particular the discrete time Lotka-Volterra equation

$$N_i(t + 1) = N_i(t)\{1 + \alpha_i - \Sigma \beta_{ij}N_j(t)\}, \tag{10.42}$$

($i = 1, ..., k$), modelling the dynamics of a community of k populations of respective sizes $N_1, ..., N_k$, has been extensively analysed. While a considerable verbal discussion exists in the literature on the relation between population genetics and ecology, rather less mathematical theory exists on this relationship. Thus the model (10.42) as it stands is free of genetic considerations.

However, if the parameters β_{ij} depend on the genetic constitutions of populations i and j, a description of the evolution in the model (10.42), concurrent with a description of the genetic evolution in the various populations, is possible in principle although no doubt normally difficult in practice. In this section we outline the outstanding analysis of Roughgarden (1976, 1977) of such a model: for analyses of related models see Fenchel and Christiansen (1977), Jayakar (1970) and Yu (1972).

Consider first the case of a single species in isolation and write (10.42) in the form

$$N(t + 1) = N(t)\, \bar{w} \tag{10.43}$$

where $\bar{w}$, the absolute mean fitness of the species, is a measure of the rate of increase in numbers in this species. Suppose that $\bar{w}$ is determined by the alleles A_1 and A_2 at a given locus: if A_1 has frequency x, then

$$\bar{w} = w_{11} x^2 + 2w_{12} x(1 - x) + w_{22}(1 - x)^2 \tag{10.44}$$

where w_{ij} is the absolute fitness of $A_i A_j$. The w_{ij} themselves are assumed to depend on the current size N of the population. Thus, quite apart from the evolution of the population size N, there will be genetic evolution at the A locus determined by the standard equation (1.25). This equation, coupled with (10.43), defines, for given functions $w_{ij}(N)$, the entire genetical-ecological evolution of the population. Anderson (1971) and more generally Roughgarden (1976) prove that this evolution is such that the equilibrium frequency x^* of A_1 is that producing the largest equilibrium population size and also maximizes mean fitness. This is the first principle of such genetical-ecological systems as enunciated by Roughgarden (1976).

Consider now a set of k co-evolving species. Here two different forms of behavior arise, the first when the fitnesses for any species are not directly functions of the allele frequencies in other species (although they may depend on the sizes of the other species which in turn are determined by these frequencies), and the second where they are. The analysis of the second case is quite complex, although Roughgarden has been able to give explicit principles governing its evolutionary behavior. Here we concentrate on the first case, for which

$$\Delta N_i(t) = N_i(t)\{\bar{w}_i(N_1, ..., N_k) - 1\}, \quad i = 1, ..., k. \tag{10.45}$$

Here, as indicated, $\bar{w}_i$ depends on $N_1 ... N_k$, as well as on the frequency x_i of A_1 in species i, but not directly on allele frequencies for species other than species i. Define the gradient matrix $A = \{a_{ij}\}$ by

$a_{ij} = \partial(\Delta N_i)/\partial N_j$ at equilibrium

$\qquad = N_i \partial \bar{w}_i/\partial N_j$ at equilibrium.   (10.46)

It is necessary to introduce the feedback F of the system, defined by

$$F = (-1)^{k+1}|A|,   \qquad (10.47)$$

where $|A|$ is the determinant of A. A subcommunity of order $k-1$ may be defined by deleting species i from the system, and in this case the feedback in the subcommunity is

$$F_i = (-1)^k|A_i|,   \qquad (10.48)$$

where A_i is obtained from A by striking out the ith row and ith column of A. For the equilibrium to be stable, $F < 0$ and $\sum_i F_i < 0$.

Apart from the difference equation (10.45), there is also a second equation describing the genetic evolution in each population, namely

$$\Delta x_i = x_i(1 - x_i)\{w_{11,i}x_i + w_{12,i}(1 - 2x_i) - w_{22,i}(1 - x_i)\}/\bar{w}_i,   \qquad (10.49)$$

where

$$\bar{w}_i = x_i^2 w_{11,i} + 2x_i(1 - x_i)w_{12,i} + (1 - x_i)^2 w_{22,i},   \qquad (10.50)$$

and the $w_{jl,i}$ depend, in a way we do not make explicit, on $N_1, \ldots, N_k$. The joint ecological-genetical evolution of the system is now determined by (10.45) and (10.48) and has the following important equilibrium properties as presented by Roughgarden (1976, 1977).

Principle 1. Under the assumptions made above suppose that, for any fixed $x_1, x_2, \ldots, x_k$, there exists a unique locally stable equilibrium for the purely ecological model. Then an equilibrium point in the coevolutionary model is locally stable if and only if $\bar{w}_i$ is maximized locally with respect to x_i at that point.

This principle concerns gene frequencies. The second principle, stated below, concerns population numbers.

Principle 2. Under the above assumptions, the equilibrium size of species i is either maximized or minimized, at a stable equilibrium, at the equilibrium value of x_i. Maximization occurs if $F_i < 0$ and minimization if $F_i > 0$. (If $F_i = 0$ then the equilibrium size of species i is not affected by genetic evolution in that species.) Further, $F_i < 0$ for at least one species in the system. (This last result follows immediately from the condition $\sum F_i < 0$ at a stable equilibrium.)

We do not prove these remarkable principles and note only the "dual" optimality of both genetic and ecological parameters at stable equilibria. Roughgarden gives particular examples of the application of these principles, together with further generalizations, but we do not consider these less mathematical analyses here.

10.7 Sociobiology

Sociobiology is the study of the biological basis of social behavior. For this study to be meaningful it must be assumed that any behavior of interest has, at least in part, a genetical basis. Some behaviors, if so genetically based, pose particular problems for evolutionary theory, the most outstanding example being that of altruism. Genes for altruism, if they exist, are at an immediate selective disadvantage in the population and should then be presumably lost from the population. Wilson (1975, p. 3) claims, indeed, that the central theoretical problem of sociobiology is to explain, assuming a genetic basis for this character, how altruism can evolve by natural selection. (For variations on this theme see Wilson, 1977.)

The sociobiological explanation for the existence of altruism, insofar as it is determined genetically, is through kin or group selection: while the behavior is disadvantageous to the individual exhibiting it, the altruistic act is sufficiently favorable to some small related or unrelated group so that the trait evolves by intergroup selection. (Note that in the *Origin of Species* Darwin also invoked intergroup selection arguments for similar traits.) In this very brief section we outline aspects of the theory on which this conclusion is based. (For an excellent survey of group selection models see Wade, 1978).

The major quantitative construct necessary for kin selection arguments is some measure of the degree of relatedness between two individuals. Various measures are available for this, the most commonly used perhaps being the "coefficient of kinship" defined as the probability that for a given locus a gene drawn at random from one individual is identical by descent to a gene drawn at random from the second individual. For any given degree of relatedness, this coefficient may be calculated by a standard path analysis method due to Wright (1921). The quantification for this theory goes back to Haldane (1932a) and Fisher (1958, p. 178): see in particular Hamilton (1964). The altruistic act is favored, for relatives of a given degree, if the number of relatives of this degree who survive as a result of it exceeds the reciprocal of the coefficient of kinship between them and the altruistic individual.

It is important to note that models for the evolution of altruism can be constructed, using group selection methods, where no concept of kin selection is invoked. Perhaps the most interesting such model is that of Matessi and Jayakar (1976), which we now briefly describe. Consider an infinite population subdivided into finite groups of fixed size N. These groups are founded anew each generation in the following way. First, the entire population breeds at random in a common mating area and the splits up into groups of size N, the membership of each group being entirely random. Suppose within a given group there are i "altruists" and $N - i$ "non-altruists" and that the fitness of each altruist is then $\phi_A(i)$ and that of each non-altruist is $\phi_N(i)$. The altruism assumption is that

$$\phi_A(i) \leqslant \phi_N(i), i = 1, 2, ..., N - 1. \tag{10.51}$$

The mean fitness of a group containing i altruists is now

$$\phi(i) = N^{-1} \{i\phi_A(i) + (N - i) \phi_N(i)\}. \tag{10.52}$$

If it happens that $\phi(i + 1) > \phi(i)$, the existence of the altruists in the group favors the group as a whole, and if this advantage is sufficiently large compared to the disadvantage of altruists within each group, altruism will in some circumstances be favored. This argument is, at the moment, non-genetic but can be placed on a genetic basis by assuming certain genotypes for altruism.

The simplest possible forms for the functions $\phi_A(i)$, $\phi_N(i)$ are

$$\phi_A(i) = \beta_0 + \alpha_0 i, \quad \phi_N(i) = \beta_1 + \alpha_1 i. \tag{10.53}$$

The conditions (19.51) and $\phi(i + 1) > \phi(i)$ jointly require

$$0 \leqslant \beta \leqslant -(2N - 1)\alpha + N, \tag{10.54a}$$

$$-N\alpha \leqslant \beta \leqslant -\alpha + N, \tag{10.54b}$$

where $\alpha = (\alpha_1 - \alpha_0)/\alpha_1$, $\beta = (\beta_1 - \beta_0)/\alpha_1$. For any given value of N, the conditions (10.54) define a set of α and β values within which, under this model, altruism can be expected to evolve. This set is a convex region in the (α, β) plane, and the smallest rectangle enclosing this set is

$$-N/(N - 1) \leqslant \alpha \leqslant N/(2N - 1), \quad 0 \leqslant \beta \leqslant N^2/(N - 1). \tag{10.55}$$

The area of the convex set relative to that of the rectangle defined by (10.55) defines a very crude measure of the "likelihood of altruism". Despite the obvious limitations of this definition it is perhaps of interest to note that the measure so defined decreases as a function of N from about 0.72 at $N = 2$ to 0.09 at $N = 20$. Clearly, even though kin selection is not involved, this form of altruism can arise in small, albeit temporarily formed, groups.

10.8 Concluding Remarks

As stated in the Introduction, this book does not aim to give an account of population genetics theory, still less of population genetics itself. Our aim is to discuss the purely mathematical aspects of the theory, and this causes a sometimes severe bias in favor of topics that although perhaps of lesser overall interest, nevertheless admit a considerable mathematical discussion. Thus, on the one hand, many topics, particularly those mentioned in the last chapter, are far more important to the general theory than our discussion of their mathematical properties would suggest. (On the other hand, several topics, particularly perhaps the neutral theory discussed at length in Chap. 9, receive more than their fair share of attention because of the richness of the mathematical theory associated with them.) In the broad context of population genetics, this book is thus intended mainly as an adjunct to those accounts that centre on describing the results of field and laboratory investigations and which discuss, with greater or lesser mathematical support, the implications of the observations so obtained from them.

What is the use of mathematical component to population genetics theory? It may be claimed, for example, that a mathematical argument merely puts in quantitative form conclusions that are already clear qualitatively from general reasoning. Although in some part such a claim is true, there are many aspects of the theory for which general reasoning, without the aid of a mathematical theory, has provided incorrect, sometimes completely incorrect, conclusions. For example, the "reasonable" view sometimes expressed that under a neutral theory approximately equal allele frequencies should be observed is shown, at least for the frequently used models (3.78), to be incorrect: under (3.78) equal frequencies are the *least* likely configuration. Similarly, many "reasonable" views concerning multi-locus principles, in particular in connection with components of variance and the Fundamental Theorem, are also incorrect. Arguments concerning genetic load often appear convincing if not scrutinized carefully by a mathematical investigation. Many further examples can be brought forward: it is clear that the non-mathematical theory is sometimes far from being a reliable guide to the principles of population genetics.

Even where general arguments are correct in principle, they often cannot provide quantitative values essential to a full understanding of evolutionary and other arguments. While perhaps the rates of change of gene frequency exemplified in Table 1.2 could be guessed at non-mathematically, this is probably not true of the corresponding values when stochastic fluctuations are taken into account. In general the effect of stochastic fluctuations, particularly their influence on fixation probabilities, is particularly difficult to assess non-quantitatively. The same is also true of interactive systems, in particular with many loci: here, for example, the correlation between relatives for quantitative traits cannot be approached other than mathematically. Many further examples can be brought forward of the great increase in richness afforded a general theory by a mathematical examination of it.

Of course, every mathematical conclusion depends on some model assumed, sometimes with little justification, to describe reasonably a population's behavior. Thus mathematically derived conclusions must be accepted only to the extent that the model on which they are based is realistic. Despite this reservation, there is little doubt that important principles such as the Hardy-Weinberg law, the Fundamental Theorem, the uneven distribution of gene frequencies under certain neutral models, and the general effects of linkage on evolutionary processes, although discussed here on the basis of specific models, are rather robust to the model chosen and can be taken to apply broadly in real genetic populations.

Much of the mathematics considered in this book is deductive, following the general lines of population genetics theory and its mathematical basis as laid down by Fisher, Haldane and Wright. The original need to show, deductively, that genetic evolution could occur under certain assumptions, is, however, now no longer felt. The deductive theory, indeed, as this book should make clear, now suffers from an embarrassment of riches: a number of theories can be invoked to explain any observed phenomenon. This has been noted in particular by Darlington (1971), who observes:

"Our difficulty now ... has been reversed. It is that we can find more than one solution for each problem. There is much evidence that a choice of solutions has also been Nature's difficulty in the course of evolution ... . Nature's difficulty has proved to be Nature's opportunity."

The answer to this problem perhaps lies in a change in direction of the theory towards a largely inductive approach. Vast quantities of data are becoming available, particularly at the molecular level, concerning the genetic make-up of contemporary populations which are, of course, the present end result of evolution. The theory must now move to consider inferential procedures, based on these data, on the course evolution has taken and the likelihood of the various processes (considered hitherto deductively) in this evolution. The neutral theory has been considered in part at such length in this book because if affords the first major mathematical inferential procedure of this kind. The construction of phylogenetic trees and the information afforded by these is another major inferential process. The mathematical concept of the reversibility of a stochastic process, discussed in this book, provides perhaps one useful tool in this endeavor. While mathematical population genetics theory is surely not yet sufficiently strong to assist decisively in such inductions, it is not too optimistic to hope that it will grow to play a key part in reaching an increasingly complete knowledge of biological evolution as inferred from the nature of contemporary populations.

Appendix A

Let $\{w_{ij}\}$ be a symmetric $k \times k$ matrix of non-negative constants, $\mathbf{x} = (x_1, \ldots, x_k)^1$ a vector of non-negative constants such that $\Sigma x_i = 1$, and $\bar{w} = \Sigma \Sigma w_{ij} x_i x_j$. Then, subject to the constraint $\Sigma \alpha_i x_i = 0$, the values $\alpha_1, \ldots, \alpha_k$ minimizing

$$S = \sum_i \sum_j x_i x_j (w_{ij} - \bar{w} - \alpha_i - \alpha_j)^2 \tag{A.1}$$

are given by

$$\alpha_j = w_j - \bar{w}, \tag{A.2}$$

where

$$w_j = \sum_i x_i w_{ij}. \tag{A.3}$$

Further, the sum of squares removed by $\alpha_1, \ldots \alpha_k$ is

$$S - S_{\min} = 2 \sum x_i (w_i - \bar{w})^2. \tag{A.4}$$

This result is required in a number of places in population genetics theory, and we here give a formal proof of it.

Differentiation of the Lagrangian function $S + 2\lambda(\Sigma \alpha_i x_i)$ with respect to $\alpha_1, \ldots, \alpha_k$ leads to

$$-\sum_i x_i x_j (w_{ij} - \bar{w} - \alpha_i - \alpha_j) + \lambda x_j = 0, \quad i = 1, \ldots, k,$$

or

$$\lambda = w_j - \bar{w} - \alpha_j.$$

Multiplying this equation by x_j and summing over all j gives $\lambda = 0$ and hence (A.2). It remains to prove (A.4): this is equivalent to showing that if α_i is defined by (A.2),

$$\sum \sum x_i x_j (w_{ij} - \bar{w})^2 - \sum \sum x_i x_j (w_{ij} - \bar{w} - \alpha_i - \alpha_j)^2 = 2 \sum x_i (w_i - w)^2. \tag{A.5}$$

This follows immediately since

$$\sum \sum x_i x_j (w_{ij} - \bar{w} - \alpha_i - \alpha_j)^2$$

$$= \sum \sum x_i x_j (w_{ij} - \bar{w})^2 - 2 \sum \sum x_i x_j (\alpha_i + \alpha_j)(w_{ij} - \bar{w}) + \sum \sum x_i x_j (\alpha_i + \alpha_j)^2$$

$$= \sum \sum x_i x_j (w_{ij} - \bar{w})^2 - 4 \sum_i x_i \alpha_i \sum_j x_j (w_{ij} - \bar{w}). + 2 \sum_i x_i \alpha_i^2 \sum_j x_j$$

$$= \sum \sum x_i x_j (w_{ij} - \bar{w})^2 - 2 \sum x_i \alpha_i^2$$

and (A.2) shows that this is equivalent to (A.5).

Appendix B

Let X_t ($t = 0, 1, 2, \ldots$) be a (possibly vector) Markovian random variable with state space $\{0, 1, \ldots, M\}$ and transition matrix P. Suppose $p_{00} = p_{MM} = 1$, that the states $\{1, 2, \ldots, M - 1\}$ are transient, and that there exists an integer m such that $p_{ij}^{(m)} > 0$ for $1 \leqslant i \leqslant M - 1$ and all j.

Suppose further a function $f(X)$ exists such that $f(0) = f(M) = 0$, $f(i) > 0$ otherwise, and for which

$$E \{f(X_{t+1})|X_t\} = \lambda_2 f(X_t) \tag{B.1}$$

for some constant λ_2. Then λ_2 is real and positive and is the leading non-unit eigenvalue of P.

The proof is elementary. Clearly P has at least two unit eigenvalues and, if the first and last rows and columns of P are removed, the remaining eigenvalues of P are those of the resultant matrix Q. Denoting $(f(1), \ldots, f(M - 1))$ by $\mathbf{f}$, Eq. (B.1) and the assumptions $f(0) = f(M) = 0$ show that

$$Q\mathbf{f} = \lambda_2 \mathbf{f}.$$

Since the matrix Q satisfies the conditions of Theorem 2.2 of Karlin and Taylor (1975, p. 545), the Frobenius theory of their Theorem 2.1 proves the desired result.

Appendix C

Empirical [a] significance levels (2.5%, 5%, 97.5%) of the test statistic $\hat{F}$ for given values of k and r (see p. 263 for details). "N.S." means significance is not possible at the probability level indicated.

r	Prob	k							
		3	5	7	10	15	20	25	30
100	2.5%	0.36	0.27	0.20	0.15	0.11	0.08	0.06	0.05
	5%	0.40	0.29	0.21	0.16	0.11	0.08	0.07	0.05
	97.5%	N.S.	0.87	0.71	0.48	0.33	0.22	0.15	0.12
200	2.5%	0.37	0.28	0.22	0.17	0.12	0.09	0.08	0.06
	5%	0.41	0.30	0.23	0.18	0.13	0.10	0.08	0.07
	97.5%	N.S.	0.89	0.78	0.63	0.41	0.29	0.23	0.17
300	2.5%	0.38	0.29	0.23	0.17	0.12	0.10	0.08	0.07
	5%	0.43	0.31	0.24	0.19	0.13	0.11	0.08	0.07
	97.5%	N.S.	0.93	0.83	0.68	0.48	0.34	0.26	0.20
400	2.5%	0.41	0.29	0.23	0.17	0.13	0.10	0.08	0.07
	5%	0.45	0.31	0.25	0.19	0.14	0.11	0.09	0.08
	97.5%	0.99	0.93	0.86	0.71	0.51	0.35	0.28	0.21
500	2.5%	0.40	0.28	0.24	0.18	0.13	0.11	0.09	0.07
	5%	0.45	0.31	0.25	0.20	0.15	0.11	0.09	0.08
	97.5%	0.99	0.93	0.86	0.74	0.52	0.41	0.31	0.24

[a] Based on 1,000 independent drawings for each (k, r) combination from the distribution (3.78). Values by courtesy of R. Anderson.

Appendix D

Values of $E(\hat{F}|k)$ and var $(\hat{F}|k)$ for various k, r values (see p. 265 for details). Values by kind courtesy of R. Anderson.

					k				
		3	5	7	10	15	20	25	30
						$E(\hat{F}|k)$				
	100	0.671	0.490	0.376	0.271	0.176	0.125	0.094	0.073
	200	0.705	0.532	0.421	0.313	0.212	0.156	0.120	0.096
r	300	0.722	0.554	0.444	0.336	0.232	0.173	0.135	0.110
	400	0.732	0.568	0.459	0.351	0.245	0.185	0.146	0.119
	500	0.740	0.579	0.470	0.362	0.255	0.193	0.153	0.126
					var $(\hat{F}|k)$					
	100	0.0325	0.0254	0.0169	0.0089	0.0033	0.0013	0.0006	0.0003
	200	0.0350	0.0306	0.0224	0.0133	0.0058	0.0028	0.0014	0.0008
r	300	0.0359	0.0331	0.0253	0.0159	0.0075	0.0038	0.0021	0.0012
	400	0.0364	0.0346	0.0272	0.0176	0.0087	0.0046	0.0026	0.0015
	500	0.0366	0.0356	0.0286	0.0190	0.0096	0.0052	0.0030	0.0018

References

Abramowitz, M., Stegun, I. A.: Handbook of mathematical functions. New York: Dover Publ. Inc. 1965

Anderson, R.: Unpublished M. Sc. Thesis, Monash University (1978)

Anderson, W. W.: Genetic equilibrium and population growth under density-regulated selection. Amer. Nat. *105*, 489–498 (1971)

Avery, P. J.: The effect of random selection coefficients on populations of finite size – some particular models. Gen. Res. Camb. *29*, 97–112 (1977)

Avery, P. J., Hill, W. G.: The effect of linkage disequilibrium on the genetic variance of a quantitative trait. Adv. App. Prob. *10*, 4–6 (1978)

Ayala, F., Powell, J. R., Tracey, M. L., Mouras, C. A., Perez-Salas, S.: Enzyme variability in the *Drosophila willistoni* group, IV. Genetic variation in natural populations of *Drodophila willistoni*. Genetics *70*, 113–139 (1972)

Barker, W. C., Ketcham, L. K., Dayhoff, M. O.: A comprehensive examination of protein sequences for evidence of internal gene duplication. J. Mol. Evol. *10*, 265–281 (1978)

Bennett, J. H.: On the theory of random mating. Ann. Eugenics *18*, 311–317 (1954)

Bennett, J. H.: Selectively balances polymorphism at a sex-linked locus. Nature *180*, 1363–1364 (1957)

Berger, E. M.: Pattern and chance in the use of the genetic code. J. Mol. Evol. *10*, 319–323 (1978)

Blundell, T. L., Wood, S. P.: Is the evolution of insulin Darwinian or due to selectively neutral mutation? Nature *257*, 197–198 (1975)

Bodmer, W. F.: Differential fertility in population genetics models. Genetics *51*, 411–424 (1965)

Bodmer, W. F., Edwards, A. W. F.: Natural selection and the sex ratio. Ann. Hum. Genet. *24*, 239–244 (1960)

Bodmer, W. F., Felsenstein, J.: Linkage and selection – theoretical analysis of the deterministic two locus random mating model. Genetics *57*, 237–265 (1967)

Bulmer, M. G.: The effect of selection on genetic variability: a simulation study. Gen. Res. *28*, 101–117 (1976)

Burt, C.: Quantitative genetics in psychology. Brit. Jour. of Math. & Stat. Psych. *24*, 1–21 (1971)

Cannings, C.: Equilibrium, convergence and stability at a sex-linked locus under natural selection. Genetics 56, 613–618 (1967)

Cannings, C.: Equilibrium under selection at a multi-allelic sex-linked locus. Biometrics *24*, 187–189 (1968)

Cannings, C.: The latent roots of certain Markov chains arising in genetics: a new approach 1. Haploid models. Adv. Appl. Prob. *6*, 260–290 (1974)

Castle, W. E.: The laws of Galton and Mendel and some laws governing race improvement by selection. Proc. Amer. Acad. Arts and Sci. *39*, 233–242 (1903)

Cavalli-Sforza, L. L., Edwards, A. W. F.: Phylogenetic analysis: models and estimation procedures. Amer. J. Hum. Genet. *19*, 233–257 (1974)

Chakraborty, R., Nei, M.: Hidden genetic variability within electromorphs in finite populations. Genetics *84*, 385–393 (1976)

Charlesworth, B.: Selection in populations with overlapping generations. I. The use of Malthusian parameters in population genetics. Theor. Pop. Biol. *1*, 352–370 (1970)

Charlesworth, B.: Selection in density-regulated populations. Ecology *52*, 469–474 (1971)

Charlesworth, B.: Selection in populations with overlapping generations. III. Conditions for genetic equilibrium. Theor. Pop. Biol. *3*, 377–395 (1972)

Charlesworth, B: Selection in populations with overlapping generations. V. Natural selection and life histories. Amer. Natur. *107*, 303–311 (1973)

Charlesworth, B.: Selection in populations with overlapping generations. VI. Rates of change of gene frequency and population growth rate. Theor. Pop. Biol. *6*, 108–133 (1974)

Charlesworth, B.: Natural selection in age-structured populations. Lectures on Mathematics in the Life Sciences *8*, 69–87 (1976)

Christiansen, F. B., Frydenberg, O.: Geographical patterns of four polymorphisms in *viviperus* as evidence of selection. Genetics *77*, 765–770 (1974)

Cockerham, C. C.: An extension of the concept of partitioning hereditary variance for analysis of covariances among relatives when epistasis is present. Genetics *39*, 859–882 (1954)

Cockerham, C. C.: Effects on linkage on the covariances between relatives. Genetics *41*, 138–141 (1956)

Cockerham, C. C.: Higher order probability functions of identity of alleles by descent. Genetics *69*, 235–246 (1971)

Cockerham, C. C., Weir, B. S.: Descent measures for two loci with some applications. Theor. Pop. Biol. *4*, 300–330 (1973)

Conley, C. C.: Unpublished lecture notes. University of Wisconsin, Madison (1972)

Cornette, J. L.: Some basic elements of continuous selection models. Theor. Pop. Biol. *8*, 301–313 (1975)

Cornish-Bowden, A., Marson, A.: Evolution of the non-randomness of protein composition. J. Mol. Evol. *10*, 231–240 (1977)

Coyne, J. A.: Lack of genetic similarity between two sibling species of *Drosophila* as revealed by varied techniques. Genetics *84*, 593–607 (1976)

Crow, J. F., Felsenstein, J.: The effect of assortative mating on the genetic composition of a population. Eugenics Quart. *15*, 85–97 (1968)

Crow, J. F., Kimura, M.: Evolution in sexual and asexual populations. Am. Nat. *99*, 439–450 (1965)

Crow, J. F., Kimura, M.: Evolution in sexual and asexual populations: a reply. Am Nat. *103*, 89–91 (1969)

Crow, J. F., Kimura, M.: An introduction to population genetics theory. New York: Harper and Row 1970

Crow, J. F., Kimura, M.: The effective number of a population with overlapping generations – a correction and further discussion. Amer. J. Hum. Genet. *24*, 1–10 (1972)

Dayhoff, M.: Atlas of protein sequence and structure, Vol. 5. Washington, D. C.: Nat. Biomed. Res. Foundn. 1972

Demetrius, L.: Primitivity conditions for growth matrices. Math. Biosci. *12*, 53–58 (1971)

Demetrius, L.: Demographic parameters and neutral selection. Proc. Nat. Acad. Sci. *71*, 4645–4647 (1974)

Demetrius, L.: Reproductive strategies and natural selection. Am. Nat. *109*, 243–249 (1975)

Demetrius, L.: Measures of variability in age-structured populations. Journ. Theor. Biol. *63*, 397–404 (1976)

Demetrius, L.: Measures of fitness and demographic stability. Proc. Natl. Acad. Sci. *74*, 384–386 (1977)

Darlington, C. D.: Evolution. New Scientist *49*, 329–330 (1971)

Eaves, L. J., Last, K., Martin, N. G., Jinks, J. L.: A progressive approach to non-additivity and genotype-environmental covariance in the analysis of human differences. Br. J. Math. Statist. Psychol. *30*, 1–42 (1977)

Edwards, A. W. F.: The population genetics of "sex-ratio" in *Drosophila pseudoobscura*. Heredity *16*, 291–304 (1961)

Elliott, J.: Eigenfunction expansions associated with singular differential operators. Trans. Amer. Math. Soc. *78*, 406–425 (1955)

Eshel, I.: Selection on sex ratio and the evolution of sex-determination. Heredity *34*, 351–361 (1975)

Eshel, I., Feldman, M. W.: On the evolutionary effect of recombination. Theor. Pop. Biol. *1*, 88–100 (1970)

Ethier, S. N., Norman, M. Frank: An error estimate for the diffusion approximation to the Wright-Fisher model. Proc. Nat. Sci. *74*, 5096–5098 (1977)

Ewens, W. J.: The pseudo-transient distribution and its uses in genetics. J. Appl. Prob. *1*, 141–156 (1964)

Ewens, W. J.: A note on the mathematical theory of the evolution of dominance. Amer. Nat. *101*, 532–540 (1967)

Ewens, W. J.: A genetic model having complex linkage behaviour. Theor. and Appl. Gen. *38*, 140–143 (1968)

Ewens, W. J.: A generalized fundamental theorem of natural selection. Genetics *63*, 531–537 (1969a)

Ewens, W. J.: Mean fitness increases when fitnesses are additive. Nature *221*, 1076 (1969b)

Ewens, W. J.: Population genetics. London: Methuen 1969c

Ewens, W. J.: The transient behavior of stochastic processes, with applications in the natural sciences. Bull. 36th session I. S. I. 603–622 (1969d)

Ewens, W. J.: Remarks on the substitutional load. Theor. Pop. Biol. *1*, 129–139 (1970)

Ewens, W. J.: The sampling theory of selectively neutral alleles. Theor. Pop. Biol. *3*, 87–112 (1972)

Ewens, W. J.: Testing for increased mutation rate for neutral alleles. Theor. Pop. Biol. *4*, 251–259 (1973)

Ewens, W. J.: Mathematical and statistical problems arising in the non-Darwinian theory. Lectures on Mathematics in the Life Sciences. *7*, 25–42 (1974)

Ewens, W. J.: Remarks on the evolutionary effect of natural selection. Genetics *83*, 601–607 (1976)

Ewens, W. J., Feldman, M. W.: The theoretical assessment of selective neutrality. In: Population Genetics and Ecology. Karlin, S., Nevo, E. (eds.), pp. 303–337. New York: Academic Press 1976

Ewens, W. J., Kirby, K.: The eigenvalues of the neutral alleles processes. Theor. Pop. Biol. *7*, 212–220 (1975)

Ewens, W. J., Thomson, G.: Heterozygote selective advantage. Ann. Hum. Gen. *33*, 365–376 (1970)

Ewens, W. J., Thomson, G.: Properties of equilibria in multi-locus genetic systems. Genetics *87*, 807–819 (1977)

Fahady, K. S., Quine, M. P., Vere-Jones, D.: Heavy traffic approximations for the Galton-Watson process. Adv. Appl. Prob. *3*, 282–300 (1971)

Feldman, M. W.: On the offspring number distribution in a genetic population. J. Appl. Prob *3*, 129–141 (1966)

Feldman, M. W.: Selection for linkage modification – I. Random mating populations. Theor. Pop. Biol. *3*, 324–346 (1972)

Feldman, M. W., Christiansen, F. G.: The effect of population subdivision on two loci without selection. Genet. Res. Camb. *25*, 151–162 (1975)

Feldman, M. W., Franklin, I., Thomson, G. J.: Selection in complex genetic system I. the symmetric equilibria of the three-locus symmetric viability model. Genetics *76*, 135–162 (1974)

Feldman, M. W., Karlin, S.: The evolution of dominance – a direct approach through the theory of linkage and selection. Theor. Pop. Biol. *2*, 482–492 (1971)

Feldman, M. W., Krakauer, J.: Genetic modification and modifier polymorphisms. Population Genetics and Ecology. Karlin, S., Nevo, E. (eds.), pp. 547–583. New York: Academic Press 1976

Feldman, M. W., Lewontin, R. C., Franklin, R. I., Christiansen, F. B.: Selection in complex genetic systems III. An effect of allele multiplicity with two loci. Genetics *79*, 333–347 (1975)

Feller, W.: Diffusion processes in genetics. Proc. 2nd Berkeley Symp. on Math. Stat. and Prob. Neyman, J. (ed.), pp. 227–246. Berkeley: University of California Press 1951

Feller, W.: Diffusion processes in one dimension. Trans. Amer. Math. Soc. *77*, 1–31 (1954)

Felsenstein, J.: Inbreeding and variance effective numbers in populations with overlapping generations. Genetics *68*, 581–597 (1971)

Felsenstein, J.: Maximum likelihood and minimum-steps methods for estimating evolutionary three from data on discrete characters. Sys. Zoology *22*, No. 3, 240–249 (1973)

Felsenstein, J.: The evolutionary advantage of recombination. Genetics *78*, 737–756 (1974)

Felsenstein, J., Yokoyama, S.: The evolutionary advantage of recombination. II Individual selection for recombination. Genetics *83*, 845–859 (1976)

Fenchel, T. M., Christiansen, F. B.: Measuring selection in natural populations. In: Lecture notes in Biomathematics 19. Christiansen, F. B., Fenchel, T. M. (eds.), Berlin: Springer 1977

Fisher, R. A.: The correlation between relatives on the supposition of Mendelian inheritance. Trans. of the Roy. Soc. of Edinburgh *52*, 399–433 (1918)

Fisher, R. A.: On the dominance ratio. Proc. Roy. Soc. Edin. *42*, 321–431 (1922)

Fisher, R. A.: The arrangements of field experiments. Jour. of the Min. of Agr. of Great Brit. *33*, 503–513 (1926)

Fisher, R. A.: The possible modification of the response of the wild type to recurrent mutation. Amer. Nat. *62*, 115–226 (1928a)

Fisher, R. A.: Two further notes on the origin of dominance. Amer. Nat. *62*, 571–574 (1928b)

Fisher, R. A.: The evolution of dominance – reply to Professor Sewall Wright. Amer. Nat. *63*, 553–556 (1929)

Fisher, R. A.: The genetical theory of natural selection. Oxford: Clarendon Press 1930a

Fisher, R. A.: The evolution of dominance in certain polymorphic species. Amer. Nat. *64*, 385–406 (1930b)

Fisher, R. A.: The evolution of dominance. Biol. Reviews 6, 345–368 (1931)

Fisher, R. A.: Prof. Wright on the theory of dominance. Amer. Nat. *68*, 370–374 (1934)

Fisher, R. A.: The wave of advance of advantageous genes. Ann. Eug. *7*, 355–369 (1937)

Fisher, R. A.: Population genetics. The Croonian lecture. Proc. Roy. Soc. B. *141*, 510–523 (1953).

Fisher, R. A.: The genetical theory of natural selection (second revised edit.). New York: Dover 1958

Fisher, R. A.: Heredity (Address to Camb. Uni. Eug. Soc.) Notes and Records of the Roy. Soc. of London *31*, 155–162 (1976)

Fitch, W. M.: Toward defining the course of evolution: minimum change for a specific tree topology. Sys. Zoology *20*, No. 4, 406–416 (1971)

Fitch, W. M., Margoliash, E.: Construction of phylogenetic trees. Science *155*, 279–284 (1967)

Franklin, I. R., Feldman, M. W.: Two loci with two alleles: linkage equilibrium and linkage disequilibrium can be simultaneously stable. Theor. Pop. Biol. *12*, 95–113 (1977)

Franklin, I., Lewontin, R. C.: Is the gene the unit of selection? Genetics *65*, 701–734 (1970)

Freedman, D.: Brownian motion and diffusion. San Francisco: Holden-Day 1971

Freeling, M.: Allelic variation at the level of intragenic recombination. Genetics *89*, 211–224 (1978)

Frydenberg, O.: Population studies of a lethal mutant in *Drosophila melanogaster*. I. Behaviour in populations with discrete generations. Hereditas *50*, 89–116 (1963)

Gallais, A.: Covariances entre apparantés quelconques avec linkage et épistasie. I-Expression générale. Ann. Génét. Sél. Anim. *2*, 281–310 (1970)

Gallais, A.: Covariances between arbitrary relatives with linkage disequilibrium. Biometrics *30*, 429–446 (1974)

Geiringer, H.: On the probability theory of linkage in Mendelian heredity. Ann. Math. Statist. *15*, 15–50 (1944)

Gillespie, J.: Polymorphism in patchy environments. The Amer. Nat. *108*, 145–151 (1974a)

Gillespie, J.: The role of environmental grain in the maintenance of genetic variation. Amer. Nat. *108*, 831–836 (1974b)

Gillespie, J. H.: The role of migration in the genetic structure of populations in temporally and spatially varying environments. II. Island models. Theor. Pop. Biol. *10*, 227–238 (1976a)

Gillespie, J. H.: A general model to account for enzyme variation in natural populations. II. Characterization of the fitness function. Amer. Nat. *110*, 809–821 (1976b)

Gillespie, J. H.: A general model to account for enzyme variation in natural populations. III. Multiple Alleles. Evolution *31*, 85–90 (1977a)

Gillespie, J. H.: A general model to account for enzyme variation in natural populations. IV. The quantitative genetics of fitness traits. In: Lecture notes in Biomathematics 19. Christiansen, F. B., Fenchel, T. M. (eds.), pp. 301–314. Berlin: Springer 1977b

Gillespie, J. H.: A general model to account for enzyme variation in natural populations. V. The SAS-CFF model. Theoret. Pop. Biol. *14*, 1–45 (1978)

Gillespie, J. H., Langley, C.: A general model to account for enzyme variation in natural populations. Genetics *76*, 837–848 (1974)

Gillespie, J. H., Langley, C.: Multi-locus behavior in random environments I. Random Levene models. Genetics *82*, 123–137 (1976)

Gillespie, J. H., Langley, C. H.: Are evolutionary rates really variable? (To appear) 1979

Gladstien, K.: The characteristic values and vectors for a class of stochastic matrices arising in genetics. SIAM J. of Appl. Math. *34*, 630–642 (1978)

Goldberger, A. S.: Models and methods in the I. Q. debate: Part 1. University of Wisconsin. SSRI series, 1978a

Goldberger, A. S.: Pitfalls in the resolution of IQ inheritance. In: Genetic Epidemiology. Morton, N. E., Yung, C. S. (eds.). New York: Academic Press 1978b

Griffiths, R. C.: On the distribution of allele frequency in a diffusion model. (To appear) (1978)

Griffiths, R. C.: Exact sampling distributions from the infinite neutral alleles model. (To appear in Adv. Appl. Prob.) (1979a)

Griffiths, R. C.: A transition density expansion for a multi-allele diffusion model. (To appear in Adv. Appl. Prob.) (1979b)

Hadeler, K. P., Liberman, U.: Selection models with fertility differences. J. Math. Biol. *2*, 19–32 (1975)

Haigh, J., Maynard Smith, J.: The hitch-hiking effect – a reply. Genet. Res. Camb. *27*, 85–87 (1976)

Haldane, J. B. S.: A mathematical theory of natural and artificial selection, Part II, the influence of partial self-fertilisation, inbreeding, assortative mating, and selective fertilisation on the composition of Mendelian populations, and on natural selection. Proc. Camb. Phil. Soc., Biol. Sci. *1*, 158–163 (1924)

Haldane, J. B. S.: A mathematical theory of natural and artificial selection, Part III, Proc. Camb. Phil. Soc. *23*, 363–372 (1926)

Haldane, J. B. S.: A mathematical theory of natural and artificial selection, IV, Proc. Camb. Phil. Soc. *23*, 607–615 (1927a)

Haldane, J. B. S.: A mathematical theory of natural and artificial selection, V, selection and mutation. Proc. Camb. Phil. Soc. *23*, 838–844 (1927b)

Haldane, J. B. S.: A mathematical theory of natural and artificial selection, VII, selection intensity as a function of mortality rate. Proc. Camb. Phil. Soc., *27*, 131–136 (1930a)

Haldane, J. B. S.: A mathematical theory of natural and artificial selection, VI, isolation. Proc. Camb. Phil. Soc. *26*, 220–230 (1930b)

Haldane, J. B. S.: The causes of evolution. London: Longmans Green 1932a

Haldane, J. B. S.: A mathematical theory of natural and artificial selection, Part IX, rapid selection. Proc. Camb. Phil. Soc. *28*, 244–248 (1932b)

Haldane, J. B. S.: The effect of variation on fitness. Amer. Nat. *71*, 337–349 (1937)

Haldane, J. B. S.: The cost of natural selection. J. Gen. *55*, 511–524 (1957)

Hamilton, W. D.: The genetical evolution of social behavior. Jour. of Theor. Biol. *7*, 1–52 (1964)

Hardy, G. H.: Mendelian proportions in a mixed population. Science *28*, 49–50 (1908)

Hedrick, P. W.: A new approach to measuring genetic similarity. Evolution *25*, 276–280 (1971)

Hill, W.: Effective size of populations with overlapping generations. Theor. Pop. Biol. *3*, 278–289 (1972)

Hill, W. G., Robertson, A.: The effect of linkage on limits to artificial selection. Genet. Res. Camb. *8*, 269–294 (1966)

Hill, W. G., Robertson, A.: Linkage disequilibrium in finite populations. Theor. and Appl. Gen. *38*, 226–231 (1968)

Hoppensteadt, F. C.: A slow selection analysis of two-locus, two-allele traits. Theor. Pop. Biol. *9*, 68–81 (1976)

Ito, K., McKean, H. P.: Diffusion processes and their sample paths. Berlin: Springer 1965

Jayakar, S. D.: A mathematical model for interaction of gene frequencies in a parasite and its host. Theor. Pop. Biol. *1*, 140–164 (1970)

Jinks, J. L., Eaves, L. J.: IQ and inequality. Nature *248*, 287–289 (1974)

Jukes, T. H.: Evolutionary changes in insulin. Nature *259*, 250–251 (1976)

Karlin, S.: Rates of approach to homozygosity for finite stochastic models with variable population size. Amer. Nat. *102*, 443–455 (1968)

Karlin, S.: Sex and infinity: a mathematical analysis of the advantages and disadvantages of recombination. In: The mathematical theory of the dynamics of biological populations. Bartlett, M. S., Hiorns, R. W., (eds.), pp. 155–194. London and New York: Academic Press 1973

Karlin, S.: Selection with many loci and possible relations to quantitative genetics. In: Proc. Int. Conf. Quart. Gen. Pollak, E., Kempthorne, O., Bailey, T., (eds.), pp. 207–226. Ames: Iowa State U. Press (1977a)

Karlin, S.: Gene frequency patterns in the Levene subdivided population model. Theor. Pop. Biol. *11*, 356–385 (1977b)

Karlin, S.: Mathematical population genetics. (To appear) 1979

Karlin, S., Carmelli, D.: Numerical studies on two-loci selection models with general viabilities. Theor. Pop. Biol. *7*, 399–421 (1975)

Karlin, S., Feldman, M. W.: Linkage and selection – two locus symmetric viability model. Theor. Pop. Biol. *1*, 39–71 (1970a)

Karlin, S., Feldman, M. W.: Convergence to equilibrium of the two locus additive viability model. J. Appl. Prob. *7*, 262–271 (1970b)

Karlin, S., Liberman, U.: Random temporal variation in selection intensities: case of large population size. Theor. Pop. Biol. *6*, 355–382 (1974)

Karlin, S., McGregor, J.: On some stochastic models in genetics. In: Stoch. models in Med. and Biol. Gurland, J., (ed.), pp. 245–279. Madison: University of Wisconsin Press 1964

Karlin, S., McGregor, J.: Direct product branching processes and related induced Markoff chains. I. Calculations of rates of approach to homozygosity. In: Bernoulli (1723), Bayes (1773), Laplace (1813): Anniv. Vol. LeCam, L., Neyman, J., (eds.), pp. 111–145. Berlin, Heidelberg, New York: Springer 1965

Karlin, S., McGregor, J.: The number of mutant forms maintained in a population. Proc. Fifth Berk. Symp. of Math. Stat. and Prob. *4*, 403–414 (1966).

Karlin, S., McGregor, J. L.: Rates and probabilities of fixation for two locus random mating finite populations without selection. Genetics *58*, 141–159 (1968)

Karlin, S., McGregor, J. L.: Addendum to a paper of W. Ewens. Theor. Pop. Biol. *3*, 113–116 (1972)

Karlin, S., McGregor, J.: Towards a theory of the evolution of modifier genes. Theor. Pop. Biol. *5*, 59–105 (1974)

Karlin, S., Taylor, H.: A first course in stochastic processes. New York: Academic Press 1975

Keeler, C.: Some oddities in the delayed appreciation of "Castle's Law". J. Heredity *59*, 110–112 (1968)

Kelly, F. P.: On stochastic population models in genetics. J. Appl. Prob. *13*, 127–131 (1976)

Kelly, F. P.: Exact results for the Moran neutral allele model. J. Appl. Prob. 9, 197–201 (1977)

Kemeny, J. G. Snell, J. L.: Finite Markov chains. New York: Van Nostand 1960

Kempthorne, O.: The design and analysis of experiments. New York: John Wiley & Sons 1952

Kempthorne, O.: The correlation between relatives in a random mating population. Proc. Roy. Soc. B *143*, 103–113 (1954)

Kempthorne, O.: The theoretical values of correlations between relatives in random mating populations. Genetics *40*, 153–167 (1955)

Kempthorne, O.: An introduction to genetic statistics. New York: Wiley 1957

Kempthorne, O., Pollak, E.: Concepts of fitness in Mendelian populations. Genetics *64*, 125–145 (1970)

Kesten, H., Ney, P., Spitzer, F.: The Galton-Watson process with mean one and finite variance. Theor. of Prob. and Applns. *11*, 513–540 (1966)

Kidwell, J. F., Clegg, M. T., Stewart, F. M., Prout, T.: Regions of stable equilibria for models of differential selection in the two sexes under random mating. Genetics *85*, 171–183 (1977)

Kimura, M.: Process leading to quasi-fixation of genes in natural populations due to random fluctuation of selection intensities. Genetics *39*, 280–295 (1954)

Kimura, M.: Solution of a process of random genetic drift with a continuous model. Proc. Natl. Acad. Sci. *41*, 144–150 (1955a)

Kimura, M.: Random drift in a multi-allelic locus. Evolution *9*, 419–435 (1955b)

Kimura, M.: Stochastic processes and distribution of gene frequencies under natural selections. Cold Spring Harbor Symp. on Quant. Biol. *20*, 33–53 (1955c)

Kimura, M.: Random genetic drift in a tri-allelic locus – exact solution with a continuous model. Biometrics *12*, 57–66 (1956a)

Kimura, M.: A model of a genetic system which leads to closer linkage under natural selection. Evolution *10*, 278–287 (1956b)

Kimura, M.: Some problems of stochastic processes in genetics. Ann. Math. Stat. *28*, 882–901 (1957)

Kimura, M.: On the change of population fitness by natural selection. Heredity *12*, 145–167 (1958)

Kimura, M.: A probability method for treating inbreeding systems especially with linked genes. Biometrics *19*, 1–17 (1963)

Kimura, M.: Attainment of quasi linkage equilibrium when gene frequencies are changing by natural selection. Genetics *52*, 875–890 (1965)

Kimura, M.: On the evolutionary adjustment of spontaneous mutation rates. Genet. Res. *9*, 25–34 (1967)

Kimura, M.: Evolutionary rate of the molecular level. Nature *217*, 624–626 (1968)

Kimura, M.: The number of heterozygous nucleotide sites maintained in a finite population due to steady flux of mutation. Genetics *61*, 893 (1969)

Kimura, M.: The length of time required for a selectively neutral mutant to reach fixation through random frequency drift in a finite population. Gen. Res. *15*, 131–134 (1970)

Kimura, M.: Theoretical foundations of population genetics at the molecular level. Theor. Pop. Biol. *2*, 174–208 (1971)

Kimura, M.: Preponderance of synonymous changes as evidence for the neutral theory of molecular evolution. Nature 267–275 (1977)

Kimura, M., Crow, J. F.: The number of alleles that can be maintained in a finite population. Genetics *49*, 725–738 (1964)

Kimura, M., Ohta, T.: Theoretical aspects of population genetics. Princeton: Princeton University Press 1971a

Kimura, M., Ohta, T.: On the rate of molecular evolution. J. Mol. Evol. *1*, 1 (1971b)

Kimura, M., Ohta, T.: The age of a neutral mutant persisting in a finite population. Genetics *75*, 199–212 (1973)

King, J. L., Jukes, T. H.: Evolutionary loss of ascorbic acid synthesizing ability. J. Hum. Evol. *4*, 85–88 (1975)

Kingman, J. F. C.: A matrix inequality. Quart. J. Math. *12*, 78–80 (1961a)

Kingman, J. F. C.: A mathematical problem in population genetics. Proc. Camb. Phil. Soc. *57*, 574–582 (1961b)

Kingman, J. F. C.: Random discrete distributions. J. Roy. Stat. Soc. B *37*, 1–22 (1975)

Kingman, J. F. C.: Coherent random walks arising in some genetical models. Proc. R. Soc. Lond. A *351*, 19–31 (1976)

Kingman, J. F. C.: A note on multi-dimensional models of neutral mutation. Theor. Pop. Biol. *11*, 285–290 (1977a)

Kingman, J. F. C.: The population structure associated with the Ewens sampling formula. Theor. Pop. Biol. *11*, 274–283 (1977b)

Kojima, K., Kelleher, T. M.: Changes of mean fitness in random-mating populations when epistasis and linkage are present. Genetics *46*, 527–540 (1961)

Kolman, W.: The mechanism of natural selection for the sex ratio. Amer. Natur. *94*, 373–377 (1960)

Langley, C. H., Fitch, W. M.: An examination of the constancy of the rate of molecular evolution. J. Mol. Evol. *3*, 161–177 (1974)

Last, K.: Genetical aspects of human behaviour. Unpublished Ph. D. thesis. Dept. of Genetics., University of Birmingham, England (1976)

Leigh, E. G. Jr.: The evolution of mutation rates. Genetics Supp. *73*, 1–18 (1973)

Levene, H.: Genetic equilibrium when more than one ecological niche is available. Amer. Nat. *87*, 331–333 (1953)

Levikson, B.: The age distribution of Markov processes. J. Appl. Prob. *14*, 492–506 (1977)

Levikson, B., Karlin, S.: Random temporal variation in selection intensities acting on infinite diploid populations: diffusion method analysis. Theor. Pop. Biol. *8*, 292–300 (1975)

Lewontin, R. C.: The interaction of selection and linkage. I. General considerations; heterotic models. Genetics *49*, 49–67 (1964a)

Lewontin, R. C.: The interaction of selection and linkage. II. Optimum models. Genetics *50*, 757–782 (1964b)

Lewontin, R. C.: The apportionment of human diversity. Evol. Biol. *6*, 381–398 (1973)

Lewontin, R. C.: The genetic basis of evolutionary change. New York: Columbia University Press 1974

Lewontin, R. C., Kojima, K.: The evolutionary dynamics of complex polymorphisms. Evolution *14*, 458–472 (1960)

Lewontin, R. C., Krakauer, J.: Distribution of gene frequency as a test of the theory of the selective neutrality of polymorphisms. Genetics *74*, 175–195 (1973)

Li, W.-H.: A mixed model of mutation for electrophoretic identity of proteins within and between populations. Genetics *83*, 423–432 (1976)

Li, W.-H.: Maintenance of genetic variability under mutation and selection pressures in a finite population. Proc. Nat. Acad. Sci. *74*, 2509–2513 (1977)

Li. W.-H., Nei, M.: Stable linkage disequilibrium without epistasis in subdivided populations. Theor. Pop. Biol. *6*, 173–183 (1974)

Li, W.-H., Nei, M.: Persistence of common alleles in two related populations or species. Genetics *86*, 901–914 (1977)

Littler, R. A.: Linkage disequilibrium in two-locus, finite, random mating models without selection or mutation. Theor. Pop. Biol. *4*, 259–275 (1973

Littler, R. A.: Loss of variability at one locus in a finite population. Math. Bio. *25*, 151–163 (1975)

Littler, R. A., Fackerell, E. D.: Transition densities for neutral multi-allele diffusion models. Biometrics *31*, 117–123 (1975)

Ludwig, D.: Stochastic population theories. Lecture notes in Biomathematics 3. Berlin: Springer 1974

McKean, H. P.: Elementary solutions for certain parabolic partial differential equations. Trans. Amer. Math. Soc. *82*, 519–548 (1956)

McNaughton, S. J.: Natural selection at the enzyme level. Am. Nat. *108*, 616 (1974)

Malécot, G.: Les mathématiques de l'héredité. Paris: Masson 1948

Mandl, P.: Analytical treatment of one-dimensional Markov processes. Berlin: Springer 1968

Maruyama, T.: On the fixation probability of mutant genes in a subdivided population. Gen. Res. *15*, 221–226 (1970)

Maruyama, T.: An invariant property of a structured population. Gen. Res. *18*, 81–84 (1971)

Maruyama, T.: The rate of decay of genetic variability in a geographically structured finite population. Math. Biosci. *14*, 325–335 (1972)

Maruyama, T.: The age of an allele in a finite population. Genetics. Res. Camb. *23*, 137–143 (1974)

Maruyama, T.: Stochastic problems in population genetics. Lecture notes in Biomathematics 17. Berlin: Springer 1977

Maruyama, T., Kimura, M.: A note on the speed of gene frequency changes in reverse directions in a finite population. Evolution *28*, 161–163 (1974)

Maruyama, T., Yamazaki, T.: Analysis of heterozygosity in regard to the neutrality theory of protein polymorphisms. J. Mol. Evol. *4*, 195 (1974)

Mather, K., Jinks, J. L.: Introduction to biometrical genetics. London: Chapman & Hall 1977

Matessi, C., Jayakar, S. D.: Conditions for the evolution of altruism under Darwinian selection. Theor. Pop. Biol. *9*, 360–387 (1976)

May, R. M.: Stability and complexity in model ecosystems. (Second ed.) Princeton: Princeton University Press 1975

May, R. M.: Theoretical ecology. Philadelphia: W. B. Saunders 1976

Maynard Smith, J.: Evolution in sexual and asexual populations. Am. Nat. *102*, 469–673 (1968)

Maynard Smith, J.: What use is sex? J. Theoret. Biol. *30*, 319–335 (1971)

Maynard Smith, J., Haigh, J.: The hitch-hiking effect of a favourable gene. Genet. Res. Camb. *23*, 23–35 (1974)

Miller, G. F.: The evaluation of eigenvalues of a differential equation arising in a problem in genetics. Proc. Camb. Phil. Soc. *58*, 588–593 (1962)

Mitton, J. B., Koehn, R. K.: Population genetics of marine pelecypods. III. Epistatis between functionally related isoenzymes of *Mytilus edulis.* Genetics *73*, 487 (1973)

Moran, P. A. P.: Random processes in genetics. Proc. Camb. Phil. Soc. *54*, 60–71 (1958)

Moran, P. A. P.: The statistical processes of evolutionary theory. Oxford: Clarendon Press 1962

Moran, P. A. P.: Wandering distribution and the electrophoretic profile. Theor. Pop. Biol. *8*, 318–330 (1975)

Moran, P. A. P.: Wandering distributions and the electrophoretic profile II. Theor. Pop. Biol. *10*, 145–149 (1976)

Moran, P. A. P., Watterson, G. A.: The genetic effects of family structure in natural populations. Aust. J. Biol. Sci. *12*, 1–15 (1958)

Muller, H. J.: Some genetic aspects of sex. Am. Nat. *66*, 118–138 (1932)

Nagylaki, T.: The moments of stochastic integrals and the distribution of sojourn times. Proc. Nat. Acad. Sci. *71*, 746–749 (1974a)

Nagylaki, T.: The decay of genetic variability in geographically structured populations. Proc. Nat. Acad. Sci. *71*, 2932–2936 (1974b)

Nagylaki, T.: Continuous selective models with mutation and migration. Theor. Pop. Biol. 5, 284–295 (1974c)

Nagylaki, T.: Conditions for the existence of clines. Genetics *80*, 595–615 (1975)

Nagylaki, T.: The evolution of one- and two-locus systems. Genetics *83*, 583–600 (1976)

Nagylaki, T.: The evolution of one- and two-locus systems. II. Genetics *85*, 347–354 (1977a)

Nagylaki, T.: Selection in one- and two-locus systems. Lecture notes in Biomathematics. Berlin: Springer 1977b

Nagylaki, T.: Decay of genetic variability in geographically structured populations. Proc. Nat. Acad. Sci. *74*, 2523–2525 (1977c)

Nagylaki, T.: The correlation between relatives with assortative mating. Ann. Hum. Genet. *42*, 131–137 (1978a)

Nagylaki, T.: A diffusion model for geographically structured populations. (To appear) (1978b)

Nagylaki, T.: The island model with stochastic migration. (To appear) (1978c)

Nagylaki, T.: Clines with asymmetric migration. Genetics *88*, 813–827 (1978d)

Nagylaki, T., Crow, J. F.: Continuous selection models. Theor. Pop. Biol. *5*, 257–283 (1974)

Nei, M.: Effective population size when fertility is inherited. Gen. Res. *8*, 257–260 (1966)

Nei, M.: Modification of linkage intensity by natural selection. Genetics *57*, 625–641 (1967)

Nei, M.: Linkage modification and sex difference in recombination. Genetics *63*, 681–699 (1969)

Nei, M.: Genetic distance between populations. Amer. Nat. *106*, 283–292 (1972)

Nei, M.: Analysis of gene diversity in subdivided populations. Proc. Nat. Acad. Sci. *70*, No. 12, Part I, 3321 -3323 (1973)

Nei, M.: Molecular population genetics and evolution. Amsterdam, Oxford: North-Holland Publishing Company 1975

Nei, M.: Mathematical models of speciation and genetic distance. In: Population Genetics and Ecology. Karlin, E., Nevo, E. (ed.), pp. 723–765. New York: Academic Press 1976

Nei, M., Maruyama, T.: Lewontin-Krakauer test for neutral genes. Genetics *80*, 395 (1975)

Norman, M. F.: A central limit theorem for Markov processes that move by small steps. Annals. of Prob. *2*, 1065–1074 (1974)

Norman, M. F.: Approximation of stochastic processes by Gaussian diffusions, and applications to Wright-Fisher genetic models. SIAM. J. Appl. Math. *29*, 225–242 (1975a)

Norman, M. F.: Limit theorems for stationary distributions. Adv. Appl. Prob. *7*, 561–575 (1975b)

Norman, M. F.: Diffusion approximation of non-Markovian processes. The Annals. of Prob. *3*, 358–364 (1975c)

Norman, M. F.: personal communication (1978)

Norton, H. T. J.: Natural selection and Mendelian variation. Proc. London Math. Soc. (series 2) *28*, 1–45 (1928)

Ohta, T.: Fixation probability of a mutant influenced by random fluctuation of selection intensity. Gen. Res. *19*, 33–38 (1972)

Ohta, T.: Mutational pressure as main cause of molecular evolution. Nature *252*, 351–354 (1974)

Ohta, T.: Role of very slightly deleterious mutations in molecular evolution and polymorphism. Theor. Pop. Biol. *10*, 254–275 (1976)

Ohta, T., Kimura, M.: Linkage disequilibrium due to random genetic drift. Gen. Res. *13*, 47–55 (1969a)

Ohta, T., Kimura, M.: Linkage disequilibrium at steady state determined by random genetic drift and recurrent mutation. Genetics *63*, 229–238 (1969b)

Ohta, T., Kimura, M.: Development of associative overdominance through linkage disequilibrium in finite populations. Gen. Res. *16*, 165–177 (1970)

Ohta, T., Kimura, M.: A model of mutation appropriate to estimate the number of electrophoretically detectable alleles in a finite population. Gen. Res. Camb. *22*, 201–204 (1973)

Ohta, T., Kimura, M.: The effect of selected linked locus on heterozygosity of neutral alleles (the hitch-hiking effect). Gen. Res. Camb. *25*, 313–326 (1975)

Ohta, T., Kimura, M.: Hitch-hiking effect – counter reply. Gen. Res. Camb. *28*, 307–308 (1976)

Olby, R. C.: Francis Galton's derivation of Mendelian ratios in 1875. Heredity *20*, 636–638 (1965)

Owen, A. R. G.: A genetical system admitting of two distinct stable equilibria under natural selection. Heredity *7*, 97–102 (1953)

Pearson, E. S., Hartley, H. O.: Biometrika tables for statisticians. Cambridge: Cambridge University Press 1962

Pearson, K.: Mathematical contributions to the theory of evolution. XII. on a generalized theory of alternative inheritance, with special reference to Mendel's laws. Phil. Trans. of the Roy. Soc. A. *203*, 53–86 (1904)

Pielou, E. C.: Population and community ecology: principles and methods. New York: Gordon and Breach 1974

Prohorov, Y., Rozanov, Y.: Probability theory, Berlin: Springer 1969

Provine, W. B.: The origins of theoretical population genetics. Chicago: University of Chicago Press 1971

Punnett, E. C.: Eliminating feeblemindedness. J. Heredity *8*, 464–465 (1917)

Robertson, A.: Selection for heterozygotes in small populations. Genetics *47*, 1291–1300 (1962)

Robertson, A.: Artificial selection in plants and animals. Proc. Roy. Soc. B *164*, 341–349 (1966)

Robertson, A.: The spectrum of genetic variation. In: Population Biology and evolution. By Lewontin, R. C. (ed.), pp. 5–16. Syracruse: Syracruse University Press 1968

Robertson, A.: Remarks on the Lewontin-Krakauer test. Genetics *80*, 396 (1975)

Rogers, J. S.: Measures of genetic similarity and genetic distance. Studies in Genetics VII (Univ. Texas Publ. No. 7213), 145–153 (1972)

Roughgarden, J.: Resource partitioning among competing species – a coevolutionary approach. Theor. Pop. Biol. *3*, 388–424 (1976)

Roughgarden, J.: Coevolution in ecological systems: results from "loop analysis" for purely density-dependent coevolution. In: Lecture Notes in Biomathematics 19. Christiansen, F. G., Fenchel, T. M. (ed.). Berlin: Springer 1977

Roux, C. Z.: Hardy-Weinberg equilibria in random mating populations. Theor. Pop. Biol. *5*, 393–416 (1974)

Sawyer, S.: On the past history of an allele now known to have frequency p. J. Appl. Prob. *14*, 439–450 (1977)

Schaeffer, H. E., Johnson, F. M.: Isozyme allelic frequencies related to selection and gene-flow hypothesis. Genetics 77, 163 (1974)

Schnell, F. W.: The covariance between relatives in the presence of linkage. In: Statistical Genetics and Plant Breeding. Nat. Acad. Sci. Nat. Res. Council. Publication 982, 468–483 (1963)

Seneta, E.: Quasi-stationary distribution and time-reversion in genetics. J. Roy. Stat. Soc. B 28, 253–277 (1966)

Seneta, E.: On a genetic inequality. Biom. 29, 810–813 (1973)

Serant, D.: Linkage and inbreeding coefficients in a finite random mating population. Theor. Pop. Biol. 6, 251–263 (1974)

Serant, D., Villard, M.: Linearization of crossing-over and mutation in a finite random-mating population. Theor. Pop. Biol. 3, 249–257 (1972)

Shaw, R. F.: The theoretical genetics of the sex-ratio. Genetics 43, 149–163 (1958)

Shaw, R. F., Mohler, J. D.: The selective significance of the sex ratio. Amer. Nat. 87, 337–342 (1953)

Simmons, M. J., Crow, J. F.: Mutations affecting fitness in Drosophila populations. In: Annual Review of Genetics. 11. Roman, H. L., (ed.), pp. 49–48. Palo Alto: Annual Review Inc. 1977

Slatkin, M.: On treating the chromosome as the unit of selection. Genetics 72, 157–168 (1972)

Sokal, R. R., Sneath, P. H. A.: Principles of numerical taxonomy. San Francisco: Freeman 1963

Sprott, D. A.: The stability of a sex-linked allelic system. Ann. Hum. Gen. 22, 1–6 (1957)

Strobeck, C.: The two-locus model with sex differences in recombination. Genetics 78, 791–797 (1974)

Strobeck, C., Morgan, K.; The effect of intragenic recombination on the number of alleles in a finite population. Genetics 88, 829–844 (1978)

Tanaka, K.: On limiting distributions for one-dimensional diffusion processes. Bull. of Math. Stat. 7, 84–91 (1957)

Thomson, G.: The effect of a selected locus on linked neutral loci. Genetics 85, 753–788 (1977)

Tier, C., Keller, J. B.: Asymptotic analysis of diffusion equations in population genetics. SIAM J. Appl. Math. 34, 549–576 (1978)

Trajstman, A. C.: On a conjecture of G. A. Watterson. Adv. Appl. Prob. 6, 489–493 (1974)

Uyenoyama, M. K., Bengtsson, B. O.: Towards a genetic theory for the evolution of the sex ratio. (To appear) (1979)

van Aarde, I. M. R.: Covariances of relatives in random mating populations with linkage. Unpublished Ph. D. thesis, Iowa State University, Ames (1963)

van Aarde, I. M. R.: The effect of linkage on the mean value of inbreds derived from a random mating population. Genetics 78, 1245–1249 (1974)

van Aarde, I. M. R.: The covariance of relatives derived from a random mating population. Theor. Pop. Biol. 8, 166–183 (1975)

Verner, J.: Selection for sex ratio. Amer. Nat. 99, 419–420 (1965)

Veronka, R., Keller, J.: Asymptotic analysis of stochastic models in population genetics. Math. Biosci. 25, 331–362 (1975)

Wade, M. J.: A critical review of the models of group selection. Quart. Rev. Biol. 53, 101–114 (1978)

Watterson, G. A.: Markov chains with absorbing states: a genetic example. Ann. Math. Stat. 32, 716–729 (1961)

Watterson, G. A.: Some theoretical aspects of diffusion theory in population genetics. Ann. Math. Stat. 33, 939–957 (1962)

Watterson, G. A.: The effect of linkage in a finite random-mating population. Theor. Pop. Biol. 1, 72–87 (1970)

Watterson, G. A.: Errata to the effects of linkage in a finite random-mating population. Theor. Pop. Biol. 3, 117 (1972)

Watterson, G. A.: Models for the logarithmic species abundance distributions. Theor. Pop. Biol. 6, 217–250 (1974a)

Watterson, G. A.: The sampling theory of selectively neutral alleles. Adv. App. Prob. 6, 463–488 (1974b)

Watterson, G. A.: On the number of segregating sites in genetic models without recombination. Theor. Pop. Biol. *7*, 256–276 (1975)

Watterson, G. A.: Reversibility and the age of an allele I. Moran's infinitely many neutral alleles model. Theor. Pop. Biol. *10*, 239–253 (1976a)

Watterson, G. A.: The stationary distribution of the infinitely-many neutral alleles diffusion model. J. Appl. Prob. *13*, 639–651 (1976b)

Watterson, G. A.: Heterosis or neutrality? Genetics *85*, 789–814 (1977a)

Watterson, G. A.: Reversibility and the age of an allele II. two-allele models, with selection and mutation. Theor. Pop. Biol. (1977b)

Watterson, G. A.: The homozygosity test of neutrality. Genetics *88*, 405–417 (1978)

Watterson, G. A., Guess, H. A.: Is the most frequent allele the oldest? Theor. Pop. Biol. *11*, 141–160 (1977)

Wehrhahn, C. F.: The evolution of selectively similar electrophoretically detectable alleles in finite natural populations. Genetics *80*, 375–394 (1975)

Weinberg, W.: On the detection of heredity in man (in German). Jh. Ver. Vaterl. Naturk. Wurttemb. *64*, 368–382 (1908)

Weir, B. S., Brown, A. H. D., Marshall, D. R.: Testing for selective neutrality of electrophoretically detectable protein polymorphisms. Genetics *84*, 639–659 (1976)

Weir, B. S., Cockerham, C. C.: Two-locus theory in quantitative genetics. In: Proc. of Int. Conf. on Quant. Gen. Pollak, E., Kempthorne, O., Bailey, T. B., (ed.), pp. 247–269. Ames: Iowa State University Press 1977

Weir, B. S., Cockerham, C. C.: Group inbreeding with two linked loci. Genetics *63*, 711–742 (1969)

Weir, B. S., Cockerham, C. C.: Mixed self and random mating at two loci. Gen. Res. *21*, 247–262 (1973)

Weir, B. S., Cockerham, C. C.: Behavior of pairs of loci in finite moneocious populations. Theor. Pop. Biol. *6*, 323–354 (1974a)

Weir, B. S., Cockerham, C. C.: Behavior of pairs of loci in finite moneocious populations. Theor. Pop. Biol. *6*, 323–354 (1974b)

White, M. J. D.: Modes of speciation. San Francisco: Freeman 1978

Williams, G. C., Mitton, J. B.: Why reproduce sexually? J. Theor. Biol. *39*, 545–554 (1961)

Wilson, E. O.: Sociobiology: the new synthesis. Camb. Mass: Belknap Press 1975

Wilson, E. O.: Animal and human sociobiology. In: The changing scenes in the natural sciences. Goulden, C., (ed.), pp. 273–283. Philadelphia: Academy of Natural Sciences 1977

Wilson, S. R.: The correlation between relatives under the miltifactorial model with assortative mating. Ann. Hum. Gent. *37*, 189–204, 205–215 (1973)

Wilson, S. R.: Two-sided assortative mating for a single locus. Ann. Hum. Gen. *40*, 225–229 (1976)

Wright, S.: Systems of mating. III. Assortative mating based on somatic resemblance. Genetics *6*, 144–161 (1921)

Wright, S.: The evolution of dominance. Amer. Nat. *63*, 556–561 (1929a)

Wright, S.: Fisher's theory of dominance. Amer. Nat. *63*, 274–279 (1929b)

Wright, S.: Evolution in Mendelian populations. Genetics *16*, 97–159 (1931)

Wright, S.: Physiological and evolutionary theories of dominance. Amer. Nat. *68*, 25–53 (1934)

Wright, S.: The analysis of variance and the correlation between relatives with respect to deviations from an optimum. J. Gen. *30*, 243–256 (1935)

Wright, S.: Isolation by distance. Genetics *28*, 114–138 (1943)

Wright, S.: On the role of directed and random changes in gene frequency in the genetics of populations. Evolution *2*, 279–294 (1948)

Wright, S.: Adaption and selection. In: Genetics, palaeontology, and evolution. Simpson, G. G., Jepsen, G. L., Mayr, E., (eds.), pp. 365–389. Princeton: Princeton University Press 1949

Wright, S.: The genetical structure of populations. Ann. Eug. *15*, 323–354 (1951)

Wright, S.: The genetics of quantitative variability. In: Quantitative Inheritance. Reeve, E. C. R., Waddington, C. H. (eds.), pp. 5–41. Her Majesty's Stat. office: London 1952

Wright, S.: Modes of selection. Amer. Nat. *90*, 5–24 (1956)

Wright, S.: Genetics and twentieth century Darwinism – a review and discussion. Amer. J. Hum. Gen. *12*, 365–372 (1960)

Wright, S.: The interpretation of population structure by F-statistics with special regard to systems of mating. Evolution *19*, 395–420 (1965a)

Wright, S.: Factor interaction and linkage in evolution. Proc. Roy. Soc. B *162*, 80–104 (1965b)

Wright, S.: Polyallelic random drift in relation to evolution. Proc. Nat. Acad. Sci. *55*, 1074–1081 (1966)

Wright, S.: Evolution and the genetics of populations, Vol. 1. Genetics and Biometric Foundations. Chicago, London: University of Chicago Press 1969a

Wright, S.: Evolution and the genetics of populations, Vol. 2. The theory of gene frequencies. Chicago: University of Chicago Press 1969b

Wright, S.: Evolution and the genetics of populations, Vol. 3. Experimental results and evolutionary deductions. Chicago: The University of Chicago Press 1977

Wright, S.: Evolution and the genetics of populations, Vol. 4. Variability within and among natural populations. Chicago: The University of Chicago Press 1978

Yu, P.: Some host parasite genetic interaction models. Theor. Pop. Biol. *3*, 347–357 (1972)

Yule, G. U.: Mendel's laws and their probable relation to intraracial heredity. New Phyt. *1*, 193–207, 222–238 (1902)

Yule, G. U.: On the theory of inheritance of quantitative compound characters on the basis of Mendel's law – a preliminary note. Rep. of third Int. Conf. on Gen. 140–142 (1906)

Author Index

Subject Index

A. T. Winfree

The Geometry of Biological Time

1979. Approx. 290 figures. Approx. 580 pages
(Biomathematics, Volume 8)
ISBN 3-540-09373-7

Contents: Introduction. – Circular Logic. – Phase Singularities (Screwy Results of Circular Logic). – The Rules of the Ring. – Ring Populations. – Getting off the Ring. – Attracting Cycles and Isochrons. – Measuring the Trajectories of a Circadian Clock. – Populations of Attractor-Cycle Oscillators. – Excitable Kinetics & Excitable Media. – The Varieties of Phaseless Experience. – Bestiary: The Firefly Machine. Energy Metabolism in Cells. The Malonic Acid Reagent ("Sodium Geometrate"). Electrical Rhythmicity and Excitability in Cell Membranes. The Aggregation of Slime Mold Amoebae. Growth & Regeneration. Arthropod Cuticle. Pattern Formation in the Fungi. Circadian Rhythms in General. The Circadian Clocks of Insect Eclosion. The Flower of Kalanchoe. The Cell's Mitotic Cycle. The Female Cycle.

From cell division to heartbeat, the widespread appearance of periodic patterns in nature reveals that many living organisms are communities of biological clocks. This landmark text draws on recent discoveries in 20 different fields of science to give an integrated explanation of periodic processes in living systems and in their non-living analogues. The language used is that of systems theory, emphasizing phase singularities, waves, and mutual synchronization in tissues composed of many clocklike units. However, no theoretical background is assumed: the notions required are introduced with copious pictures and examples. The entire second half of the book is devoted to descriptions of the experimental systems on which most examples are based. The book's lively presentation, timely perspective and unique bibliography will make it rewarding reading for students in applied mathematics entering research in the life sciences, as well as biologists, physiologists, physicists and chemists with interests ranging from contraception to chemical waves and oscillations.

Springer-Verlag
Berlin
Heidelberg
New York

Journal of

mathematical Biology

ISSN 0303-6812   Title No. 285

Editorial Board: H. J. Bremermann, Berkeley, CA;
F. A. Dodge, Yorktown Heights, NY; K. P. Hadeler,
Tübingen; S. A. Levin, Ithaca, NY; D. Varjú, Tübingen

Advisory Board: M. A. Arbib, Amherst, MA;
E. Batschelet, Zürich; W. Bühler, Mainz; B. D. Coleman,
Pittsburgh, PA; K. Dietz, Tübingen; W. Fleming, Provi-
dence, RI; D. Glaser, Berkeley, CA; N. S. Goel, Bing-
hamton, NY; J. N. R. Grainger, Dublin; F. Heinmets,
Natick, MA; H. Holzer, Freiburg i. Br.; W. Jäger, Heidel-
berg; K. Jänich, Regensburg; S. Karlin, Stanford, CA;
S. Kauffman, Philadelphia, PA; D. G. Kendall, Cam-
bridge; N. Keyfitz, Cambridge, MA; B. Khodorov,
Moscow; E. R. Lewis, Berkeley, CA; D. Ludwig, Van-
couver; H. Mel, Berkeley, CA; H. Mohr, Freiburg i. Br.;
E. W. Montroll, Rochester, NY; A. Oaten, Santa Barbara,
CA; G. M. Odell, Troy, NY; G. Oster, Berkeley, CA;
A. S. Perelson, Providence, RI; T. Poggio, Tübingen;
K. H. Pribram, Stanford, CA; S. I. Rubinow, New York,
NY; W. v. Seelen, Mainz; L. A. Segel, Rehovot;
W. Seyffert, Tübingen; H. Spekreijse, Amsterdam;
R. B. Stein, Edmonton; R. Thom, Bures-sur-Yvette;
Jun-ichi Toyoda, Tokyo; J. J. Tyson, Blacksburgh, VA;
J. Vandermeer, Ann Arbor, MI.

Springer-Verlag
Berlin
Heidelberg
New York

The **Journal of Mathematical Biology** publishes papers
in which mathematics leads to a better understanding of
biological phenomena, mathematical papers inspired by
biological research and papers which yield new experimental data bearing on mathematical models. The scope
is broad, both mathematically and biologically and
extends to relevant interfaces with medicine, chemistry,
physics and sociology. The editors aim to reach an
audience of both mathematicians and biologists.

Subscription information and sample copy upon request.